Drilbert Engineering Inc.

You can find this and other great drilling books at:

www.Drilbert.com
70 E. Evangeline Oaks Circle
The Woodlands, TX 77384
Phone: +1 (936) 321-3534 office

www.iadc.org
IADC Publications
PO Box 4287
Houston, Texas 77210-4287
Phone: +1/713-292-1945
Fax: +1/713-292-1946
Email: publications@iadc.org

Drilbert Engineering Inc.

Notice

Drilbert Engineering, Inc.
The Woodlands, TX
Phone: 936-321-3534
Drilbert@aol.com
WWW.Drilbert.com

Acknowledgement

The Author, John Mitchell PE, of Drilbert Engineering Inc., can also be reached at: www.Drilbert.com

Dedication

This book is dedicated to my Grandfather, John C. Mitchell Sr. PE, (March 22, 1909 - September 13, 2001), who inspired me to pursue an engineering career in the drilling industry. Grandad was a true pioneer in the drilling industry who endeavored to apply his scientific understanding of physics, metallurgy, and mining to practical applications and innovations. He taught me the value of practical experience and the need to use my understanding of science to interpret my experiences. More importantly, he taught me the value of sharing my understanding with others.

Preface

This book is intended to help drilling personnel improve their understanding of the down hole mechanics of drilling. **The focus of this book is the physics of stuck pipe**. The mechanics of what causes stuck pipe is explained in great detail. Warning signs, preventive measures, and various freeing procedures are also discussed. Some of this material is based on personal observations and experiments and has never been presented before. Most of the information, however, is based on the hundreds of technical papers I researched while writing this book.

The drilling industry has devoted a lot of effort to the problem of stuck pipe in research facilities around the world. The first technical paper on this subject was written in 1937; today, the topic of stuck pipe appears in more than 8,000 technical papers and books. Unfortunately, most of the men and women working on rigs are not even aware that this research exists. This information is of no use if it does not reach the people in the field. **The primary goal of this book is to get this information to rig-based engineers and supervisors in an easy-to-understand format**.

The people on the rig drill the well. They are the ones who need to understand stuck pipe prevention. In particular, drillers and drilling supervisors must have this understanding. A thorough understanding is not easy to obtain. Most technical papers focus on a very narrow portion of the whole picture, and are written in a manner that is difficult even for an engineer to comprehend.

I have attempted to capture the information that is most useful to the driller and present it clearly and simply, with strong visual support from pictures and graphs. Unfortunately, non-technical people may still find some of this material to be difficult to comprehend, especially the material on hole cleaning and well bore instability. Drilling is an engineering subject. It simply cannot be fully understood without the language of physics and math. I have tried to keep the physics and math as simple as possible, but the more educated the reader is, the easier this book will be to comprehend.

Education is a tool to help us interpret our experiences. Experience and education work together; one is worthless without the other. The more education we get up front, the better we can interpret our experiences. However, we will not benefit from our experiences if we do not carefully observe and analyze them. We must study the cause and effect relationship of everything we see and do. If we analyze our experiences, we will learn from them. Experience is thus an education, and education is part of our experience.

An education in engineering gives an individual a huge advantage toward interpreting and understanding his experiences on the rig. It is crucial to pass this understanding along to other rig hands. The people on the rig drill the well. They are the ones who must understand the science and art of drilling if we are to perform well.

Unfortunately, there are those who believe the men who work on rigs are not capable of understanding the material presented in this book. I strongly disagree with this sentiment. Seventeen years at the rotary table taught me that the men I rubbed elbows with are every bit as intelligent as the engineers in the office.

This book will be challenging for most rig-based personnel. It is, however, a challenge they are up to. Like me, they will probably need to read it many times. Not everyone will fully understand every concept presented—but every reader will increase their understanding of downhole mechanics.

A fundamental responsibility of any engineer or scientist is to build a bridge for those who follow behind. That is exactly what this book is intended to be—a bridge to help others obtain the deeper understanding of drilling problems that I have devoted most of my career pursuing.

Table of Contents

Chapter 1
Stuck Pipe and Other Unscheduled Events

Introduction

Recently, the drilling industry has focused a lot of attention toward "unscheduled events." An unscheduled event is an unplanned incident that costs time and money. Stuck pipe is a typical unscheduled event. *Roughly 25% to 33% of the cost of drilling an average well is due to unscheduled events.*

This chapter focuses on the chain of events leading up to an unscheduled event, and two links that are found in any incident.

The Chain of Events Leading to Trouble

It seems that when an accident such as stuck pipe occurs, the first thing that management wants to know is what caused it. After investigating a number of unscheduled events, such as stuck pipe, blowouts, and injuries, I have found that they seldom happen because of a single cause. There is usually a sequence or chain of events that leads up to the incident (Fig 1-1). One cause alone is not enough; several causes must be in place in order for an accident to happen. For example, a defective blowout preventor cannot cause our rig to burn down unless we first take a large enough kick to go under-balanced. We normally don't take a large kick unless we lower our guard or use poor practices.

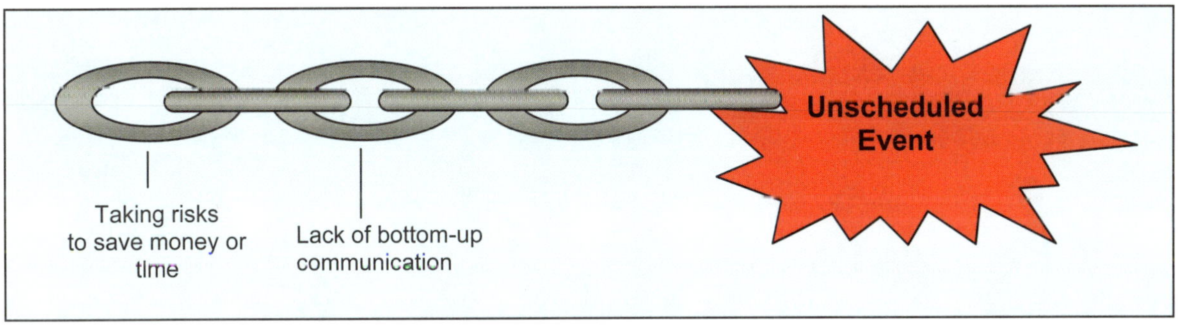

Fig 1-1 The chain of events leading to unscheduled events

As I investigate unscheduled events, I find that two links are present in practically every unscheduled event—**reckless risk taking to save money**, and a **lack of bottom-up communication**. Every disaster during the last century, from the Hindenburg to the Challenger space shuttle disaster, reinforces this belief.

Reckless Risk Taking

The Hindenburg Zeppelin was intended to be filled with helium. However, helium was considered too expensive after the U.S. stopped selling it to Germany, so the Nazis begin filling Zeppelins with hydrogen gas. When the design engineers learned of this, they were quoted as saying, "We are tempting fate." We tempt fate every time we take a risk.

There is nothing wrong with risk taking, but if the consequences and payoff of the risk are not carefully analyzed, the risk taking is reckless.

Bottom-Up Communication

"Bottom-up communication" refers to communication that is initiated by subordinates. It could be a question, a comment, or even a criticism. This is the opposite of "top-down communication," in which the communication is initiated by management.

Some managers are too interested in their own opinions and tend to stymie attempts from their workforce to contribute opinions or "waste their time" by asking questions. These managers are frequently "blind sided" with costly unscheduled events.

We will discuss risk taking and communication in more detail in the next two chapters.

Key Lessons

There is never just one "root cause" for an unscheduled event. There is a chain of events that leads up to it. All we must do to avoid the disaster is break the chain of events that leads up to it by identifying and removing one of those links.

Two links are present in every unscheduled event: ***reckless risk taking****, and a* ***lack of bottom-up communication****.*

Bibliography

1. BP Amoco "Training to Reduce Unscheduled Events" course.

Chapter 2
Saving Money—The Root of all Evil

Introduction

This chapter addresses reckless risk taking to gain positive recognition. Emphasis is placed on why we take these risks and how the corporate culture unwittingly encourages reckless risk taking. There is some discussion on how to use good judgment before taking risks, and because "silence is consent", why we need to reward reckless risk taking with negative recognition. Several amusing case studies are offered to reinforce this concept.

The fastest, and cheapest, way to drill is to drill trouble free!

The root of all evil

Saving money is the root of almost all costly disasters. Stuck pipe is no exception. When I investigate stuck pipe incidents, or any other unscheduled event for that matter, I almost always find reckless cost cutting in the chain of events leading up to the incident (Fig 2-1).

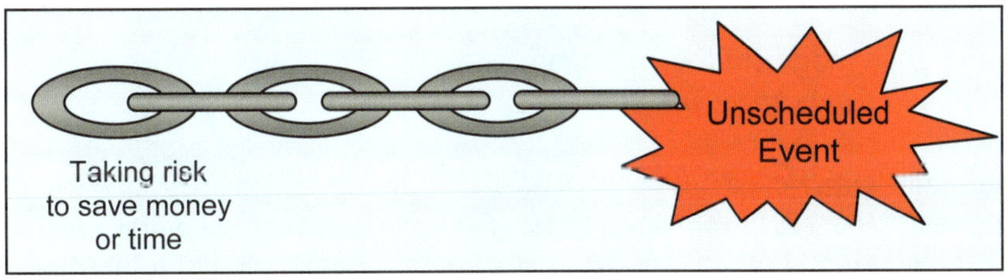

Fig 2-1 Reckless risk taking

I began my study of stuck pipe prevention as a means of optimizing penetration rates. Early in my career, I endeavored to gain recognition in my family's drilling company by achieving the fastest penetration rates possible. We drilled strictly on footage contracts, and I occasionally found myself paid on a footage rate instead of an hourly rate. Obviously, the more footage I gained, the more money I made for my company and myself.

It was also a matter of pride. My ego demanded that I be recognized as the fastest driller in every area I drilled. I turned the operation into a race toward total depth.

I soon learned however, that to be the first driller to reach TD, we must first reach TD! If our tools became stuck, we lost valuable time. Neither the company nor I made any income until we freed our pipe and continued drilling. More importantly, we also ran the risk of losing our tools and the well we just drilled! On deeper wells, such a loss could potentially bankrupt the company! Becoming stuck was an emotional trauma in such instances. I learned to temper my reckless desire for optimization with caution and prudence.

As I pursued a better understanding of drilling practices by working with other companies, I saw this same lesson repeated in a wide variety of applications. It seems everyone wants to achieve recognition through extraordinary performance. With most companies, performance is measured in dollars—revenue earned and costs saved. Those who save the most or produce the most are rewarded with recognition; those who don't are punished with either no recognition or negative recognition.

We are too willing to take risks to get positive recognition!

Saving Money

It has been my observation that one of the root causes of almost all stuck pipe incidents can be traced to an endeavor to save money or time. In fact, *saving money is one of the root causes of practically all the accidents and disasters we refer to as unscheduled events, in any industry*!

In 1989, I became so exasperated with the reckless pursuit of saving money that I began writing a satire called "101 Ways To Save Money by Cutting Corners On Quality and Safety." It contained examples of how millions of dollars were lost by taking extreme risks to save only a few dollars. It seems we must encounter a disaster before we learn our lesson, and then we need only to shuffle people around with promotions or layoffs to forget the lesson and revert to the reckless pursuit of saving money.

Saving money in itself is not evil. In fact, one of the "Drilbert Principles" is "It's not how much one earns, but how much one saves that ultimately determines his financial strength." It is *unnecessary* and *reckless* risk taking to save or make money that must be addressed. My goal with this chapter is to encourage you to *use good judgment while taking risks to save money*.

Decisions and Good Judgment

How does one make a good decision regarding risk taking to save money?
The answer is simple; *good decisions require good judgment.*

How does one acquire good judgment for making decisions?
Good judgment comes from experience.

How does one acquire this experience?
From poor judgment! *Our best experience comes from the mistakes we make when using poor judgment.* It seems we must get burned before we learn how close we can safely get to the fire.

Case Studies

I have a couple of interesting "stories" that I offer here to reinforce this point. I will not reveal rig or company names, dates, or specific locations in an effort to avoid embarrassment to those who may have earned it.

The Great Train Wreck

"The great train wreck" involves a land rig in the North Western United States. In order to save time, this rig was typically moved in big chunks rather than being broken down into highway-sized loads. The mast was moved in one piece, with the crown resting on the back of one truck and the base resting on the back of another truck. One truck would drive forward and the other would back up. This is a common method of moving land rigs.

(Please be aware that although this is an acceptable practice, there is a cost incurred by excessive wear at the pins. The pins are stressed more by the bending loads of the mast in a horizontal position than by the hook loads in a vertical position.)

Unfortunately, the next move of this rig would take it along an interstate highway where large loads like this are not permitted. The rig move was scheduled over the weekend and the tool pusher thought he could get away with moving the larger chunks between 2:00 AM and 3:00 AM on a Sunday morning. There would be very little traffic, if any, and they could post guards with CB radios to alert them of any approaching highway patrol cars.

There was a railroad overpass across the highway that was too low for the larger chunks to pass under. They had to exit the highway, cross the tracks, and get back onto the highway. Everything went well until the 125-foot-long mast high-centered on the railroad tracks and the lead truck drove out from underneath it.

The crew was struggling to get the mast off the tracks before they were discovered by the highway patrol when they heard a train approaching. The train was traveling too fast to stop in time and ran into the mast, carrying it several hundred feet down the track! The train was damaged, several people suffered minor injuries, and the mast was destroyed. The rig was now unable to drill, there were fines and lawsuits to pay, and much damage to repair. I am told it nearly bankrupted this major contractor.

The goal behind this risk was to save time by not having to disassemble and reassemble the mast. I doubt that the combined time they saved on every rig move in their fleet could pay for the damage done that night.

The Jinxed Semi-Submersible

Here is another fine example of Murphy's Law—Whatever can go wrong, will go wrong.

A semi-submersible I once worked on was at a shipyard in Europe undergoing a periodic inspection. They were about to inspect the ballast tanks and opted to remove the hatch covers to all 12 tanks at once. The hatch covers protruded a few inches above the deck of the pontoons and were nearly awash. The deck of the pontoon was only a few inches above sea level.

The Marine Operations Manual for this vessel calls for only two hatches to be opened at a time because only two tanks could be accidentally flooded without causing the rig to capsize. However, the crew opted to remove all the hatches at once in order to save time packing tools up and down between the main deck and the pontoon deck. It would also allow them to air out all the tanks at once, which would allow the inspectors to get through the job a lot quicker.

The goal was to save time and trouble, and the crew felt they could safely take this risk because they were at shipyard. Even though the open hatches to the enormous ballast tanks were only a few inches above seawater, there seemed to be little risk that they could become flooded. After all, the vessel was in a protected harbor on a clear day, with no more than 3-inch high waves splashing against the hull.

A passing speedboat sped alongside to get a closer look at the odd looking vessel. The bow waves from this boat were a couple of feet high and were high enough to splash into the open hatches. After a few waves had spilled into the ballast tanks, the rig began to list to that side. This allowed the 3-inch high waves to spill into the ballast tanks as well. Soon, the hatch covers were slightly below sea level and the ballast tanks were being down flooded!

The only thing that prevented the vessel from capsizing was the shallow depth of the water it was in. The port side of the vessel came to rest on the bottom, severely damaging its thrusters. The thrusters had to be removed and other damage repaired, so the vessel remained in the shipyard for a considerable time. The rig was unable to obtain a day rate until the damage was fixed. The rig then had to work for a lower day rate and incurred higher insurance premiums for several years, as it no longer had thrusters.

When new thrusters were available, the rig was brought into a different shipyard to install them. The economics of getting a higher day rate and lower insurance premiums justified the cost of returning to shipyard to install the new thrusters.

The rig was brought into the shipyard with the aid of a pilot (a legal requirement), who also navigated the vessel back out of the shipyard. With four new thrusters, the semi-submersible now had 15 feet more draft leaving the shipyard than it had coming in. Apparently, this was not adequately communicated to the pilot. The brand new thrusters were raked off on a sand bar on the tow out! The rig was returned immediately to shipyard so the damaged thrusters could be removed and repairs to the hull could be made.

The rig then worked for several more years at a lower day rate while paying higher insurance rates because it still was not "self propelled."

Finally, an opportunity came to put on a third set of thrusters. Unfortunately, while installing these thrusters, a storm came up and blew a neighboring vessel away from her moorings. This errant vessel careened into the semi and punched a hole in her hull. The thrusters were in place, but had not been welded up. A ¾-inch gap existed all around the 3-foot diameter crawlway between the large open hull and the thruster. As the damaged ballast tank quickly flooded, it became apparent the thrusters would soon be under water and that the open area of the hull would flood. Quick stability calculations revealed that this would cause the vessel to capsize, and this time the water was deep enough to allow it!

Fortunately, a resourceful night pusher crammed a one-inch air hose into the gap around the 3-foot crawlway and pressured it up. This created enough of a seal for the small bilge pumps and a bucket brigade to keep the vessel afloat. This semi now has four working thrusters, but it lost eight, suffered three unnecessary shipyard repairs, and was nearly capsized twice! All because of an effort to save a little time and effort by opening 12 hatches instead of just two (as recommended in the operations manual).

Dynamite Factories

I want to leave this chapter with the story of dynamite, Dupont, and the emergence of the safety movement.

Dupont is recognized for pioneering the safety movement during the industrial revolution. Their interest in safety began with the manufacturing of explosives, such as black powder and dynamite.

Dupont made its fortune manufacturing black powder. Because the manufacturing process is inherently dangerous, they had to take unprecedented safety precautions to be successful. The grinding of black powder could create sparks, causing the whole factory to go up in a chain reaction. Dupont quickly learned to isolate the grinding mills from the rest of the factory to limit the damage caused by a mill explosion.

They also took unprecedented steps toward making their managers accountable for factory safety, and are credited with the first written safety policy. This policy required senior managers to operate each piece of equipment before it could be brought into service and operated by anyone else. The number of accidents and lost factories was drastically reduced by this action.

Now here is a more interesting story about dynamite manufacturing. Although it cannot be wholly substantiated with written documentation, I have no reason to doubt it.

Alfred Nobel invented dynamite in 1866 and sold the rights to manufacture it to Dupont. Dupont was one of only a few companies in the world making dynamite at this time. It was relatively cheap to make and was in huge demand. They could sell as much dynamite as they could make for practically any price they asked. However, they could not make money with dynamite. In fact, they were losing money because they were blowing up their dynamite factories. The gains they made from the sale of dynamite were overshadowed by the loss of their factories.

Dupont's upper management implored their managers to find a way to make dynamite more safely. (Remember, Dupont's management was already credited with being the world's foremost champion of safety.) The science of management at this time however, was focused on optimizing efficiency. Managerial specialists would study the layout and movements in the factory and try to find ways to squeeze out more productivity with less labor and cost. The regional managers insisted there just was no safe way to make dynamite. If one person were to trip and drop a case of dynamite, the whole plant would go up in a chain reaction!

According to the legend, Dupont was on the verge of giving up and was considering selling the right to make dynamite in order to regain some of their losses. Before giving up, they tried one last, desperate act—they encouraged their managers to move their families into the dynamite factories!

The manufacturing of dynamite could not proceed until the managers and their families were on the premises! If the manager's family wanted to leave the factory for any reason, all production was halted and the laborers had to leave the factory as well. Upper management told them they didn't care if it took a week to make one stick of dynamite, as long as they made it safely. They never blew up another factory!

Dupont made a fortune selling dynamite. The lesson they learned about optimization and safety spread to all of their operations. It became obvious that *the avoidance of catastrophes was far more important than optimization.*

Dupont still leads the safety movement today. They are clearly recognized as having the safest operations in the world. They still manufacture dangerous chemicals, but they do so without taking unnecessary risks. This is what keeps them profitable.

Here is an interesting fact. Alfred Nobel became distraught over the use of dynamite for destructive purposes. Prior to his death, he established the Nobel peace and science awards with funds that came largely from the money he made from dynamite.

Key Lessons

Most major companies have been guilty of "blowing up their factories" in an attempt to be more efficient.

Our rigs are our factories for making hole.

We have to be more concerned with avoiding catastrophes, such as stuck pipe and blowouts, than we are with optimization in order to achieve success. The fastest way to drill is to drill trouble-free.

If management rewards a successful endeavor to save money while overlooking any reckless risk taking in the process, they are, in effect, rewarding the reckless risk taking.

Silence is consent! *Praising success without criticizing carelessness sends a dangerous message.*

Chapter 3
Communication and Morale

Introduction

Poor communication may well be the number one cause of stuck pipe around the world, and is one of the root causes in almost every stuck pipe incident.

This chapter explains how communication is linked to morale, and how morale is a measure of trust and confidence between team members. Emphasis is placed on how to cultivate trust, morale, and effective communication.

This chapter also illustrates the importance of bottom-up communication. Communication that works its way up the chain of command is the most effective way to break the chain of events leading to an unscheduled event!

Morale

The morale on the rig determines the quantity and quality of the communication taking place. Everyone recognizes the importance of morale, but we don't really understand something unless we can explain it with one sentence, just as Newton explained physics with his three fundamental laws of nature. I struggled for many years to understand and explain exactly what "morale" is.

I have worked on dozens of rigs and visited more than a hundred others. On some of these rigs, everyone was smiling and happy. There was a lot of friendly communication taking place and people looked forward to coming to work. We say that the morale on these rigs is high.

On other rigs however, the opposite was true. Few people were smiling and there was very little communication. Everyone was either frustrated, fearful, or both. No one was talking, either because they were afraid to or because they didn't like the people around them. The morale on these rigs was low.

On the rigs with high morale, everything just seemed to go right. There were very few accidents, they seldom got stuck, and they generally didn't have many unscheduled events. On the other hand, on rigs with low morale, nothing seemed to go right. These rigs had a lot of accidents and seemed to get stuck more often.

One can immediately sense the level of morale on a rig. If morale is high, there is a lot of communication taking place. People are eager to talk and share information. They are more alert and always looking for an opportunity to talk with someone. When morale is low, there is very little communication taking place. People are withdrawn and detached. They are not as attentive and feel less motivated to make an impact on their work environment.

Morale represents the level of confidence that exists on a team. An individuals morale reflects his or hers confidence in, individual and team success, in their own abilities and knowledge, in their teammates abilities and knowledge, and most importantly, in their ability to communicate intimately with their teammates.

I believe *you can measure morale by the amount of communication taking place.* This includes communication between rig hands, between subordinates and supervisors, between contractor and operator, and between the rig and the office. This communication doesn't have to pertain to work. If there is an abundance of communication, I know the morale is high. It is this ease of communication that prevents unscheduled events.

For this ease of communication to exist, there must be a high level of earned trust amongst the crew. By earned trust, I mean they are confident they can approach each other with questions, comments, criticisms, or admissions of mistakes without any fear of negative consequence. They can admit weaknesses and mistakes, knowing that it won't be used against them. They also know they can point out their co-workers weaknesses and mistakes without upsetting or angering them. Some would say they feel they are amongst family or friends.

When morale is high, the level of earned trust is also high. **Morale is thus a barometer that measures the level of earned trust**. When people trust each other, such that they feel confident they can speak their minds without negative consequence, they communicate freely. It becomes very easy to admit you don't know something or that you've made a mistake that needs to be corrected. It is also easy to criticize or make comments about things you think should be corrected. In other words, it becomes very easy for anyone to speak up about, and thus remove, one of the links in the chain of events that leads to an unscheduled event.

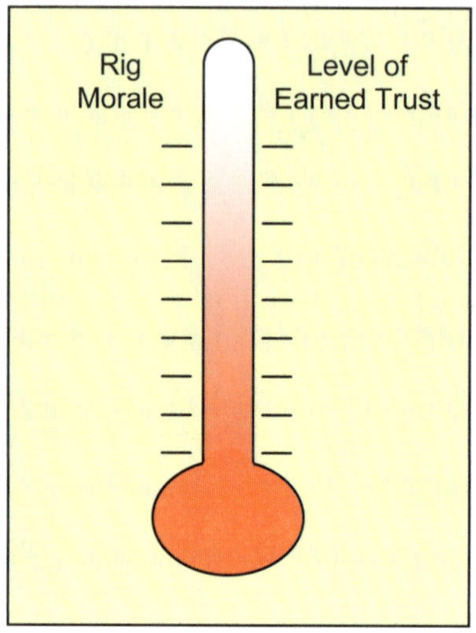

Fig 3-1 Morale barometer

Bottom-Up Communication

This brings us back to the second link that appears in almost every unscheduled event—a lack of bottom-up communication (Fig 3-2). "Bottom-up communication" refers to a communication that is initiated by the subordinate. This communication could be a question about something he doesn't understand or just a comment about something he does or doesn't like.

It seems that during every unscheduled event I ever investigated, someone has come up to me and said, "I knew this was going to happen." When I ask them what they did to try to prevent it, or more specifically, what they said to try to prevent it, I always get the same answer. They said nothing. They believed their supervisors weren't interested in their opinions or they felt too intimidated to speak up. In other words, they felt communication from the bottom up wasn't tolerated or encouraged.

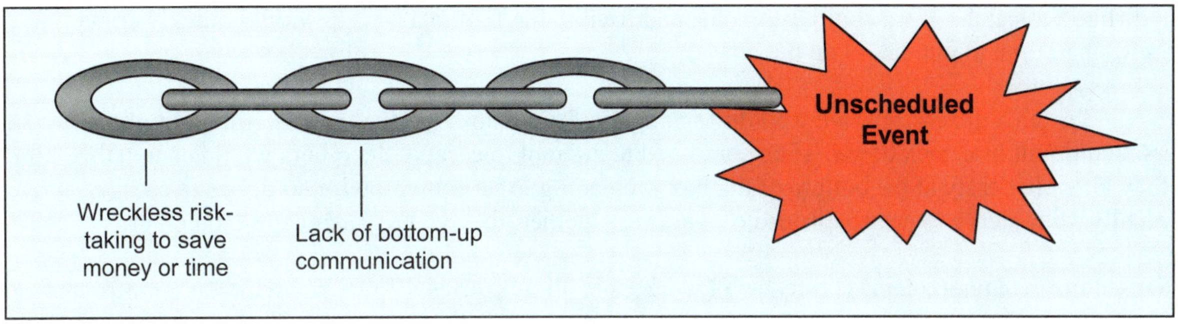

Fig 3-2 Bottom-up communication

All we need to do to avoid any unscheduled event is to <u>break the chain of events</u> that lead up to it. We only need to remove one of the links to do this. Usually, somebody saw something that alerted him to the impending disaster but didn't say anything. We can usually blame low morale for his silence.

All it takes to break the chain of events leading to an unscheduled even is for one person to say one thing to one other person. When morale is high and there is an abundance of communication taking place, it is easy for someone to be that one person who speaks up and triggers the process toward removing one of the links from that chain of events.

The most important form of communication for preventing stuck pipe is upward (bottom-up) communication. Subordinates must feel comfortable approaching superiors with questions and comments. It is unfortunate that on many rigs the foreman, tool pushers, or drillers are too insecure to encourage this type of communication. Insecure supervisors often endeavor to intimidate subordinates to prevent them from asking questions or making comments that might expose the supervisor's weaknesses or lack of understanding.

When investigating stuck pipe incidents and looking for the root causes leading to them, I often find that a misunderstanding (or lack of understanding) about certain drilling procedures contributed to the problem. Harder to uncover, but ever present, is uncertainty or indifference about observations someone made but didn't bring up with his supervisor. A lack of upward communication cultivates an environment where uncertainty, indifference, and misunderstandings prevail. An environment like this is ripe with conditions that can lead to unscheduled events.

One goal of this book and of every course I teach is to promote the upward communication that can break the chain of events leading to an unscheduled event! Providing enough information to the rig hand so that he feels comfortable enough to ask questions is one way to promote this upward communication.

More importantly, *the supervisor must earn the trust of his subordinates.* Subordinates must know that it is safe to admit mistakes and weaknesses, and that they can look to their supervisors for help with such matters.

Horseplay and the "Box of Crabs" Mentality

Horseplay can destroy earned trust amongst coworkers, even when it is good-natured. When it is less than good-natured, it is disastrous to team morale. The drilling supervisor must never allow horseplay to exist, especially with new employees. Traditionally, we enjoyed teasing new employees by sending them to look for the key to the "V-door." Although it is intended in good fun, the new employee learns not to trust his fellow employees in the process. It takes a long time to regain his trust.

There is an unfortunate analogy between the little rock crabs we see in the surf, and many workers in entry-level positions. These little crabs are excellent climbers, with a remarkable grip. They can hang upside down on slippery rocks and cannot be dislodged by the surf. They are also incredibly quick, which makes them difficult to catch. If we could catch one, and tried to put him in a box, he would quickly crawl out before we had time to put the lid on the box.

If we could catch a dozen crabs we wouldn't need a lid for the box—if one tried to climb out of the box, the others would pull him back down. If he continued to attempt to escape, the others would cannibalize him. So, they all sit listlessly at the bottom of the box pretending to be content, with no desire to escape. Eventually, they starve to death. Instead of working together, they work to sabotage each other.

I often see this behavior amongst entry-level workers in all industries. If this behavior isn't confronted early on, it quickly becomes accepted—with disastrous consequences to team morale. The supervisor must be alert to any oppressive behavior that resembles the box of crabs mentality and put a stop to it. The driller and drilling supervisors must take leadership positions in promoting and cultivating earned trust.

Key Lessons

Supervisors must promote bottom-up communication to reduce the occurrence of unscheduled events.

High morale leads to abundant communication.

The level of earned trust between co-workers, and especially between supervisor and subordinates, determines the level of morale. *The supervisor must work to cultivate this trust!*

Supervisors must encourage the development of their subordinates. They cannot allow a "box of crabs" mentality to keep the climbers in the box.

All we need to do to avoid any unscheduled event is to <u>break the chain of events</u> that lead up to it.

Chapter 4
Problem Solving

Introduction

Stuck pipe is a problem for the drilling industry. For any problem, there is problem-solving approach that can be followed to efficiently and effectively solve the problem. This chapter explains the importance of the "Scientific Method", and it's impact on the ascent of mankind. A methodical approach to problem solving is introduced and the importance of a "process" is emphasized in order to learn from our experiences.

The Scientific Approach

The scientific approach to problem solving is a tried and proven method of successfully solving problems. The scientific method can be credited for delivering mankind from the dark ages into the age of enlightenment and information. Every technological marvel we enjoy today owes its success to this modern way of thinking.

Prior to the scientific method, kings and religious leaders ruled the world. They believed themselves to be omniscient and imposed their ideas and decisions on everyone beneath them. Opposing views and disagreement were simply not tolerated. If a problem developed, the king chose someone to blame and then thought up a solution. If his solution failed, he found someone to blame and thought up another solution. Nobody dared to forward their own opinion or disagree with the king for fear of a negative performance appraisal and subsequent torture and death.

The modern scientific method owes its roots to Copernicus and Galileo. Prior to Copernicus, the earth was believed to be the center of the universe. Copernicus reasoned that the sun was the true center of the universe and the earth revolved around it. Galileo later proved this with his invention of the telescope. Religious leaders were offended that someone dared disagree with their assertion of the truth and accused Galileo of heresy. They summoned him to an inquisition in 1633 and forced him to watch a fellow scientist be stretched in half on a rack. He was offered an opportunity to recant his statements about the sun being the center of the universe in exchange for his life. (If he did not have powerful friends within the church he would not have been given this choice). He wisely recanted and spent the rest of his life in prison.

The scientific community was horrified and world was forever changed. Mankind recognized the value of finding the real truth and accepted the fact that opposing views are not necessarily evil.

A forum was developed where scientists could present their findings and opinions to their peers. These opinions would be argued and defended in a highly dignified manner. No one was ridiculed or intimidated for offering new ideas or for disagreeing with someone else's idea. The truth was reached through a process of logic known as the scientific method, rather than by a ruling from the powers that be.

The Scientific Method

The scientific method involves inductive reasoning. This means that we formulate a hypothesis concerning some observed phenomena, and then perform experiments to confirm the validity of our hypothesis. Objectivity is important when making our observations. *To be objective, we must see the world as it really is—without falsifying our observations to agree with any preconceived views we might have.* In other words, to see things the way they really are, not the way we want them to be. Most importantly, we need to be open to the observations and suggestions of others.

Unfortunately, the scientific approach to problem solving is often discouraged when it comes to stuck pipe. The drilling industry has its own kings who identify problems as they see them, then implement solutions without accepting objective observations from subordinates. Opposing views and their supporting arguments are not always permitted. Only the views of the loudest are heard. The analytical people who are best suited for solving problems are often suppressed and pushed aside.

The best hope for solving stuck pipe problems is to develop an environment in which everyone involved can freely offer objective observations and suggestions. In other words, everyone needs to be involved in analyzing the problem and finding a cure. We must involve the entire team and embrace the systematic and objective scientific method, as opposed to a hierarchical, top-down approach.

The Problem Solving Process

There are many books and courses about problem solving, but all approaches follow a simple *think-do-think* process.

The first step in any problem-solving process is to define the problem or prepare a plan. We must sit on our hands until we are convinced that we have adequately defined the problem and come up with the best plan to solve it. Then, we put the plan to work. Finally, we must learn from the process, so that we can improve. The final step is always to analyze and compile the lessons learned while solving the problem.

A Five-Step Problem Solving Process

The *Training to reduce Unscheduled Events* manual by BP Amoco presents a 5-step problem solving process that has been presented to tens of thousands of rig hands.

Step 1: Define the Problem

The first step is to correctly define the problem. This is a step most people try to bypass in their hurry to find a solution to the problem. In their haste, they will identify one of the causes leading to the problem rather than defining the problem itself. We will eventually want to identify all the causes leading to the problem but we must first define the problem itself. Otherwise, we end up focusing on a cure for one of the causes, rather than a cure to the problem itself.

As far as stuck pipe is concerned, the problem is defined when the mechanism that is sticking the pipe is identified.

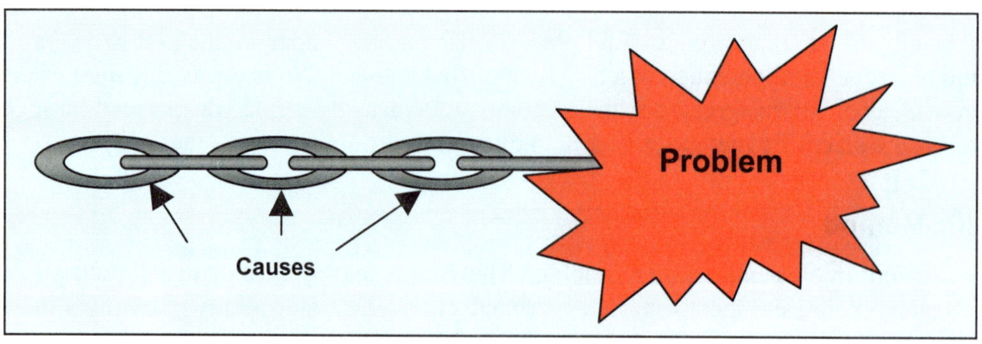

Fig 4-1 Problem Solving

Step 2: Identify Causes

The next step is to identify all the likely causes of this problem. There is seldom just one cause of any problem; there is usually a chain of events leading to the incident. Each link in this chain has a cause and effect relationship with other links. As mentioned earlier, two likely links in the chain of events leading up to stuck pipe include (1) a reckless attempt to reduce costs, and (2) poor bottom-up communication.

If the problem turns out to be differential sticking, one of the causes could be that the pipe was left static too long. This, in turn, may have been caused by an equipment breakdown. Something must have caused the equipment breakdown. The driller may not have been aware that the BHA was across permeable sand. If he did not know of this potential danger, then weak communication is certainly one of the causes.

It is not necessary to eliminate all the causes leading up to the incident in order to prevent its reoccurrence. Normally, we need eliminate only one link in the chain of events leading up to the incident to break the chain and prevent the stuck pipe. Some links are bound to be in place from time to time. Our goal is to eliminate as many links as possible to prevent the chain from completing.

Step 3: Formulate a Solution

The next step in the problem solving process is to conjure up a variety of solutions for solving the problem. This is the easy part. Solutions generally present themselves as we identify the causes contributing to the problem. However, we must avoid the temptation to jump at the first solution proposed. The more alternative solutions we come up with the more likely we are to find one that best fits our needs.

Step 4: Implement a Solution

The fourth step is to choose a solution and implement it.

Step 5: Evaluation of the Solution and Process

The final step is to evaluate the effectiveness of the plan. Like the first step, this step is often skipped. Once a problem is solved, the tendency is to go on to the next one. If we don't gather the team for a project "postmortem," we pass up an excellent opportunity to learn from it. As an industry, we recognize the value of getting together for a pre-spud meeting, but how often do rig personnel and engineers get together after the well to discuss what went right and what went wrong?

Professional chess players always go over their games with their opponents at the end of a match. They explain their strategies and analyze what worked and what didn't. This post-game analysis helps chess players improve their game more than any other activity. A post game analysis is also carried out with every professional, competitive sport I'm aware of. It is surprising that we can see the value of post game analysis in sports, but have trouble justifying it where it matters the most—with our professional careers!

Rig personnel, drilling engineers, and drilling managers will not improve their understanding of drilling substantially if they do not invest that small amount of time at the end of a project to review its successes and failures. It seems we are too concerned with saving money to invest time to analyze our operations and performance.

Don't skip the post-project review meeting to save money! The information gained in this meeting can help prevent future accidents and mistakes that can cost many times more than what is spent on the meeting.

Expect Resistance to a Process

Don't be surprised if attempts to initiate post-project analysis meetings are met with strong resistance, even hostility. Engineers and managers are often unwilling to have their own failures and mistakes revealed. They may fear poor performance appraisals or losing face in front of co-workers or subordinates.

Part of the problem is just plain old laziness. It takes a substantial amount of mental work to analyze our performance.

A few oil companies are now using a process known as "Technical Limit." It is conducted as a pre-spud workshop where the drilling crew details the work to be done and makes suggestions on how to do it smarter. This process is followed through with a post-well evaluation in order to capture the lessons learned.

Key Lessons

The Scientific Method must be embraced to deliver management from the "Dark ages" of blame and cover up, and into the age of enlightenment. *Everyone on the team must be encouraged to contribute his unique insight and perspective of the problem.* Objectivity is critical when making our observations. *We must see the world the way it really is, without falsifying our observations to agree with any preconceived views we might have.*

All formal problem-solving approaches employ a *think-do-think* process. The problem is analyzed, solved, and the solution process is evaluated. *The organization can learn and grow from this process.*

The most important part of the problem solving process is to capture what we have learned.

Chapter 5
Well Planning

Introduction

Well planning is responsible to some extent for more than half of all stuck pipe incidents. *The focus of this chapter is to encourage rig based supervisors to involve themselves in the planning stage of the well.* It is better to be proactive while planning, than reactive while drilling. If rig hands are not allowed to participate in the planning stage, they can still stay ahead of "planned stuck pipe" by being alert to the plans weaknesses.

One weakness is a well path that is too aggressive. We may be tempted to "bite off more than we can chew" when we pick our targets and select the well path. Another weakness may be in the casing program. Casing is one of the biggest costs of a well, so we try to minimize the number of casing strings to the point that open hole sections become over exposed. Sometimes it is the bit or BHA selection that is to blame. Perhaps the biggest weakness is a lack of communication during the planning phase. Insufficient research, poor record keeping (including records that are inaccurate, falsified or incomplete), and lack of involvement from the field prevent us from averting an otherwise avoidable stuck pipe incident.

To better understand the impact of well planning on down hole problems, we will briefly review the basic principles of well planning and see how they impact stuck pipe. A more detailed analysis of how these issues affect stuck pipe will be covered in later chapters.

Communication Issues

Effective communication is necessary for any successful endeavor. The first step in the planning phase is getting the well design team together and communicating the goals and objectives of the well. The design engineers will require as much useful information as possible in order to anticipate potential problems and optimize the drilling program.

Offset well information provides clues to hole problems and potential drilling rates. It might also provide a working well program that can be improved on. One thing to be aware of is that the offset well information may not be complete or accurate. Important information is sometimes omitted or falsified in morning reports to hide mistakes made by the drilling crew. A major effort is underway today, through "TRUE" and "Technical Limit" courses, to learn from mistakes rather than hiding them. The more complete and accurate the offset well information is, the more prepared our design engineers will be for designing around potential hazards.

During the design phase, all potential hazards should be identified and anticipated. One goal is always to drill the well in record time with the lowest possible cost. However, we must never overlook the potential hazards. If any hazards are anticipated, they must be communicated to the entire drilling team. A large percentage of stuck pipe could have been avoided if the driller had been aware of, and had prepared for, the potential hazards. It is the drilling manager's responsibility to ensure this communication has taken place.

The pre-spud meeting is one forum for discussing potential drilling hazards. The main objective of the pre-spud meeting is to clearly define the well plan and its objectives. It is also used to identify who is responsible for the various tasks and components of the plan.

Many operators hold a **Training to Reduce Unscheduled Events** course in addition to, or in place of, the pre-spud meeting. A Training to Reduce Unscheduled Events course presents the well plan and then focuses on potential down hole problems. The attendees prepare action plans designed to prevent and deal with the anticipated problems. Most operators who have used this process have had outstanding success in minimizing down hole problems.

Another course that has gained popularity is the "**Technical Limit**" course. This course is also presented in addition to, or in place of, the pre-spud meeting. It is often held in conjunction with a Training to Reduce Unscheduled Events course.

"Trouble-Free Drilling", taught from this manual, is largely replacing the Training to Reduce Unscheduled Events course. It is also taught in conjunction with the Technical Limit course.

Attendees of the Technical Limit course are actively involved in fine-tuning the well plan to optimize it for their rig. The goal is to work smarter, with more pre-planning, which results in less non-productive time. The Technical Limit course is intended to be an ongoing learning process with post-job analysis. As mentioned before, post job analysis the most important step of the learning process—it allows us to learn from both our successes and our failures.

A post-well analysis meeting would be beneficial, but is seldom done. I recommend gathering the same group of people who attended the pre-spud meeting and then reviewing the well plan after the well has been drilled. Design engineers and drilling crew can both learn from this process.

As mentioned before, the most important type of communication is "bottom-up" communication. Those on-site individuals who are actually drilling the well must communicate their observations, objections, concerns, and questions up the chain of command to the ultimate decision makers. The information gained by our experienced "hands on" people can improve our well design and execution decisions—but only if we encourage their input.

The Drilling Foreman, Tool Pusher, and their immediate supervisors are key players to successful communication. Promoting and maintaining effective communication should be their primary responsibility, from the design phase to the post-well analysis. The success or failure of any well can be traced to the effectiveness of their ability to communicate and promote bottom-up communication.

Basic Well Planning Principles

Well planning begins with some end or goal in mind, such as achieving oil or gas production from a geologic target. Therefore, the first step is to identify and choose a geologic target and a well path to reach the target.

Well Path Trajectory

The choice of our well path is one of our earliest opportunities to avoid becoming stuck. Choosing a well path that is too aggressive for the rig or formations increases the chance of stuck pipe. Several factors must be considered when choosing a well path, including:

- Location of the target(s)
- Position of the rig
- Capability of the rig and drill string
- Inclination and direction
- Wellbore stability
- Bedding planes and natural bit walk
- Horizontal production
- MWD tools and other down hole measurements

Location of the Target(s)

A frequent cause of stuck pipe is *choosing too many targets* to reach with a single well path. A main objective is chosen, but along the way, our geologists want to evaluate other possible targets. The well is turned first one way, then another. Higher inclinations and larger doglegs are incurred than would otherwise be necessary for reaching our prime objective.

It may actually be less expensive to drill multiple wells to evaluate multiple targets. When multiple wells don't fit into the budget, we are tempted to accept more risk than is justified by the potential benefits. The cost of evaluating multiple targets with a single well is often artificially lowered by downplaying the risk of becoming stuck.

Another problem that occurs is *selecting a target that is barely within reach of an existing rig location.* This frequently occurs with offshore platforms and mountain or jungle locations. Because a new location cannot be budgeted, high inclinations and long open hole sections are risked in hopes that the target can be reached without creating a new location.

The bit should be allowed to follow a natural path to the target with as little correction as possible. Adhering too strictly to the well path can cause too many direction corrections and excessive open hole exposure. Reaching the target is the goal. We must be able to get our logs and casing to bottom. To accomplish this, we should strive to minimize well bore tortuosity rather than strictly adhere to a particular well path.

We must be careful not to abandon common sense as we rationalize our budget-trimming endeavors. Sometimes these risks pay off, other times they don't. It is in this stage of the planning that objectivity must be maintained. *We must strive to see the risks and probabilities of success and failure as they really are, not what we want them to be in order to justify the project.* There is nothing wrong with pushing the envelope, but we must be realistic and retain our objectivity as we evaluate the consequences of failure and the chances for success.

Position of the Rig

Ideally, the rig will be placed in such a way that the bit can drill naturally into the target. We also like to choose a location that allows for an easy rig up. In areas of high tectonic stress or other geological hazards, the rig may have to be placed in such a way as to allow a trajectory that maximizes well bore stability.

Occasionally, the site location is limited by environmental and economic factors. The same common sense that applies to selecting the well path applies here. The potential cost of stuck pipe should be considered when weighing the cost of potential site locations.

Inclination and Direction

The direction and inclination of the well are largely dictated by the path to, and through, the target. Sometimes however, the direction and inclination of the well are chosen because of their influence on well bore stability. Direction and inclination both influence well bore stability. Inclination also influences hole cleaning, differential sticking, and to some extent, well bore geometry issues.

The inclination of the well should be carefully considered during the design phase. It is not advisable to build angle in troublesome shale. If we are worried about differential sticking, we should avoid building angle in troublesome sands as well.

Remember that:

- As inclination and tortuosity increase, it becomes more difficult to free pipe once it becomes stuck.
- Hole drag reduces the ability to move down freely out of any stuck pipe scenario.
- Casing points may be affected by mud weights, which must be increased as angle builds.

Well Bore Stability

In regions of high tectonic stress, a well path can be chosen to minimize the difference between the principal and minor stresses. With local tectonics, such as around salt domes and faults, a path may be chosen to avoid the stressed area altogether. Other geological hazards, such as shallow gas and unconsolidated formations, are often avoided this way.

We may choose to build angle in stable formations and hold a constant angle through troublesome formations to limit our open hole exposure time. We don't want a dogleg in troublesome shale. We will want to rotate the string to disturb cutting beds on the low side in high angle wells. Rotating through the dogleg can cause additional stress and failure of the shale.

Subsidence from production of underlying formations could lead to highly fractured or locally stressed shale. This can cause trouble in shale that previously could be drilled trouble-free.

Bedding Planes and Natural Bit Walk

Natural bedding planes dictate the well path to some extent, in that the bit tends to drill up-dip into beds dipping at shallow angles and down-dip along beds dipping at steep angles. This tendency increases when there are multiple formations with differing degrees of drillability.

Bedding planes can lead to instability problems, so a well path may be chosen to avoid passing through the bedding planes at too high an angle.

Horizontal Production

Toward the end of the 20th century, horizontal drilling became the norm instead of the exception in order to increase the well productivity. Multilateral wells also became more popular. The benefits of horizontal production are irrefutable, and the challenges have proven to be quite manageable.

The approaches to horizontal and vertical drilling are different, and these differences need to be recognized and understood. It is especially important that the drilling crew and foreman understand that what worked in one well may not work in another.

MWD Tools and Other Down-Hole Measurements While Drilling

The need to shut down while taking measurements can lead to differential sticking. While planning the path and inclination, we should consider how the well path will be measured.

We must also avoid too much static time in troublesome shale and depleted sands. As sands are produced and depleted, the tendency to become differentially stuck increases. This should be considered while choosing the well path.

Casing Program

After a well path is selected, a casing program is designed. The casing program is designed from the bottom up. We choose the size of completion we want to produce through, and then select the minimum size casing that can accommodate that completion.

Occasionally, we choose the next larger size of casing in order to have a contingent string.

The next step is to decide how much open hole can be tolerated before setting this casing. Sometimes, the previous casing string will be run to the top of the pay zone so that the production zone can be drilled quickly, with minimum formation damage or hole enlargement.

The pore pressures and fracture gradients of the formations penetrated usually determine the maximum length of open hole. The mud weight in the open hole section must be heavy enough to prevent the well from flowing and to support the well bore, yet light enough to avoid lost circulation

The procedure for selecting the maximum open hole section can be summarized as follows:

1. Plot the pore pressure and fracture gradients on a chart.
2. Mark off the bottom of the production string or TD on the chart.
3. Choose the design mud weight for the open hole section drilled for the production string.
    - The design mud weight is the heaviest mud used for that section.
    - *The design mud weight must be greater than the highest pore pressure gradient, yet less than the weakest fracture gradient of the open hole section.*
    - A mud that is heavy enough for the bottom portion of the well may be too heavy for the upper sections. If so, these upper formations must be cased off to allow the use of the heavier mud. A casing point must be selected somewhere up the well to case off the weaker formations (Fig 5-1).
4. This process is repeated until the surface is reached.

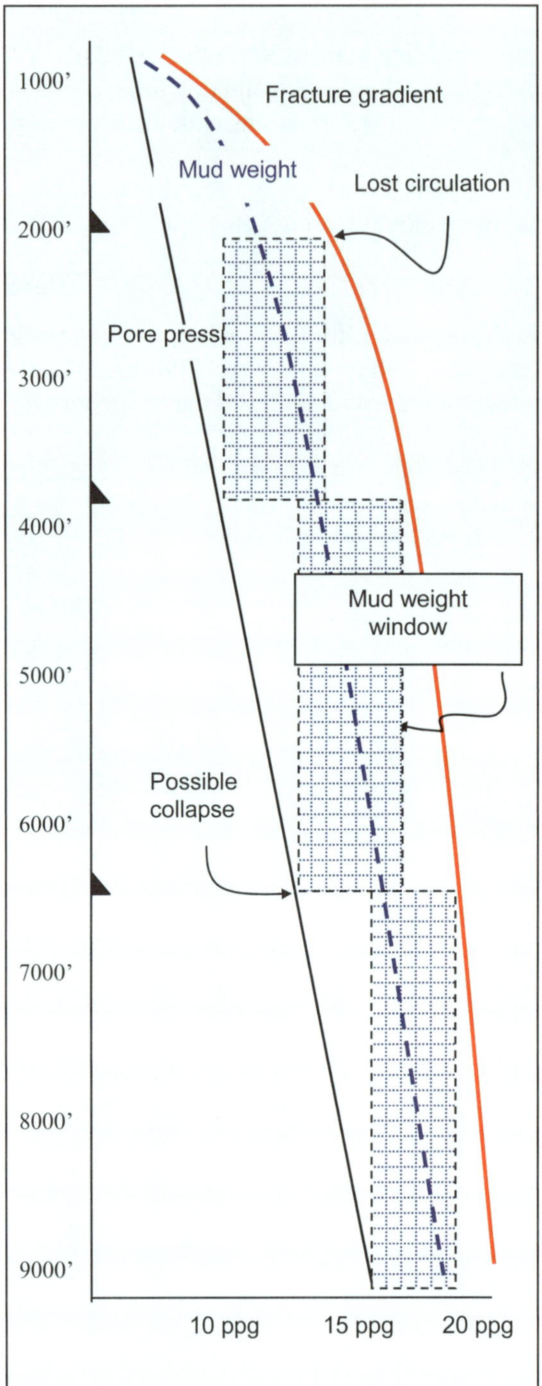

Casing programs are designed from the bottom up around acceptable mud weight windows.

Fig 5-1 Mud weight window

A similar process can be used to design around the prevention of differential sticking.[1] Casing points can be selected to maintain overbalances below a statistically critical level, such as 1,400 psi in the Gulf of Mexico.

Aggressively minimizing casing costs is a common cause of stuck pipe. If there is too much open hole section, the shale can become over-exposed. Allowing too small a window for mud weight fluctuations can also cause problems. (See chapter 8)

Sometimes it is the casing that gets stuck! This becomes more likely with less annular clearance around the casing. Well bore geometry, cutting beds in high angle holes, and differential sticking are all notorious for sticking casing. Each of these cases becomes less severe as the annular clearance around the casing increases.

Hole size

Hole size is largely dictated by the casing size. There must be sufficient clearance to run and cement the casing. There must also be enough clearance to ensure a good cement job. Too much or too little clearance allows cement channeling.

Sometimes a larger hole is drilled to allow for mobile or squeezing formations. The hole size influences *hole cleaning, well bore stability, differential sticking, and packoff.* Generally, the larger the hole, the harder the hole cleaning, but the less likely we are to get stuck.

Bits

Bits are chosen with the hope of achieving the optimum rate of penetration. This means they will drill fast, but will last long enough to minimize trips for a new bit. Ideally, the entire open hole section would be drilled with just one bit.

Bit selection has a lot to do with stuck pipe. Poor bit selection leads to unnecessary bit trips and extended open hole exposure. In directional holes, bit selection also influences well bore tortuosity.

Bit balling leads to slower penetration rates and extended well bore exposure. It also leads to swabbing and packoff. Lost cones lead to lost time and excessive trips. Erratic torque from cone failure can mask well bore problems. If the bit does not follow the intended path, additional open hole time and trips must be suffered to correct it.

Occasionally, it is the bit that gets stuck, especially when drilling with a motor.

BHA and Drill String

We have different BHA approaches for vertical and directional wells.

The BHA in a vertical well must provide sufficient weight on the bit, and have the mass and size to stabilize the bit and dampen vibrations. The BHA should also provide the correct inclination and direction while drilling. Large, stabilized BHAs provide straight, full-gauge holes. They also fill the hole and can apply great stress to the low side of the hole as inclination angle increases.

Vertical holes should be drilled with large bottom-hole assemblies. The stiffness or resistance to bending increases to the fourth power of the diameter[2]. If the diameter of the drill collars is doubled, their resistance to bending is increased by a factor of 16. This stiffness keeps the well bore straight and stabilizes the vertical load across the bearings in the bit.

The mass of the drill collars provides a flywheel effect to keep the bit rotating, and a dampening effect of vibrations between bit and drill string. Vertical, axial, and torsional vibrations from the drill string are absorbed before they can be transferred to the bit, and vice versa. When rotating off bottom however, the heavier BHA's increase string tension so side loads and drill string vibrations are higher.

Large, stabilized drill collars also provide longer bit life by preventing bit walk and maintaining constant weight on the bearings. (Fig 5-2) If the drill collars bend or buckle, the weight on one side of the bit will be higher than the other. The load across the bearing will then oscillate as the bit rotates. This causes unnecessary fatigue and shortens the life of the bearings, teeth, and shirttails of the bit. In extreme cases, it may even cause the bit shank to break.

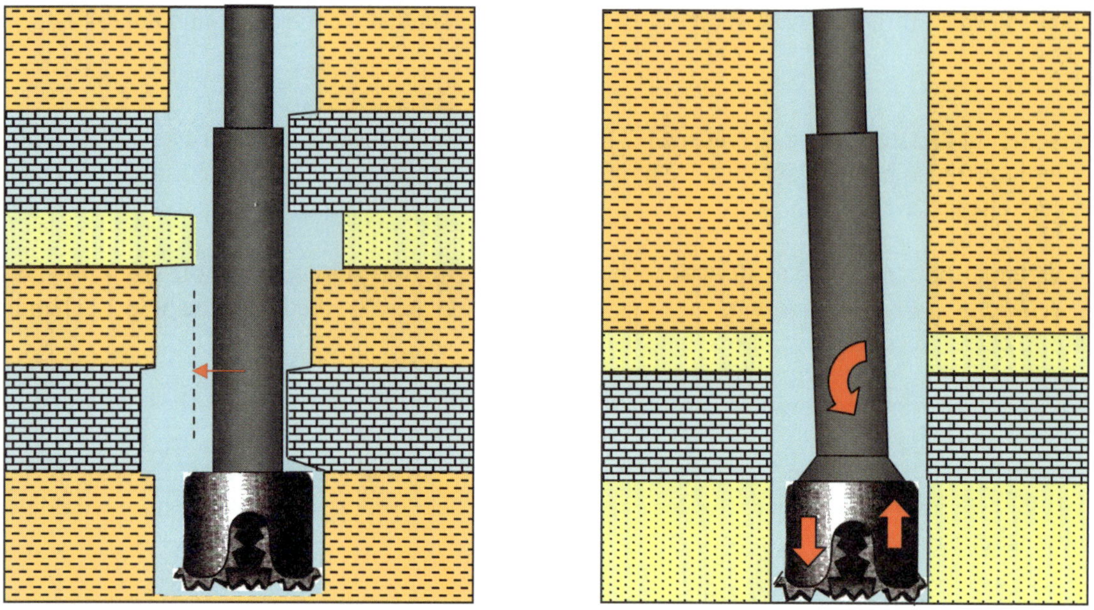

When the bit is not properly stabilized, the bit can walk or rotate off of its true axis. This causes the teeth to scrape excessively against the bottom of the well and rapidly wear down.

When the bit is not properly stabilized, the collars will bend at the bit. This causes the bit to rotate in the vertical plane, resulting in cyclical loading across the bearings.

Fig 5-2 Bit stabilization

There is often reluctance to use large collars for fear of eroding the well bore with high annular velocities. This is seldom the case. The shear force caused by high annular velocities is well below the strength of most formations.[3] Caliper logs reveal that the bottom 300 feet of almost any well is always in gauge. The "washouts" occur higher in the hole and are more the result of well bore instability than high fluid velocity.

There is also reluctance to use large collars for fear of getting them differentially stuck, but statistically it is the drill pipe that gets stuck most often, not the drill collars.[4] (See chapter 9)

At higher inclinations, large drill collars become a liability. As the angle increases, they provide less weight to the bit and more weight to the low side of the well. This increases torque and drag. Large collars provide less annular clearance for dragging through the cutting beds and tend to pack off more. Drill collars are no longer needed to absorb vibration and provide a mass for the mud motors to work against because the pipe is laying on the low side of the hole. Fewer and smaller drill collars are used as the inclination angle increases.

In a more recent trend, directional wells are now being drilled without stabilizers to further reduce the problem of packing off while pulling out of the hole.

Hole Cleaning and Hydraulics

It is necessary to both flush the cuttings from beneath the bit so they are not re-drilled, and to flush cuttings out of the well to avoid getting stuck. Insufficient jetting action at the bit can lead to bit balling and greatly reduced penetration rates. This is why we sometimes endeavor to get the most hydraulic horsepower at the bit that our pumps will allow while pumping the minimum flow rate required to clean the hole.

Unfortunately, too much emphasis is often put on cleaning beneath the bit, and not enough on cleaning the hole. It is possible to reduce penetration rate with too much jet impact at the bit by pistoning the bit off bottom. The penetration rate will also suffer if we must stop to clean the well. Unless we are limited by annular friction losses, it is better to have too much annular velocity than not enough. Unless the jet velocity is absolutely needed at the bit, it may be better to use larger nozzles so the flow rate can be increased.

High flow rates clearly help clean the well bore. However, there is some reluctance to higher flow rates because of an unnecessary fear of eroding the hole. We often talk about maximum flow rates and how we must keep the flow rate low enough to prevent the hole from washing out. In fact, *the well bore will seldom wash out with high flow rates*.

The hole enlargement we see from our caliper logs is due to other mechanisms, not hole erosion. If erosion were responsible for the hole enlargement, we would see this enlargement in those sections of the well bore where the velocity is highest, such as around the drill collars. Our caliper logs usually show the bottom part of the hole to be in gauge. This is where we have the smallest annular clearance, and thus the highest velocity. In inclined sections of the well bore, we know that the velocity is highest on the high side of the hole. Yet, the hole enlarges horizontally, not vertically.

Finally, as the hole enlarges, the velocity must decrease. The cross sectional area of the hole is a function of the diameter squared. If the hole diameter is doubled, the velocity will be reduced to 1/4th it's previous value through that section. If erosion were responsible for hole enlargement, a steady state would eventually be reached. The diameter would increase until the velocity was no longer high enough to erode the well.

There may be concern that the flow regime will reach turbulent flow. Higher pressures and more horsepower are required for turbulent flow. Turbulent flow may be able to erode the dispersed and softened surfaces of the shale formations, thus leading to some hole enlargement. Equally troublesome would be the higher circulating pressures in the annulus that could lead to lost circulation.

If lost circulation is anticipated we should be certain that our nozzles are large enough to pump lost circulation material without plugging.

Drilling Mud

Drilling mud must be fine-tuned to the well. The mud weight must be carefully balanced to fit within the mud weight window allowed by the casing program. Chemical inhibition may be required in shale. Lubricants, fluid loss additives, and other chemical additives may be required to prevent sticking. The Yield Point and Plastic Viscosity must be controlled to optimize hole cleaning. Chemicals and weighting agents are expensive, so we try to minimize their use.

Mud typically accounts for only 7% of the total well cost. However, its typically affects 50% of the total well cost. An inadequate drilling mud is one of the main contributers to down hole problems. Mud costs are also hard to predict, and often cause the project to go over-budget.

Onsite management must recognize when a prescribed mud program needs adjustment, and must be able to make those adjustments when necessary. Well design teams should build this flexibility into their program.

Mud weight is one of the biggest culprits of stuck pipe. We drill faster with lighter mud, and everyone wants to drill faster to win positive recognition. But if the mud weight is too light, we may suffer wellbore instability, which ultimately slows or ends forward progress. There have been many arguments between the rig and home office over raising and lowering mud weight. When the mud weight is repeatedly raised and lowered in a particular section of the well, instability and stuck pipe is likely to occur.

Recommendations for mud weight and chemical additives are offered in the sections on hole cleaning, instability, and differential sticking.

Solids Control

Inadequate solids control leads to thick filter cake, high surge and swab pressures, slower drilling, bit and collar balling, equipment wear, poor casing cementing jobs, and lost circulation. All of these problems can contribute to stuck pipe. It is important to have adequate solids control equipment specified for the well, and it must be properly installed and supervised. Many drilling engineers and foremen assume they have excellent solids control on their rigs but have never really investigated it.

Solids control programs on most rigs are like the volleyball that hits the sand. Everyone assumes someone else is looking after it, when in fact, nobody is. The equipment is there and running but perhaps no one on the rig knows how to optimize its performance. In fact, they may not even know if the equipment is being used correctly. It often isn't.

Down hole problems are generally reduced as solids removal is optimized.

Summary

Drillers and tool pushers were more involved in the drilling process in the 1970's than they were in the 1980's and 1990's. The trend toward the end of the 20th century, saw increasingly more drilling decisions being made by office-based engineers, rather than by rig-based supervisors. With the advent of the personal computer, and spreadsheet and word processing software, "canned" drilling programs also became more common.

A well-documented well plan and drilling program would be copied, renamed, and modified for the next well. Eventually, a well program for the Gulf of Mexico would be used to drill a well off the West coast of Africa. Casing depths, mud programs, formation names, etc., would be modified or edited to suit the new area. Unfortunately, the well might still be designed more for recently deposited shale and clay rather than older, more consolidated shale and carbonates. It became too easy to try to drill the entire world like the region the drilling engineers cut their teeth in.

What works in one region does not necessarily work in another. For example, when drilling horizontal production wells, the last casing shoe is typically at true vertical depth, no matter how far out we drill from the shoe. Influxes are typically bullheaded back into formation. This technique is not likely to be successful in a deep, vertical well. Most probably, the shoe will break down and cause an underground blowout. As another example, in North Western New Mexico, when one operator becomes differentially stuck, they call out a nitrogen truck and blow all of the mud out of the well. The hard rock formations are competent enough to stand with no mud in the well, and the oil influxes will not flow to surface. In many coastal areas, however, if the mud level drops much below the bell nipple, the well will cave in or start flowing.

The trend with canned programs and an abundance of drilling engineers has resulted in a "dumbing down" of the drill crew. They have become less likely to involve themselves in the planning stage of the well or in the day-to-day decision-making involving the well. They simply accept the drilling program "as is" and carry it out without objection or suggestions. In one sense, drilling crews have been transformed into equipment operating janitors.

One of the first things taught in management school is that if a person will be affected by a decision, it is best to involve him in that decision. It is good for morale and he will be more motivated to carry out that decision.

Rig based personnel are also best situated to make many drilling decisions. They are right on top of the well. They are the eyes and ears the well is talking to. Unfortunately, many decisions that should be delegated to the drilling crew are not.

There is some effort to reverse these trends with various training programs and initiatives around the world. The Murchinson course, the Technical Limit course, and the Training to Prevent Unscheduled Events course are some examples.

To avoid costly stuck pipe and other unscheduled events, rig-based personnel must be more involved in the planning and decision-making processes of the well. A well plan should never be taken for granted. There is always room for improvement. <u>Everyone</u> makes mistakes, including engineers. The drilling foreman, tool pusher, and drillers should scrutinize the well plan and compare it to any offset well information they may have.

Bibliography

1) Maurice I. Stewart Jr., U.S. Minerals Management Service, Metairie, LA: "A method of Selecting Casing Setting Depths to Prevent Differential Pressure Pipe Sticking"
2) Bill Garrett, & Gerald Wilson; "How To Drill A Useable Hole" World Oil (August 1, 1976)
3) Gray, George R. & Darley, H.C.H.: "Composition and Properties of Oil Well Drilling Fluids" fourth edition, Gulf Publishing Company (1980)
4) Maurice I. Stewart Jr., U.S. Minerals Management Service, Metairie, LA: "A method of Selecting Casing Setting Depths to Prevent Differential Pressure Pipe Sticking"

Chapter 6
Stuck Pipe Mechanisms

Introduction

In this chapter we explore a methodical problem solving process for preventing and freeing stuck pipe. We begin with a definition of stuck pipe, then proceed to categorize it into three broad categories. A quick and simple method of determining the type of stuck pipe known as the "Stuck Pipe Freeing Worksheet" is presented.

Once the "mechanism" of stuck pipe is identified a pre-planned freeing procedure can immediately be executed. The key to getting free from any type of stuck pipe is to take the correct first action as quickly as possible.

Stuck Pipe Defined

The first step in the problem solving process is to define the problem. If the problem is not properly defined, it will be more difficult to solve. When dealing with stuck pipe problems, we must define *when* and *how* the pipe became stuck.

A drill string is considered "stuck" when the operation is suspended because the pipe cannot be removed from the well.[1] It may be possible to move the pipe below the "stuck point" with full rotation and circulation, as is often the case with keyseats and collapsed casing. But if we can't come out of the hole when we want to, we're stuck.

The next step is to determine how the pipe became stuck. In other words, what type of stuck pipe or "mechanism" is responsible.

Stuck Pipe Categories

Historically, stuck pipe was identified as being either *mechanically stuck* or *differentially stuck*.[1] Modern thinking now breaks mechanical sticking into two separate categories, *packoff and bridging*, and *well bore geometry related sticking*. This is because the mechanisms that stick the pipe in packoff and geometry related cases are clearly different.

These three categories are often referred to as **stuck pipe mechanisms**.[2] The mechanism is defined as the down hole force that is preventing the pipe from being removed from the well.

In almost any region of the world, less than 20% of the stuck pipe incidents account for 80% of the stuck pipe costs. We must identify the mechanisms for these incidents in each area we drill in then focus our attention on them.

Pack-Off and Bridging

This type of sticking occurs when there is debris in the well that wedges itself between the drill string and the well bore. This debris is usually *cuttings*, *cavings*, or *junk*. Large pieces of debris bridge easily and can stick the pipe, even though full circulation is possible. Even though the debris is smaller than the annular clearance between the drill string and well bore, it can still bridge and prevent pipe movement. Cuttings and cavings can become packed together so tightly that they prevent circulation. This type of bridging is referred to as "packing off."

Packoff and bridging is the most frequent cause of stuck pipe around the world. It usually happens while pulling out of the hole, but is also common when the pipe has been stationary with the pumps off for a while. Occasionally, it happens while running in the hole.

Packoffs tend to be the most serious type of stuck pipe. Typically, there is a lower chance of freeing pipe from a packoff than from differential or well bore geometry related sticking. We therefore lose more tools and have to sidetrack more often because of packoffs.

Most packoff and bridge sticking occurs while pulling out of the hole.

Pack-Off and bridging is caused by inadequate hole cleaning or wellbore instability, and is covered in greater detail in chapters 7 and 8.

Differential Sticking

Differential sticking occurs when a motionless drill string is forced into the filter cake against a permeable formation by differential pressure. The friction between the drill string and the formation becomes so high that the pipe cannot be moved. Differential sticking is the second most common type of stuck pipe around the world. It is becoming increasingly prevalent as we deplete reservoirs and drill through them.

Differential sticking almost always occurs after the drill string has been stationary for a length of time.

The mechanics of differential sticking is discussed in chapter 9.

Well bore Geometry

Well bore geometry related sticking occurs when there is a conflict between the shape of the BHA and the shape of the well bore. The BHA does not want to pass through this portion of the well. In order to become stuck, the BHA must be moved into that area where the conflict occurs.

In other words, *the drill string must be moving up or down to become stuck due to a well bore geometry issue.*

The mechanics of well bore geometry related sticking is covered in chapter 10.

Other Types of Sticking

Other types of stuck pipe include equipment failure, drilling string failure, and force majure.

If the rig is shut down due to mechanical failure, the pipe cannot be removed from the well. By definition, the pipe is stuck in the hole. By the time the rig is running again, the drill string is likely to be stuck from down hole mechanisms. This same logic holds for drill string failures, evacuations, or suspensions due to weather, political unrest, labor disputes, etc.

Stuck Pipe Problem

Properly identifying the problem is the first step in the problem solving process. Solving the problem of stuck pipe therefore begins with identifying the sticking mechanism. Once the mechanism is identified the correct freeing procedure can immediately be taken. It is imperative to take the correct first action to attempt to free ourselves as quickly as possible. Whatever is causing the stuck pipe problem only gets worse with time. Statistically, 50% of all stuck pipe becomes free within 4 hours of getting stuck while less than 10% becomes free after 4 hours.[2]

The problem solving process does not end when the pipe is freed. The final step of any problem solving process is to analyze and measure our performance so we can learn and improve from the process.

Stuck Pipe Freeing Worksheet

The "Training to Reduce Unscheduled Event" book[2] presents a stuck pipe freeing worksheet to quickly find the mechanism responsible for the sticking. The table 6.1 is a slightly simplified version of this worksheet. This table is a compilation of probabilities that when taken collectively help us to determine which mechanism is at work. We need only answer the four questions on the left, circle the numbers in the row of each answer, then total the circled numbers in each column. The column with the most points indicates which sticking mechanism is responsible. (Fig 6.1)

For example, we don't become differentially stuck unless the pipe is static. Therefore there is no probability of differential sticking if the pipe was moving prior to sticking. Thus there is a "0" under the differential sticking column in the moving up and moving down cells for "Direction of pipe movement just prior to sticking". There is a "2" in the cell for "Static" because there is a high probability of differential sticking in this case. There is a "0" in the cell under Wellbore Geometry for static pipe just prior to sticking because we don't get stuck in Wellbore Geometry unless the pipe is moving. If the pipe is static there is no probability of Wellbore Geometry related sticking.

Note that there is also a "2" in the cell under Pack-off or Bridge when the pipe was static just prior to sticking. This is because we also tend to pack off when the pumps are shut off and the pipe is still during a connection. Knowing which way the pipe was moving just prior to sticking is not enough to tell us for certain how we have become stuck. We need to answer the other three questions as well, and then total all the values in each column to determine the sticking mechanism.

Table 6.1 Stuck Pipe Freeing Worksheet (from the BP Amoco "Training to Reduce Unscheduled Events" book)[2]

Direction of pipe movement just prior to sticking?	Pack-off or Bridge	Differential Pressure	Wellbore Geometry
Moving up	2	0	2
Moving down	1	0	2
Static	2	2	0
Downward motion of pipe after sticking?			
Down free	0	0	2
Down restricted	1	0	2
Down impossible	0	0	0
Pipe rotation after sticking?			
Rotate free	0	0	2
Rotate restricted	2	0	2
Rotate impossible	0	0	0
Circulating pressure after sticking?			
Circulation free	0	2	2
Circulation restricted	2	0	0
Circulation impossible	2	0	0
Totals			

Example of Stuck Pipe Freeing Worksheet

Direction of pipe movement just prior to sticking?	Pack-off or Bridge	Differential Pressure	Wellbore Geometry
Moving up	2	0	2
Moving down	1	0	2
Static	2	2	0
Downward motion of pipe after sticking?			
Down free	0	0	2
Down restricted	1	0	2
Down impossible	0	0	0
Pipe rotation after sticking?			
Rotate free	0	0	2
Rotate restricted	2	0	2
Rotate impossible	0	0	0
Circulating pressure after sticking?			
Circulation free	0	2	2
Circulation restricted	2	0	0
Circulation impossible	2	0	0
Totals	**2**	**4**	**2**

Fig 6-1 Stuck Pipe Freeing Worksheet

In the example above the pipe became stuck during a connection. We could not rotate or move the pipe downward after it became stuck but circulation was not restricted.

- The Direction of pipe movement just prior to sticking was "static" so that row is selected.
- The Downward motion of pipe after sticking is "impossible" so that row is also selected.
- The Pipe rotation after sticking is also "impossible" so that row is selected.
- Circulating pressure after sticking is "unrestricted" so that row is selected.

The numbers in each selected cell are totaled in each column.

- For "Pack-off or Bridge" the 2, 0, 0, and 0 are selected. Their total is 2.
- For "Differential Pressure" the 2, 0, 0, 2 are selected. Their total is 4.
- For "Wellbore Geometry" the 0, 0, 0, and 2 are selected. Their total is 2.

The column with the highest total is "Differential Pressure". The highest probability is that this is differential pressure sticking.

This is a chart that can be used on the rig floor to quickly suggest which mechanism is responsible for sticking the pipe, or which mechanism is causing tight hole problems. It can also be used for post mortem analysis in conjunction with a Geolograph chart.

First Action Freeing procedures

Once the sticking mechanism is identified the correct first action to free ourselves may be taken. The recommended first actions to take for each mechanism are outlined below.

Pack off and Bridge

1. Relieve trapped pressure and apply a low pressure. (To much pressure will piston the BHA into the packoff. A little pressure is required to help establish circulation if any movement is attained)
2. Apply torque and jar down. If jars are not in place, or not functioning, then apply torque and apply maximum set down weight. (We want to move the pipe in the opposite direction is was moving when we packed off. If we pull into the packoff we will only packoff harder. The goal is to establish pipe movement and then circulation in an attempt to disturb the debris and wash it up the well) (Note that if we are moving <u>down</u> in a high angle well as the packoff occurred we will want to pull or jar <u>up</u> without torque instead)
3. If any circulation can eventually be established increase the flow rate to the maximum amount without losing circulation. Circulate until the well is cleaned.
4. Ream the section and eventually return to bottom to circulate before running casing or logs.

Differential Sticking

1. Immediately apply maximum torque and work the torque to the stuck point.
2. Circulate at the maximum allowable rate. (This should be done simultaneously with the working of the torque.) (If jars are in the string reduce the pump rate to minimize the pump open force while jarring down)
3. Slump the pipe to the maximum set down weight while holding torque in the string. Never pull up! (Pulling up only pulls the pipe harder into the stuck point and tension reduces the amount of torque we can safely apply to the drill string)
4. Jar down if jars are in the string. (Remember to reduce pump pressure while jarring down to minimize the "pump open" effect on the jars)

Wellbore Geometry

1. Jar in the opposite direction the pipe was moving prior to sticking. Apply torque while jarring down but never apply torque when jarring up.
2. Do not forget to take the pump open force into account while cocking or firing the jars. Pump pressure increases the jar blow up and decreases the jar blow down. It also resists cocking the jars for an up blow and helps cock the jars for a down blow.

Secondary freeing procedures can be prepared while continuing with the first action for any of the stuck pipe mechanisms.

Bibliography

1) Bill Murchinson Drilling Practices Course, Albuquerque, New Mexico
2) BP Amoco Training to Prevent Unscheduled Events Manual

Chapter 7
Hole Cleaning

Introduction

Insufficient hole cleaning is responsible for a large portion of all stuck pipe. Some would argue that it is the number one cause of stuck pipe around the world, especially in high-angle holes. One study in the North Sea attributed 33% of the stuck pipe incidents to poor hole cleaning alone.[1] In fact, any packoff and bridge type sticking mechanism has hole cleaning as an issue. Even if insufficient hole cleaning didn't cause the packoff, the debris that is causing the packoff needs to be cleaned from the well.

Consequentially, hole cleaning has received a lot of attention by the drilling industry. The number of research and technical papers covering this subject is enormous. However, the earliest technical paper I could find on this subject wasn't written until 1950. Williams and Bruce[2] published this paper to address the cuttings-carrying capacity of drilling fluids. This paper is referred to in many subsequent papers written on the subject of hole cleaning. The drawings in Figs 7-2, 7-3, and 7-4 were redrawn from this paper. The significance of this early work was the introduction of the "laminar flow profile" and how it affects cleaning.

The focus of this chapter is on the fundamental physics that influence the "flow profile" and thus our hole cleaning effectiveness. We will begin by looking at hole cleaning in vertical wells and then move on to the broader subject of hole cleaning for extended reach drilling. *The practices that worked well in vertical wells do not work well in highly deviated wells.* For example, an increase in yield point helps hole cleaning in a vertical well, while hurting hole cleaning in a directional well.

Hole cleaning in a vertical well is examined first and then we wil look at hole cleaning in a directional well. We will see that hole cleaning in a vertical well focuses on flow rate and rheology. The goal is to produce a flat flow profile. In a directional well the focus is on flow rate, pipe rotation, and visco elastic mud properties. The best practices are summarized at the back of the chapter.

We will also explore hole cleaning with air and foam. The emphasis here is on drilling with a compressible medium. Because air and foam are compressible the volume, and thus annular velocity, are not constant along the wellbore. The annular velocity and the viscosity of foam are generally lower near the bit. *We will learn how successful hole cleaning with air or foam depends on managing the back pressure at the bell nipple to optimize down hole velocities and viscosity.*

The goal of this chapter is to bring the understanding of hole cleaning compiled from our research institutes to the rig floor where it can be used to prevent stuck pipe! Once this is accomplished we move on to the last part of this chapter to discuss preventive measures, warning signs, and freeing procedures for stuck pipe.

> **Note:** To have a proficient understanding of the science of hole cleaning, one must become familiar with its terminology. This terminology will be explained as we progress through the chapter.

Hole Cleaning Efficiency in Vertical Wells

The rig crew monitors drilling trends and the shale shakers to determine the effectiveness of hole cleaning.

There are also two mathematical methods generally used to predict and evaluate hole cleaning efficiency in vertical wells. One method is the *volumetric cuttings concentration in the annulus* and the other is the *transport ratio*.

The **volumetric cuttings concentration**[3] is the total volume of cuttings in the annulus divided by the total annular volume.

$$\text{Volumetric cuttings concentration} = \frac{\text{Volume of cuttings in annulus}}{\text{Total annular volume}} \qquad \text{eq. 7.1}$$

A lower cuttings concentration means we have better hole cleaning. To get better hole cleaning, we must be concerned with lifting the cuttings up the well. The cuttings however, are being pulled downward through the mud by gravity at a terminal velocity known as the **slip velocity**. For cuttings to have a positive upward velocity, the annular velocity must be higher than this slip velocity.

The ratio of cutting velocity to annular velocity is called the **transport ratio**[4] and is the other method used to describe cleaning efficiency.

$$\text{Transport ratio} = v_c / v_a \qquad \text{eq. 7.2}$$

Where:

v_c = velocity of cutting = $v_a - v_s$
v_a = annular velocity = flow rate/flow area

$$v_s = \text{slip velocity} = \sqrt{\frac{2gd_c(\rho_c - \rho_f)}{1.12\rho_f}} \qquad \text{eq. 7.3}$$

g = gravitational constant
d_c = diameter of cutting
ρ_c = density of cutting
ρ_f = density of fluid

Anything that increases the transport ratio increases hole cleaning efficiency in vertical wells. A reduction in slip velocity is one way that the transport ratio can be increased. The slip velocity is influenced by the density and size of the cutting, and by the viscosity and density of the fluid. The larger and heavier the cutting, and the lighter and less viscous the fluid, the faster the cutting will slip through the mud. Much of what we do to improve hole cleaning efficiency in vertical wells is aimed at reducing the slip velocity or increasing the average annular velocity.

There is a lot more to hole cleaning than having a low slip velocity. To understand hole cleaning efficiency in vertical wells, we must look at all the factors involved in the mechanics of hole cleaning.

Factors Affecting Hole Cleaning in Vertical Wells

Factors that affect hole cleaning in vertical wells include:

- Mud weight
- Annular velocity
- Fluid rheology and flow regime
- Cutting size, shape and quantity
- Rate of Penetration
- Pipe rotation and eccentricity
- Time

Mud Weight (Vertical Well Cleaning Factors)

Mud weight influences hole cleaning in three ways:

- It provides buoyancy to help lift the cuttings
- It affects the momentum of the fluid
- It affects the friction the fluid can impart on a cutting as it passes by.

The amount of lift we get from **buoyancy** can be found from the ratio of cutting density to fluid density.

$$\text{Percent lift} = \rho_f / \rho_c \qquad \text{eq. 7.4}$$

Where:

ρ_c = density of cutting
ρ_f = density of fluid

For example, the average density of drilled cuttings is about 21 ppg. When our mud weight is 8.33 ppg, the lift from buoyancy would be about 40% of the weight of the cutting. It would be 50% at 10.5 ppg and 76% at 16 ppg!

Mud Density	÷	Cutting Density	*100 =	% lift
8.33 ppg	÷	21 ppg	*100 =	40%
10.5 ppg	÷	21 ppg	*100 =	50%
16.0 ppg	÷	21 ppg	*100 =	76%

A slight increase in mud weight has a significant effect on the cuttings slip velocity, and thus improves the transport ratio.

Nothing reduces the slip velocity of a cutting more than an increase in mud weight. (See equation 7.3)

Another way that mud weight influences hole cleaning is by transferring momentum to the cutting, just as a cue ball transfers momentum to a billiard ball. Momentum is defined as mass times velocity. Momentum increases linearly with an increase in mud weight (Fig 7-1).

$$\text{Momentum} = m \bullet v \qquad \text{eq. 7.5}$$

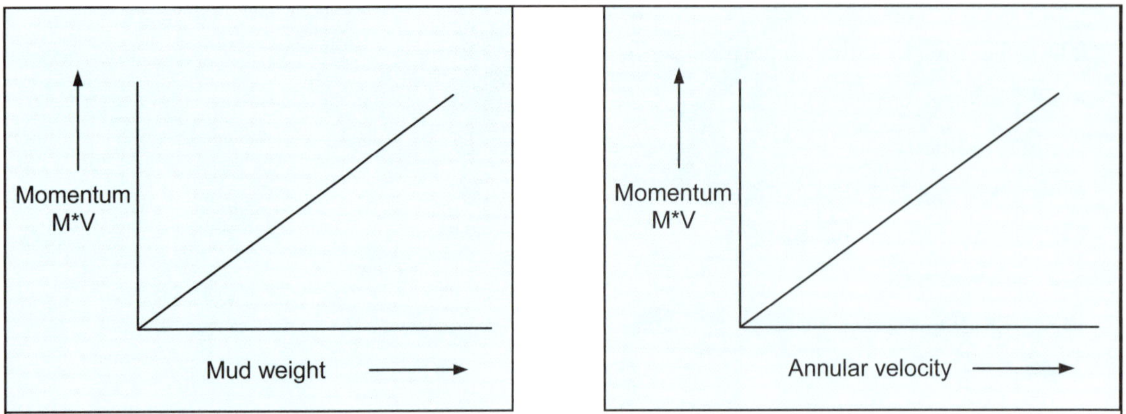

Momentum increases linearly with either an increase in mud weight or annular velocity

Fig 7-1 Momentum

A change in mud weight will influence the momentum, according to equation 7.6.

$$M_2 = M_1 \bullet (\rho_2/\rho_1) \qquad \text{eq. 7.6}$$

Where

ρ_1 = initial density M_1 = initial momentum
ρ_2 = final density M_2 = final momentum

Heavier mud transfers more momentum to the cutting. Note that momentum is equally dependent on annular velocity (Equation 7.5). Some of the lift comes from the **friction** of the mud passing by the cutting. Friction may also help drag cuttings off the wall and back into flow. Friction is also influenced by mud weight; the higher the mud weight the higher the friction.

If mud weight could be lowered to zero, it would not impart any buoyancy, momentum, or friction to the cuttings. Therefore, the hole could not be cleaned, no matter how high the annular velocity. If the mud weight could be raised higher than the cuttings density, no velocity would be needed to clean the hole. The cuttings would simply float out of the well.

Field experience tells us that as the mud weight increases, we have less trouble with hole cleaning and we can get by with lower annular velocities. This is partly because the slip velocity is reduced when the mud weight is increased. When the mud weight is low, such as with air, a much higher annular velocity is needed to clean the well because there is a much higher slip velocity.

Nothing contributes more to hole cleaning efficiency in a vertical well than an increase in mud weight. The transport ratio increases and the cuttings concentration drastically reduces as mud weight increases. Please note however, that we generally don't adjust our mud weight to improve hole cleaning. We generally try to keep the mud weight as low as hole conditions allow for economic reasons. Therefore, we adjust our annular velocity or rheological properties instead.

Annular Velocity (Vertical Well Cleaning Factors)

Annular velocity is the second most influential factor affecting hole cleaning efficiency in a vertical well. Annular velocity provides a lifting force through momentum transfer and friction as the mud strikes and slips past the cutting. The momentum transfer increases linearly with velocity in laminar flow, just as it did with mud weight (Equation 7.5). Note that the contribution from annular velocity depends on mud weight (Fig 1). If the mud weight were zero, there would be no contribution to hole cleaning from annular velocity.

Although the mechanisms of the lifting force provided by annular velocity are fairly straight forward, the total impact of velocity is more complicated because of the influence of the flow profile.

Flow Profile

When we talk about annular velocity, we generally mean **average annular velocity** because the annular velocity is not constant across the diameter of the well bore. It is zero at the wall, and becomes progressively faster away from the wall. This creates the **velocity flow profile.** (Fig 7-2) The flow profile depicts the velocity of the flow at various distances from the wall.

The flow profile of the mud in our mud ditch can be observed by "drawing" a line across the flow with powdered bentonite or LCM. This "line" assumes a flow profile in 1 second, which represents the velocity of the mud in feet/second at any point in the ditch. (Fig 7-2)

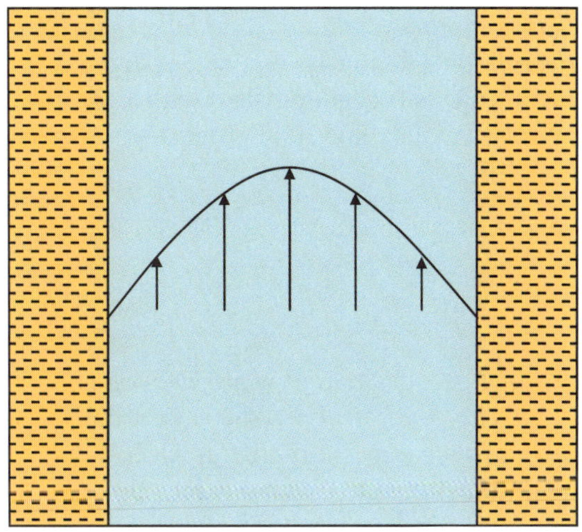

Fig 7-2 Laminar flow profile

The velocity flow profile causes an unequal distribution of force across the cuttings. (Fig 7-3) This causes the cuttings to be pushed away from the faster moving fluid and toward the wall. The larger the cutting, the more force it feels and the faster it is pushed to the wall. Even if the average annular velocity is high enough to provide a positive transport ratio, larger cuttings migrate to the wall where the velocity is slower, and then slip down the well.

Lift acts ⊥ to flow. ∴ in the vertical, the cuttings will move toward the wall of the pipe

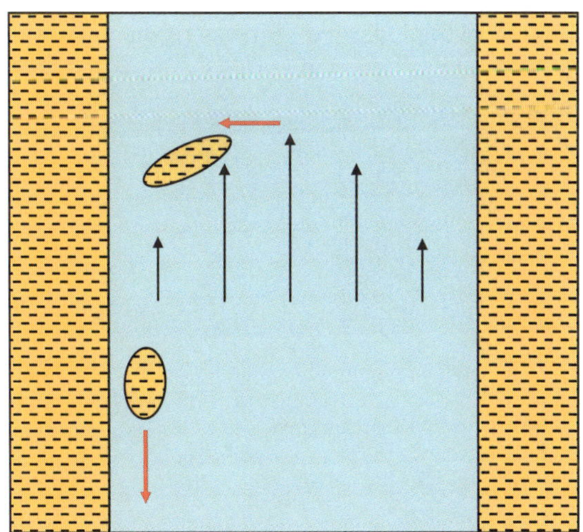

The uneven velocity distribution causes the cuttings to be pushed towards the wall.

Fig 7-3 Cuttings migration

As the downward velocity of the cutting increases, a "Bernoulli" force drags it back into the main flow where it is carried upward again. This upward and falling cycle is known as **cuttings recycling**[3] and causes the annular cuttings concentration to increase (Fig 7-4).

The severity of cuttings recycling depends largely on the flow profile. With a flat flow profile, there is less tendency for a particle to be pushed against the wall, and the velocity near the wall is also faster. (Fig 7-5) Cutting recycling is thus much lower with flatter profiles. *Cuttings recycling increases as the flow profile becomes more elongated, and hole cleaning efficiency decreases.* Obviously, our goal is to create a flat flow profile. This is accomplished with a combination of annular velocity and fluid properties.

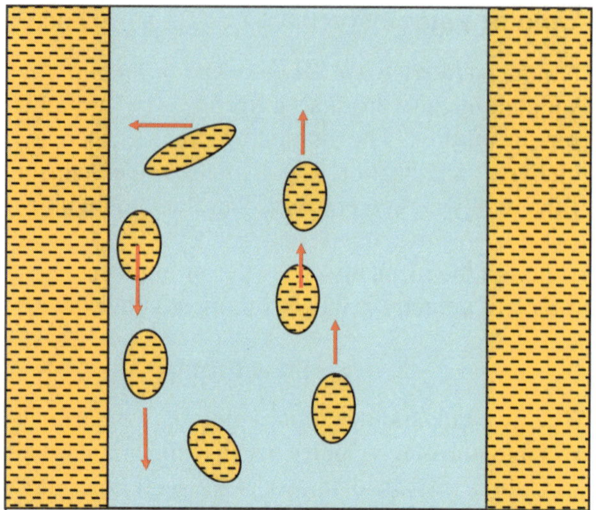

Cuttings slip down the wall where the velocity is slow until they eventually slip back into the faster moving stream where they are carried upward again.

Fig 7-4 Cuttings recycling

Annular velocity and flow profiles are influenced by hole size. The larger the hole, the slower the velocity near the wall. This is especially true in washed out sections. Thus, the problem of cuttings recycling worsens in large diameter and in enlarged sections of the hole.

Cuttings may also stick to the wall or continuously recycle in the enlarged sections of the annulus. When the pumps are shut down, these cuttings may fall back into the well bore.

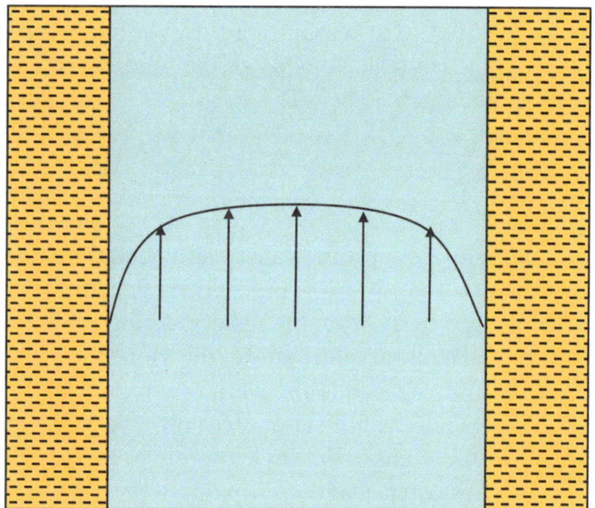

The velocity near the wall is much higher when the flow profile is flatter.

Fig 7-5 Flat flow profile

Fluid Rheology and Flow Regimes (Vertical Well Cleaning Factors)

The shape of the annular flow profile depends on the flow regime it is in. The **flow regime** is a type of relationship between *pressure* and *velocity*.

There are three flow regimes:

- Turbulent
- Laminar
- Plug flow

At low pressures and velocity, the flow is laminar.
At higher pressures and velocity, the flow is turbulent. (Fig 7-6)

With extremely viscous fluids at low velocity, a variation of laminar flow known as "plug flow" exists. (Fig 7-7C) The velocity at which flow switches from laminar to turbulent is influenced by the properties of the mud.

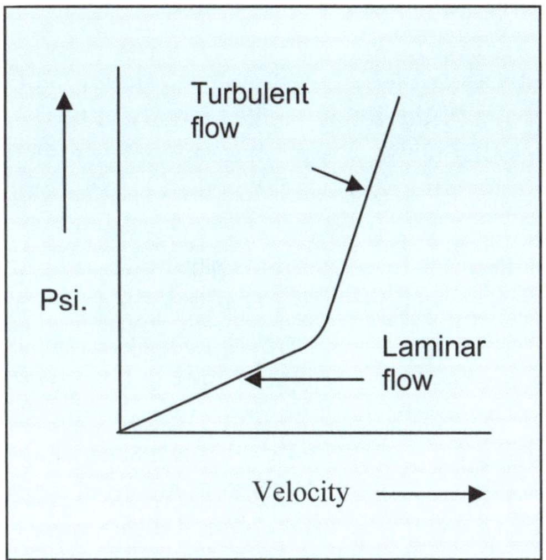

Pressure climbs more rapidly in turbulent flow

Fig 7-6 Flow rate vs. pressure

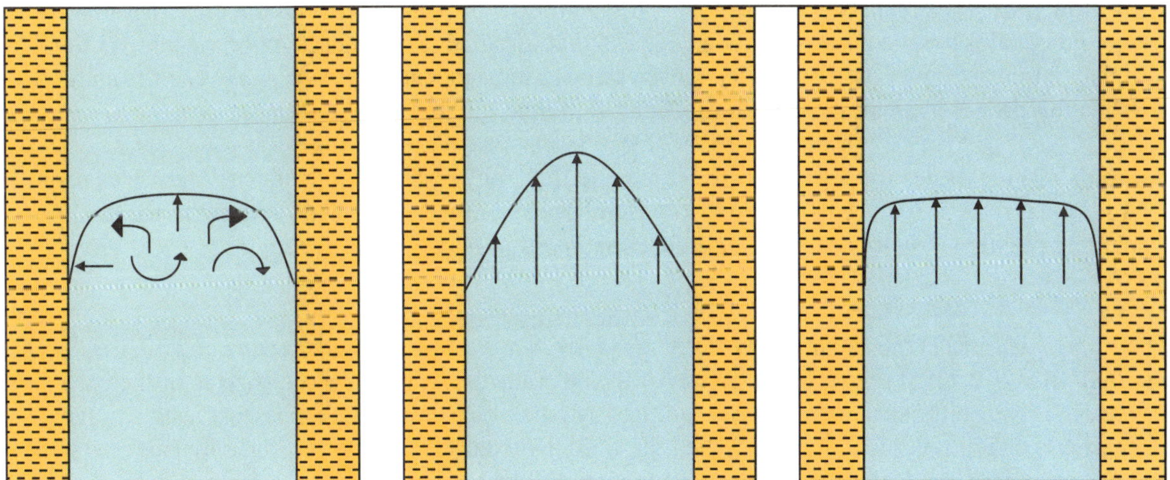

Fig 7-7A Turbulent Flow

Fig 7-7B Laminar Flow

Fig 7-7C Plug Flow

Fig 7-7 Flow regimes

Laminar flow is governed by the viscous properties of the fluid. The fluid flows smoothly, with all molecules moving in the same direction, but at different speeds.

Water flowing slowly in an irrigation ditch is an example of laminar flow. (Fig 7-7B) The water adheres to the side of the ditch, so the velocity at the wall is zero. The cohesive properties of water causes an attractive force that slows down adjacent water molecules passing by. Thus, the molecules close to the wall move very slowly and the molecules in the center of the ditch move the fastest.

The laminar flow profile is parabolic. The shape of this profile depends on the plastic viscosity of the fluid and its yield point. The profile flattens out as the ratio of Yield Point to Plastic Viscosity (YP/PV) increases.

Turbulent flow is more chaotic and is governed by the *inertial* properties of the fluid. The fluid does not flow smoothly, with all molecules moving in the same direction, as in laminar flow. Molecules stick to the wall, so flow at the wall is zero, just as in laminar flow. In the body of the stream however, the molecules are moving in all directions and at different speeds (Fig 7-7A). The average flow is in the downstream direction, with a flow profile that is much flatter than with laminar flow. A fast moving river is an example of turbulent flow.

A flat flow profile provides better hole cleaning. A long parabolic flow profile, such as water in laminar flow, is ineffective for hole cleaning. In this type of profile, the larger particles are pushed toward the wall where the flow is slower and then they slip back down the well.

Turbulent flow provides a relatively flat profile. Turbulent flow provides the best hole cleaning, but it's usually not practical for vertical wells. Note in Fig 7-6 that turbulent flow causes more annular friction pressure than laminar flow. Lost circulation, and in some cases, hole enlargement, may result from turbulent flow. When the flow is laminar, the flattest profile is with mud that has a high YP/PV ratio.

Yield point (YP) represents the force required to initiate flow, or cause the molecules to shear past each other. Additional force is required to cause the mud to flow at a higher rate. This additional force is represented by **plastic viscosity** (PV). *Yield point contributes to good hole cleaning, but plastic viscosity does not.*[5]

Yield point is a measure of the ability of the solid particles in a fluid to build a structure that resists deformation. It is a result of the repulsive forces of the electrostatic charges on the surface of the particles. The negative charge on the surface of the Bentonitic clay particles forces them apart, just as two magnets are forced apart when their poles are aligned. When the fluid is mixed with bentonite, the clay particles will try to get as far away from each other as possible, and thus form the structure shown in Fig 7-8.

Yield point represents an electrochemical resistance to flow.

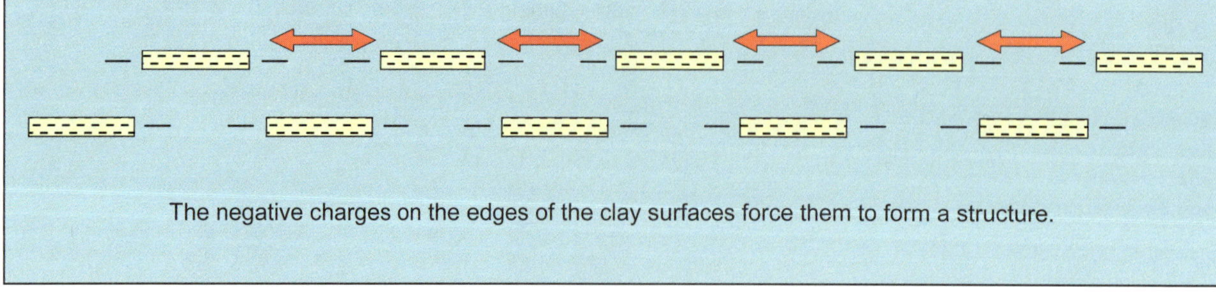

The negative charges on the edges of the clay surfaces force them to form a structure.

Fig 7-8 Yield Point

You can visualize the effect of yield point by imagining a thin sheet of mud, one molecule layer thick, magically suspended in a horizontal plane (Fig 7-9). Because the units on yield point are lb/100 ft^2, we will use an area of 100 ft^2.

If a cutting is to pass through this sheet of mud, the bentonite particles must move out of its way. This means some of the particles must get closer together. The repulsive force of the negative charge tries to keep this from happening. The repulsive force is a function of distance squared. This means that if the distance between two particles is cut in half, the force trying to separate them is increased by a factor of four. The more clay there is in the mud, the closer together these particles will be and the higher their repulsive forces.

If the yield point is 20 lbs/100 ft^2, our magically suspended sheet of mud will support 20 lbs of cuttings, if they are ground very fine and distributed uniformly across the sheet (Fig 7-9A). If we placed a 20 lb rock in the middle of this sheet, it would easily tear through it. It would, however, feel a pressure of 20 lbs/100 ft^2 as it tore through the sheet. (Fig 7-9B)

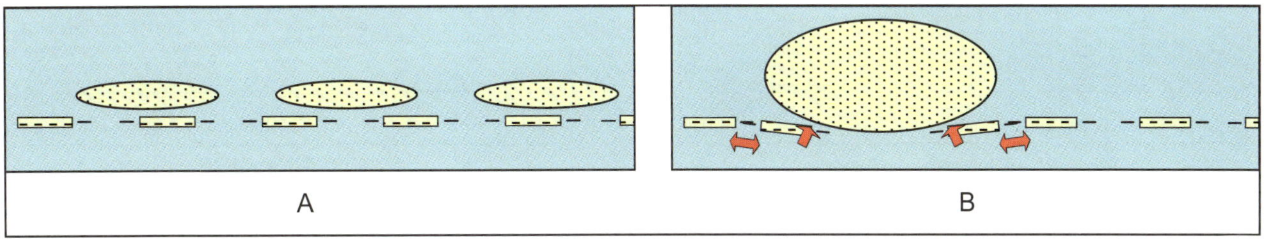

Fig 7-9 Carrying capacity

Lets return to laminar flow. Remember that the velocity of fluid at the wall is zero and it gets progressively higher as we move away from the wall. The attractive forces in the mud try to hold the molecules together. A shear stress is required to force the molecules to slip or "shear" past each other. The rate at which the molecules slip past each other is called the **shear rate**.

The shear rate is not constant across the well bore. (Fig 7-10) It is highest at the wall, and smallest in the center of the well. Plotting the shear stress and shear rate for any mud produces a graph called a **consistency curve.** A consistency curve for clear water is shown in Fig 7-10A.

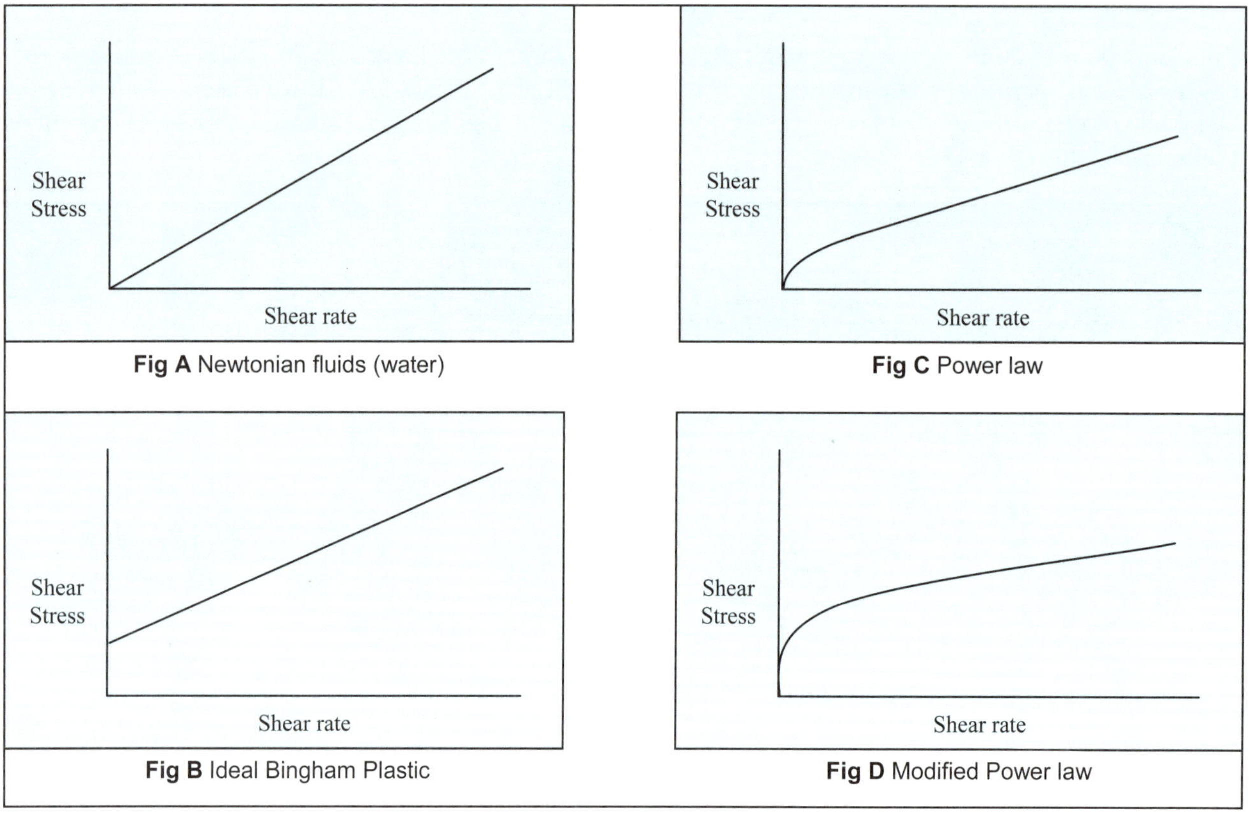

Fig A Newtonian fluids (water)

Fig C Power law

Fig B Ideal Bingham Plastic

Fig D Modified Power law

Fig 7-10 Consistency curves for typical fluids

The slope of this graph represents the viscosity of the fluid. **Viscosity** is a fluid's resistance to flow and is defined as the ratio of shear stress to shear rate.

$$\text{Viscosity} = \text{Shear stress/shear rate} \qquad \text{eq. 7.7}$$

When a fluid contains solid particles larger than molecules, it exhibits a different behavior. Fig 7-10B shows a consistency curve for a Bingham plastic. Notice that a large amount of shear stress is required to initiate flow. The shear stress required to initiate flow is called the Yield Point and is a result of the solid particles' tendency to build a structure in the fluid. This structure resists shear.

Plastic Viscosity

The slope of the straight-line part of the graphs in Fig 7-10 represent the **Plastic Viscosity.** Plastic viscosity represents the ratio of incremental changes in shear stress to shear rate ($\Delta\tau/\Delta v$). It is a result of mechanical interference between the solid particles and the fluid, and is mostly a function of the total surface area of the solids. The units of measurement for plastic viscosity are centipoises.

Plastic viscosity represents a mechanical resistance to flow.

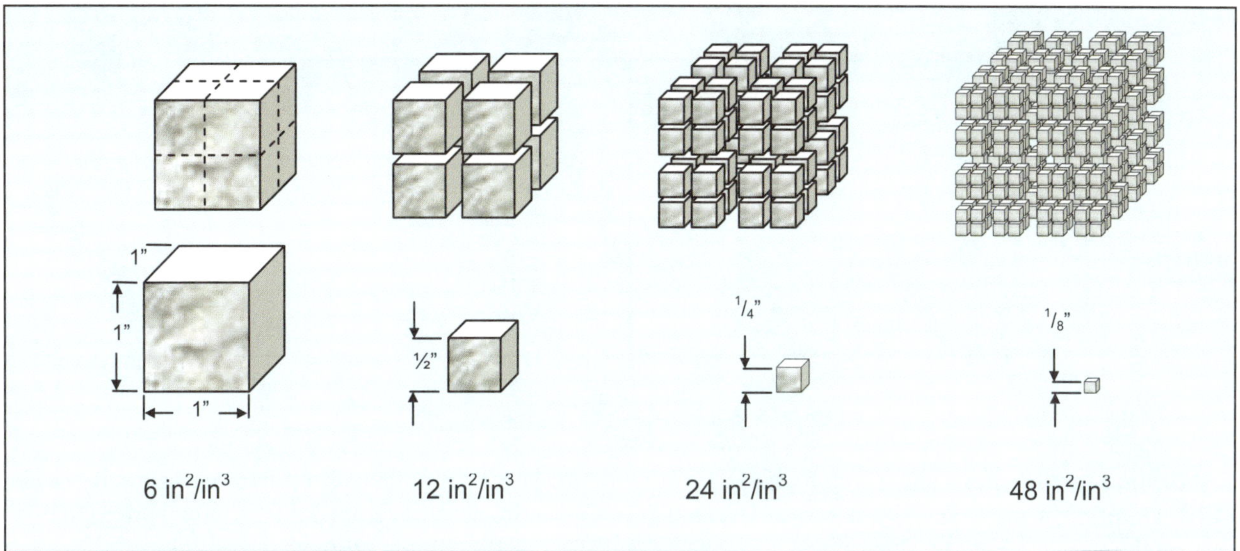

The specific area of a cutting increases as its size decreases. Specific area is the total surface area divided by the weight of the cutting. Note that the specific area doubles each time the diameter of the cutting is halved.

Fig 7-11 Surface area of cuttings

A thin film of liquid at least 2 microns thick coats every solid particle in the mud. When the particles are large, not much liquid is needed to coat them. As they are broken into smaller and smaller pieces, their total surface area increases. (Fig 7-11) More and more liquid is tied up coating the particles as they become smaller. Eventually, these immobile liquid layers constitute a sizeable percentage of the total mud volume. (Fig 7-12)

The droplets of mud begin to interfere with each other in the mud flow. The free mud has a difficult time navigating past all the immobile mud tied to the solids. The mud-coated solids thus increase the mud's resistance to flow. *They also create a tendency to elongate the flow profile.*

The plastic viscosity increases as the total surface area of cuttings increases. If the concentration of solid particles increases, or the solids in the mud are broken into smaller pieces, the plastic viscosity increases.

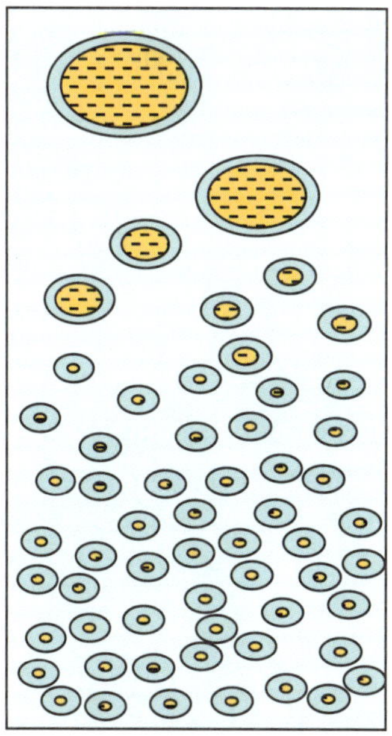

Fig 7-12 Plastic viscosity

Apparent Viscosity

Apparent viscosity represents the total pressure required to cause a certain flow rate. It is a combination of both yield point and plastic viscosity. It is the slope of a line intersecting any point on the consistency curve and the origin.

Note in Fig 7-13 that the apparent viscosity decreases as the rate of shear increases. The yield point and plastic viscosity of a fluid remains constant at moderate flow rates and gives meaningful information. The apparent viscosity does not. To provide a value of apparent viscosity, we must include the rate of shear at which it was measured.

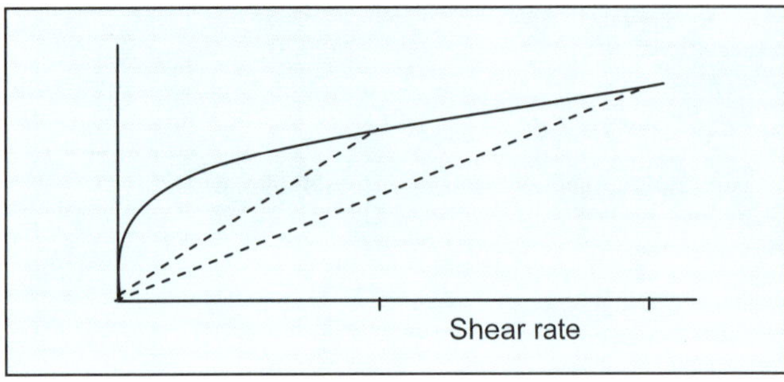

Apparent viscosity is represented by the slope of the dotted lines. Apparent viscosity decreases as the shear rate increases.

Fig 7-13 Apparent viscosity vs. shear rate

Most drilling muds behave like a Bingham plastic at moderate shear rates. However, mud will actually begin to flow at very low pressures, so the power law model in Fig 7-10C was developed. This model attempted to describe the mud at very low shear rates, such as up against the wall, where cuttings are settling.

Typical drilling mud does require some pressure just to initiate flow, so the consistency curve in Fig 7-10D is the most widely accepted model to describe mud at low shear rates. This model is called the Herschel Buckley rheological model[6] (also called the "Yield Power Law").

$$\tau = \tau_0 + K\gamma^n \qquad \text{eq. 7.8}$$

where:

τ is the shear stress at a specific shear rate
γ is the specific shear rate
τ_0 is the yield stress at zero shear rate
K is the plastic viscosity
n is the flow index
(The flow index indicates the degree of departure from Newtonian behavior.
n decreases as yield point increases.)

This model is used to predict cutting settling rates and annular velocities very near the wall or behind the eccentric drill pipe. The flow index "n" is of more concern in deviated wells.

Shear Thinning

In order to initiate flow, enough shear stress must be applied to exceed the yield point. More shear stress is required to produce flow at higher shear rates, due to the plastic viscosity. However, the shape of the solids in the mud largely affects the yield point. As the mud begins to flow, the irregularly shaped particles tend to align themselves with the flow, causing a reduction in both yield point and plastic viscosity. Thus, the higher the shear rate, the lower the viscosity. This reduction of viscosity as the shear rate increases is known as **shear thinning** and is a desired quality in drilling mud. Some muds are more shear thinning than others. The nature and shape of the solids determines how shear thinning a mud will be.

At low rates of shear, the viscosity will be relatively high and the flow will be laminar. At high rates of shear, the viscosity is lower and the flow rate may be turbulent. *The higher the yield point and the lower the plastic viscosity, the more shear thinning the mud will be.*

The shear rate is not constant across the annulus. It is highest near the wall and lowest in the center. The viscosity of shear thinning fluids is thus lower near the wall and higher in the center.

Mud that has high shear thinning properties has a flatter flow profile and is more likely to exhibit plug flow.

Measuring Yield Point and Plastic Viscosity

Mud engineers measure yield point and plastic viscosity with a Fann V-G viscometer. This instrument measures shear stress at different shear rates. The shear stress is plotted at the shear rates produced as the cup rotates at 300 and 600 rpm (Fig **7**-14).

Plastic viscosity is the slope of the line. The equation to find this slope is:

PV = (shear stress at 600 rpm - shear stress at 300 rpm)/(shear rate at 600 rpm – shear rate at 300 rpm)

We call the difference in shear rates between 600 and 300 rpm a unit value (or 1), to simplify the equation to:

PV = (shear stress at 600 rpm - shear stress at 300 rpm)

We can extrapolate the slope of this line to the point of zero shear rate to find the stress that will initiate laminar flow. We can use the mathematical trick of equal triangles to arrive at the equation for yield point.

YP = Shear stress at 300 rpm – (shear stress at 600 rpm - shear stress at 300 rpm)
or YP = Shear stress at 300 rpm – PV eq. 7.9

YP is a stress with units of pounds per 100 feet squared (lb/100 ft^2)

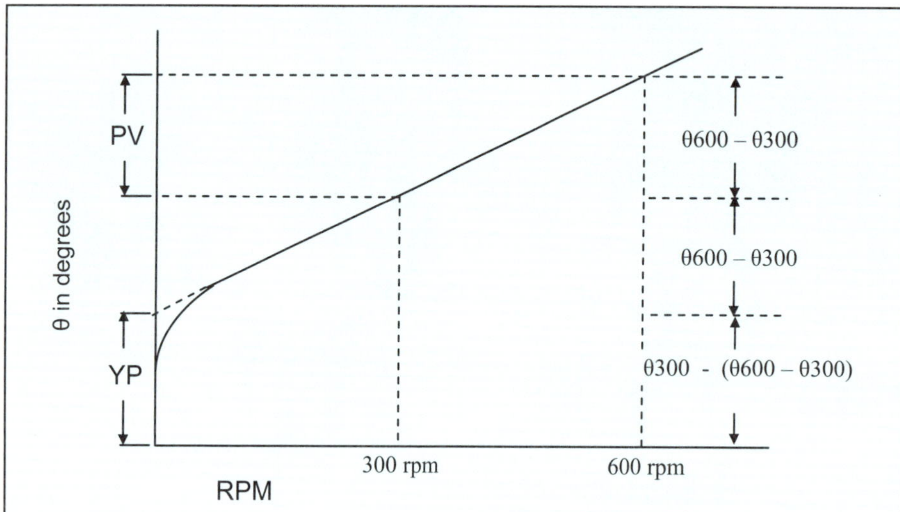

Graphical determination of YP and PV from a Fann V-G viscometer

θ = Shear stress measured in degrees

Fig 7-14 Graphical determination of YP and PV

Marsh funnel viscosity is simply a measure of how long it takes a quart of mud to run through a fixed orifice. It is a quick and simple diagnostic tool to monitor mud trends. A change in the Marsh Funnel viscosity indicates that either the PV or YP has changed, but it doesn't tell us why.

The rheology of the fluid influences the flow profile and thus the velocity of the fluid near the wall where cuttings tend to settle. *The mud engineer's goal is to produce a rheology that maximizes the velocity near the wall while minimizing the slip velocity at these low shear rates.*

Cuttings Size, Shape and Quantity (Vertical Well Cleaning Factors)

The slip velocity increases with the size and density of the cutting. It also increases as the cutting becomes more spherical. Larger cuttings have more tendency to drift to the wall, where the fluid velocity is slower. As the size of the cutting increases, hole-cleaning efficiency must increase to maintain good cleaning.

The shape and nature of the cuttings affects the shear thinning qualities of the mud. Spherical, inert solids exhibit little shear thinning. Plate-like cuttings will gradually align themselves in the direction of flow, causing a reduction of viscosity at higher flow rates. Surface-active solids such as polymers also contribute to shear thinning.

Large quantities of cuttings interfere with each other and the flow profile to frustrate the hole cleaning effort. The concentration of cuttings along the wall increases as the concentration of cuttings in the well bore increases. This causes a shifting of the flow profile inward, as shown in Fig 7-15.

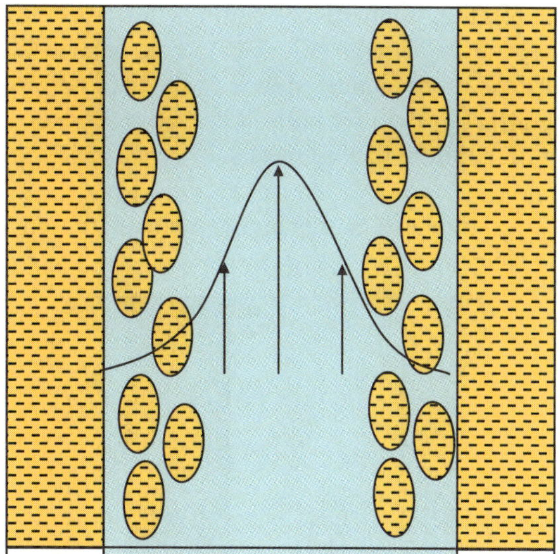

A large concentration of cuttings will force the flow profile inward. The velocity near the wall will be slower and cuttings recycling will increase

Fig 7-15 Excessive solids deform the flow profile

Solid particles in the mud also raise the plastic viscosity. This causes a decrease in the YP/PV ratio, which causes the flow profile to become more elongated. In some instances, the rate of penetration may have to be controlled to maintain good hole cleaning.

Cavings can be thought of as extremely large cuttings. They have high slip velocities and may have to be broken into smaller pieces before they can be circulated out.

Rate of Penetration (Vertical Well Cleaning Factors)

The rate of penetration controls both the size and amount of cuttings generated. At high rates of penetration, the bit is digging in deeper to produce larger cuttings. It is also producing more of them. As the rate of penetration increases, hole-cleaning efficiency may need to be increased.

An increase in torque as the rate of penetration increases suggests that the bit is digging in deeper and producing larger cuttings. If torque continues to build as penetration remains constant, it could be a sign of poor hole cleaning. The aggregation of cuttings is interfering with pipe rotation.

Pipe Rotation and Eccentricity (Vertical Well Cleaning Factor)

Pipe rotation improves the cuttings transport ratio by sweeping cuttings away from the wall and back into the faster moving stream (Fig 7-16). Pipe eccentricity in the annulus reduces the cutting transport ratio. The flow profile is more elongated on the side with no pipe, and the velocity is very low around the pipe (Fig 7-17). The effect of both rotation and eccentricity has very little impact in vertical wells, but becomes much more significant in high angle wells.[7]

Pipe rotation also causes higher annular friction losses. This effect becomes more pronounced as pipe and hole size converge.

Fig 7-16 Pipe rotation

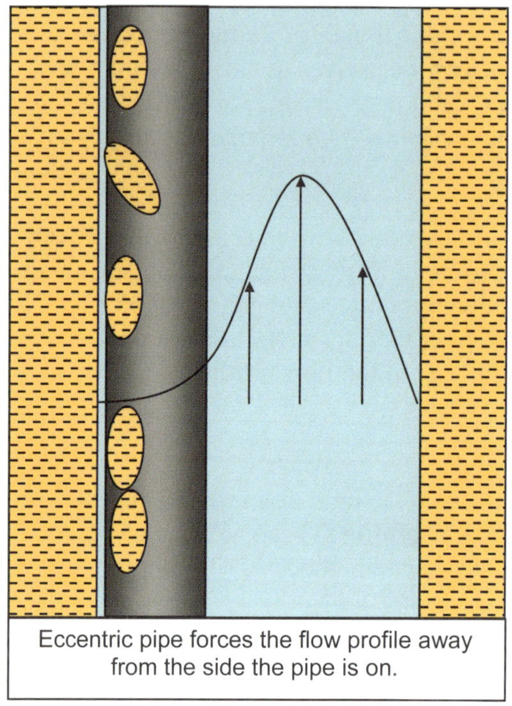

Fig 7-17 Pipe eccentricity

Time (Vertical Well Cleaning Factors)

It takes time to circulate cuttings away from the bit and BHA before making a connection. It also takes time to circulate the hole clean before tripping out of the hole.

A large number of stuck pipe incidents can be traced to not allowing enough circulating time before a connection or trip.

An estimate of the amount of time to circulate the hole clean should be established and compared against the actual observed time to clean the well before each trip.

Hole Cleaning Efficiency in Directional Wells

The same factors that influence hole cleaning in a vertical well also influence hole cleaning in a directional well. However, there are some fundamental differences in how these factors apply. This is due largely to the formation of cuttings beds and rapid settling at some angles due to a phenomenon known as "Boycott Settling." (Fig 7-21 and 7-22) The approach to assessing and predicting hole cleaning performance is also different.

In a vertical well, the effectiveness of hole cleaning is monitored by torque and drag trends and by visual inspection of the shale shakers. In a directional well however, there may be substantial cuttings beds even though the shakers are clean and the torque and drag is moderate.

In a vertical well, hole cleaning can be improved by decreasing the slip velocity of the cuttings. However, the reduction of slip velocity has a diminishing effect as hole angle increases. The transport ratio must be modified to account only for the axial component of the slip velocity, and it becomes a less effective way to predict hole cleaning efficiency.[2] (Fig 7-18)

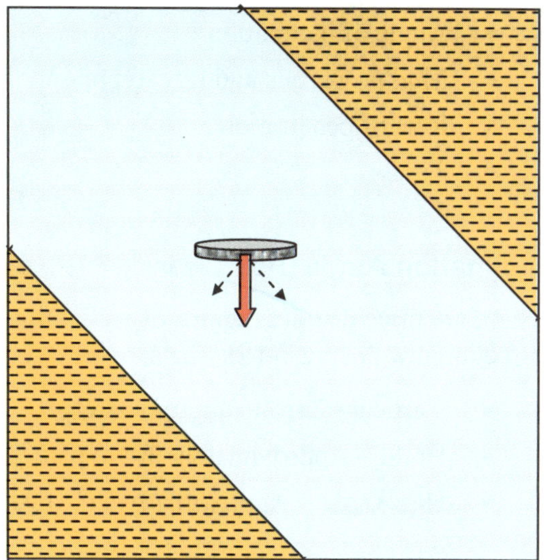

As inclination angle increases, the axial component of slip velocity decreases.

Fig 7-18 Slip velocity in inclined wells

The prediction and monitoring of hole cleaning must also take cuttings beds into consideration. The volumetric cuttings concentration increases because of these cuttings beds.

The fundamental difference between vertical and directional wells is:

- In vertical wells, hole cleaning efforts are focused on reducing the settling of cuttings
- In directional holes, the emphasis is on dragging cuttings off the cuttings bed and into suspension

In the laboratory, the average and maximum cuttings bed heights are used to measure hole cleaning performance. The average bed heights give an indication of the total volume of cuttings in the annulus, while the maximum bed height indicates where sticking is most likely to occur. The **Hole Cleaning Ratio (HCR)** proposed by Marco Rasi[8] is one example of evaluating hole cleaning efficiency using cuttings bed heights.

Another approach to assessing hole cleaning is the **Minimum Transport Velocity** (MTV), proposed by Ford, Peden, Oyeneyin, Gao, and Zarrough.[10] The MTV is defined as the *velocity required to initiate cuttings transport*.

There is one MTV for initiating the movement of cuttings beds, and another for initiating cuttings suspension. Several factors, such as rheology and pipe movement, affect hole cleaning performance. The relative impact these factors have on hole cleaning is reflected in the MTV. This makes the MTV a good indicator of how effective these other hole cleaning factors are.

Each of these hole cleaning factors, and their interaction with each other, will be discussed in the following sections.

Factors Affecting Hole Cleaning in Directional Wells

Hole cleaning in a directional well is affected by:

- Angle of inclination
- Mud properties and flow regime
- Rate of penetration
- Time
- Flow rate
- Cuttings beds
- Pipe rotation and eccentricity

Inclination Angle (Hole Cleaning Factors in Directional Wells)

There are three distinct regions of inclination in a directional well (Fig 7-19):

- Vertical - Zero to approximately 30°
- Build – Approximately 30° to 65°
- Hig Angle – Approximatly 65° to 90°

The three regions are sometimes referred to as the **vertical**, **build**, and **high angle** regions.

In the vertical section there are no cutting beds. Cuttings beds begin to form in the build section and will always slide down the well when the pumps are turned off. In the high angle section cuttings beds are stationary and will not slide down the well when pumps are turned off.

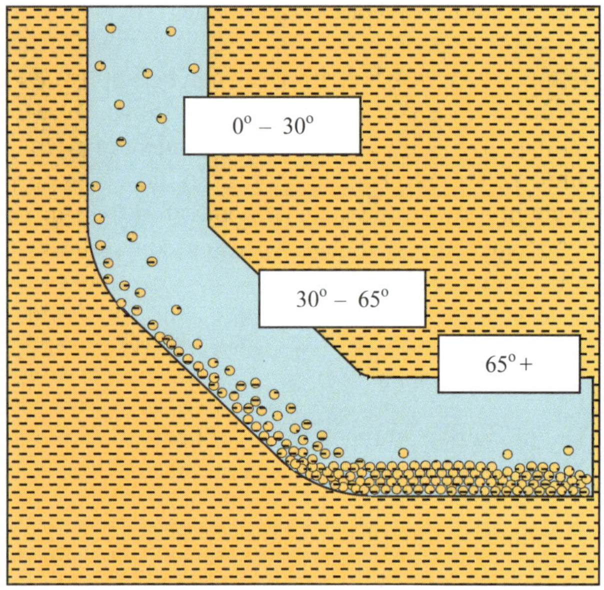

Fig 7-19 Three regions of inclination

The nature of the cuttings bed and the hole cleaning mechanisms are distinctly different in each section. In the vertical section, anything that reduces slip velocity improves hole cleaning. Actually, only the axial component of slip velocity affects hole cleaning. In the vertical section, most of the axial component is in the vertical direction. As the angle builds, the cuttings settling direction and buoyancy are still vertical, but they have a smaller axial component (Fig 7-18).

Cuttings migrate toward the low side of the well as the well begins to deviate from vertical. As the angle approaches 30°, cuttings begin to spend more time at the wall before being swept back into the main stream. The problem of cuttings recycling becomes even more severe when the angle exceeds 30°. As the angle increases from 30° to 45°, the duration of time that the cuttings spend on wall increases dramatically. As shown in Fig 7-20, the volumetric cutting concentration increases dramatically between 30° and 45° and remains relatively constant at higher angles.[3]

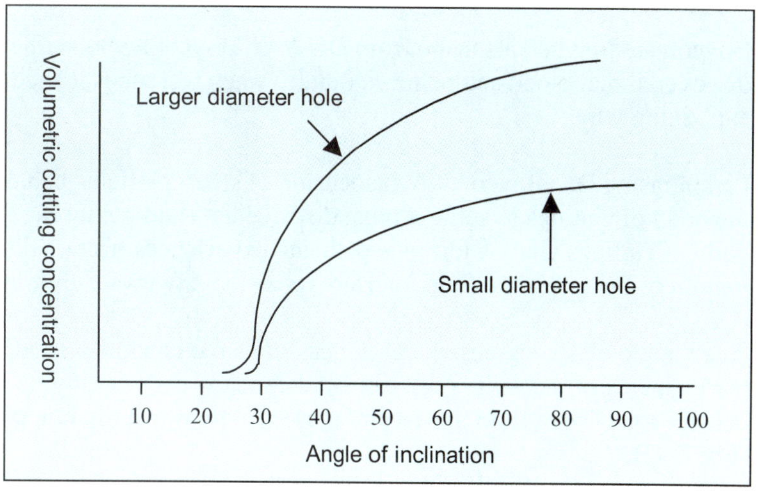

Fig 7-20 Cuttings concentration climbs rapidly after 30°

Cuttings beds begin to form at angles somewhere between 30° and 35°. These cuttings beds are loose and highly fluidized at angles less than 45°. Therefore, they are easily disturbed and slide easily. At angles less than 55°, these cuttings beds will always slide down the well when circulation is stopped.[11] Cuttings beds continually slide and tumble, even while circulating, at angles up to 65°. (This sliding is more pronounced with OBM than WBM.)

Cuttings beds become stationary at angles somewhere between 55° and 65°. They are also more packed and harder to disturb.

Cuttings tend to settle out of the 40° to 55° section more rapidly than in the other sections, due to a phenomenon known as Boycott settling.

Boycott Settling

Boycott settling gets its name from Dr. A. E. Boycott, who studied blood samples in WW1. He accidentally discovered that blood settled more rapidly when test tubes leaned at an angle of 45°. He later wrote a paper explaining why.

Paraphrasing Dr. Boycott, "A suspension of small particles behaves as a separate body of fluid. In order for the body of cuttings to settle it must displace the fluid below it. The displaced fluid must travel through the body of cuttings along a tortuous path and experiences frictional drag as it passes the particles." "The upward motion of the displaced fluid interferes with the downward motion of the cuttings" (Fig 7-21)

At an angle of 45 degrees, the particles only have to settle a small distance before a channel is exposed along the high side of the well. The clear fluid can now rush rapidly up the high side of the well, allowing the denser, solid-laden fluid to descend more rapidly down the low side in the form of loose cuttings beds." (Fig 7-22)

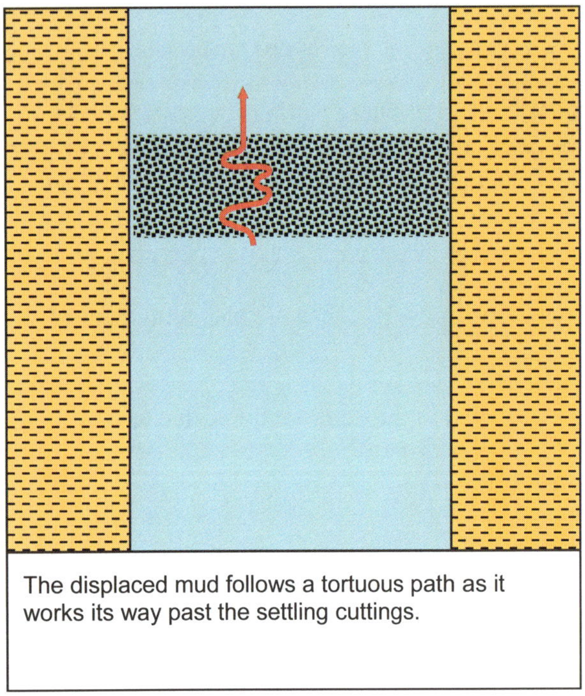

Fig 7-21 Boycott settling

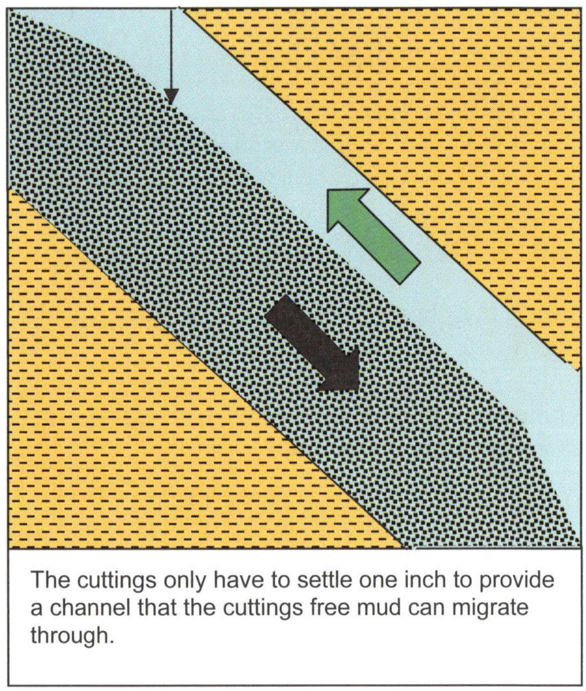

Fig 7-22 Boycott settling

Cuttings Transport Mechanisms

The cuttings transport mechanism differs in each section because of the nature of the cuttings beds. Two broad classifications of cuttings transport are **suspension** and **bed transport**. These are discussed in greater detail in the cuttings bed section.

In the vertical section, the cuttings tend to be well mixed with the drilling fluid and move in homogeneous suspension. As the angle builds, the larger cuttings migrate to the low side and move up the well in heterogeneous suspension. (Fig 7-23) At higher angles, the cuttings tend to form beds and roll along the low side of the well. At angles above 65°, the beds tend to be stationary unless disturbed by pipe movement. If the flow rate is high enough however, it may be possible to achieve homogeneous suspension, even at high angles of inclination.

The cutting transport grades from homogeneous suspension in the vertical section to a combination of heterogeneous suspension and bed transport in the build section.

Stationary beds are almost always present in the higher angle sections.

Heterogeneous suspension transport occurs on top of these beds.

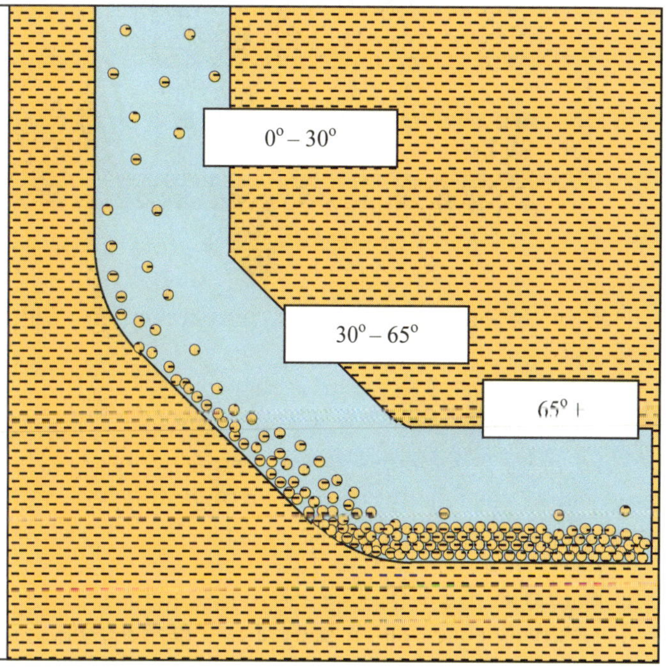

Fig 7-23 Cuttings transport at various inclination angles

Homogeneous suspension is the most efficient transport mechanism. This makes the vertical section the easiest section to clean.

Cuttings rolling or bed transport is the least efficient transport mechanism. This might lead one to believe that the horizontal section is the hardest to clean. However, cuttings will be transported in heterogeneous suspension above the cuttings bed, and will never slide back down the well at angles above 65°, as in the 30° to 65° section.

Because of the combination of sliding beds, Boycott settling, and an asymmetrical flow profile, *the moderate angle hole section is the hardest to clean.*

One problem with inclined wells is that pipe eccentricity and cuttings concentration on the low side of the hole distorts the flow profile. The fluid on the high side of the hole has a much faster velocity than the obstructed flow on the low side. (Fig 7-24) This makes it more difficult for the drilling fluid to impart the energy necessary to move the cuttings. The more elongated profile of laminar flow is much less effective in preventing or cleaning cuttings beds in high angle holes than the flatter profile of turbulent flow.

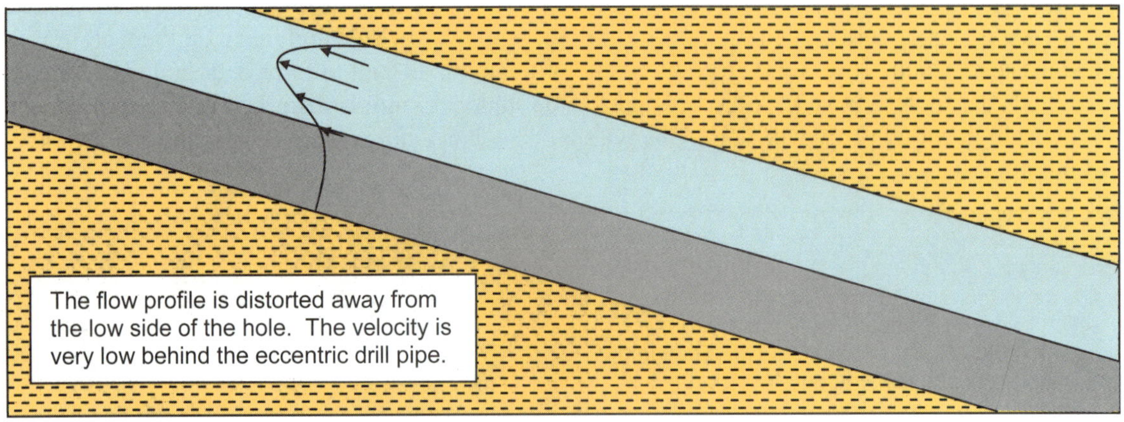

Fig 7-24 Asymmetrical flow profile

The nature of the cuttings beds and pipe eccentricity calls for different flow regimes to clean the well at different angles.

Flow Regime

In vertical wells, a laminar flow regime is preferred. The slip velocities of cuttings are lower with laminar flow and we can readily control mud properties and annular velocity to accomplish adequate hole cleaning. At high angles however, the advantage of laminar flow is nullified because of the formation of cuttings beds.

Laminar flow is less effective at scouring and dislodging the beds than turbulent flow. Also, viscous fluids do not effectively penetrate the packed cuttings beds that exist at high angles.

The studies of Sifferman[11] and Kenny[6] have led to these conclusions regarding the flow regime at various angles of inclination:

- *Laminar flow is desirable for angles less than 45°* because the reduction of slip velocity dominates in vertical wells.
- *Turbulent flow is preferred for angles above 55°* because the need to penetrate the cuttings beds and maximize velocity near the beds dominates in high angle wells. The laminar flow profile in high angle wells does not provide enough shear stress to disturb the cuttings beds unless the fluid is viscous enough for plug flow to be obtained. Turbulent flow is far more effective at disturbing the cuttings beds at angles above 55°. (Fig 7-26)
- *Laminar and turbulent flow are comparably equal in the 45° to 55° range.* A compromise must be struck between limiting the settling velocity of cuttings near the wall and maximizing the velocity near the wall.

The most challenging angles to clean are in the 45° to 55° range. The cuttings and drill pipe lay on the low side of the well. The eccentric drill pipe causes the velocity to be very low in the vicinity of these settling cuttings and cuttings beds. Our goal is to produce a mud that is effective in cleaning under the eccentric drill pipe. To do this, we need a mud with a flow profile that maximizes the velocity under the eccentric drill pipe while minimizing the slip velocity of the cuttings[6].

At these moderate angles, we are also concerned with sliding cuttings beds and, therefore, the minimum annular velocity required to prevent this. This is the minimum transport velocity (MTV) necessary to initiate cuttings rolling. Ford et al. drew the following conclusions regarding the MTV to imitate cuttings rolling[10] at various angles of inclination:

- The MTV required to initiate cuttings rolling increases as angle increases.
- The MTV reaches a maximum value, then decreases as angle increases. The hole angle where this occurs is different for different fluids, but it is approximately 65°.
- The MTV is less dependent on hole angles above 40° than below 40°.
- Annular velocity must increase to limit bed formation as angle increases.

Mud Properties (Hole Cleaning Factors in Directional Wells)

As with vertical wells, *the cuttings concentration always reduces as **mud weight** increases*. However, the effect of buoyancy has a diminishing effect on moving the cuttings in the axial direction as angle increases. Mud weight contributes to hole cleaning in higher angles by slowing down the boycott settling effect and by causing the cuttings beds to be more fluidized and less compacted. The contribution that mud weight makes to momentum transfer remains constant at any angle.

Studies by Becker and Azar at the University of Tulsa demonstrated the effect of mud weight on cuttings bed formation[3]. Here is a summary of their findings:

- The cuttings concentration increased drastically between 35° and 45° at low mud weights, but not so drastically at higher mud weights. (Fig 7-25)
- Cuttings bed height was substantially reduced with small increases in mud weight, at any angle.
- The sliding and avalanching of cuttings beds occurred less frequently with heavier mud.
- Cuttings beds are more fluidized in heavier mud and are thus more easily disturbed.
- The minimum velocity needed to initiate cuttings rolling is less with heavier mud.

In general, an increase in mud weight makes it easier to erode the cuttings bed. This means that at higher mud weights the threshold velocity that limits cuttings bed growth will be less. Thus, the annular clearance will be higher and the bed height will be lower for the same flow rate.

Cutting beds are smaller with heavier mud weight.

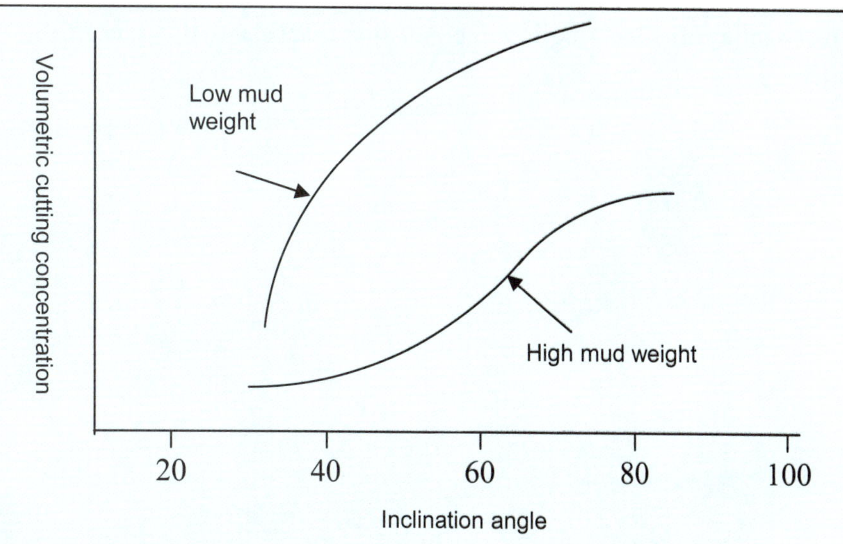

Fig 7-25 Effect of mud weight on cuttings bed height

Yield Point and Viscosity

In a vertical well, an increase in **yield point** will result in better hole cleaning. In a directional well however, an increase in yield point generally has a detrimental effect on hole cleaning. This is partly because *viscous mud cannot penetrate the cuttings beds as easily as less viscous mud. The main reason however, is the distortion of the laminar flow profile.* The eccentric drill pipe pushes the flow profile away from the cuttings beds, which produces a very low velocity across the top of the beds. (Fig 7-26)

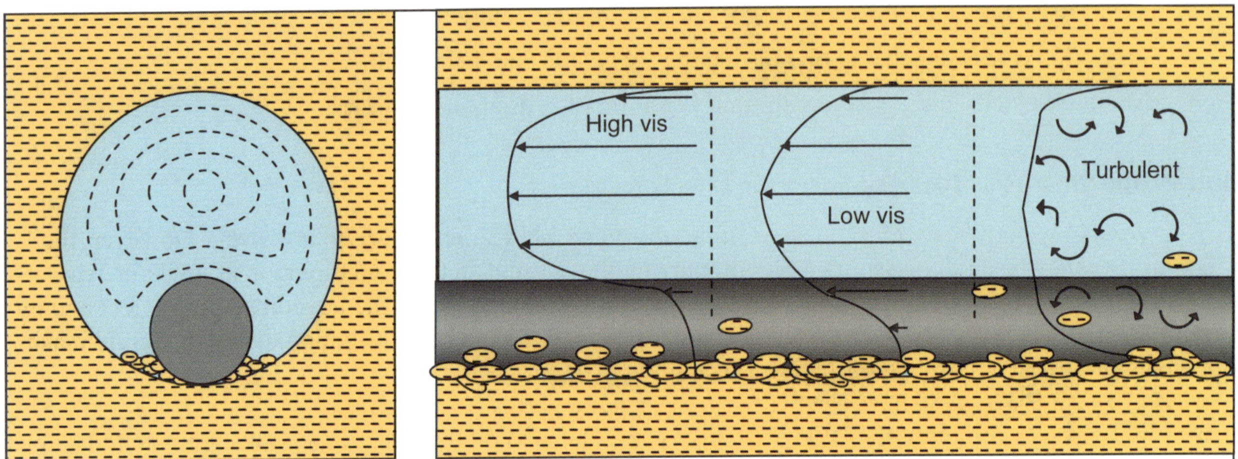

The velocity profile is shifted up away from the drill pipe and cutting beds laying on the low side of the well. This causes the velocity on the low side to be too low to erode the cutting bed. This problem is less pronounced with lower viscosity muds.

Fig 7-26 Velocity profiles in a horizontal well

Kenny, Sunde, and Hemphill[6] argue that less viscous fluids promote higher fluid velocities under the eccentric drill pipe. Referring to the Herschel Bulkley modified power law (equation 7.8), they demonstrate that hole cleaning in high angle hole sections improves as the **flow index (n)** increases, and as yield point and plastic viscosity decrease.

A mud with a high viscosity will cause the flow to divert into the open annulus above the drill pipe, thereby reducing the velocity in the vicinity of the cuttings beds. However if the mud is too thin is has no visco-elastic coupling, and the benefit of pipe rotation is substantially reduced.

Kenny, Sunde, and Hemphill also argue that increasing the fluid velocity under the eccentric pipe isn't the only consideration for hole cleaning at moderate angles. *It is equally important to limit the settling velocity at the low rate of shear that exists near the wall.* Some compromise must be struck between extremely low viscous fluids that promote high velocity near the wall and higher viscosity mud that minimizes slip velocity. (Their findings are summarized in Fig 7-27.)

In Fig 7-27A, *with no pipe eccentricity*, high viscosity mud provides a better cuttings transport ratio than low viscosity mud. The flow profile provides enough velocity near the cuttings beds to disturb the beds and the high viscosity limits cuttings settling. High viscosity mud provides good hole cleaning, even at high angles.

When pipe eccentricity is introduced, such as in Fig 7-27B and 7-27C, the flow is displaced away from the cuttings bed. High viscous mud provides lower velocities above the cuttings bed than low viscosity mud. However, low viscosity mud does not provide a good transport ratio behind the eccentric pipe because the slip velocity is too high at the low rate of shear near the wall.

Moderate viscosity mud provides the best combination of mud velocity behind the eccentric pipe and low slip velocity at low rates of shear.

When the flow rate is too low, as in Fig 7-27C, the high viscosity mud performs even worse. This is because the velocity is reduced even more behind the eccentric drill pipe.

The intermediate viscosity mud performs the best at any angle and at all laminar flow rates when drill pipe eccentricity exists.

Note: The graphs in Fig 7-27 apply to laminar flow only. These calculations do not hold for turbulent or plug flow. <u>These graphs also assume no pipe rotation.</u>

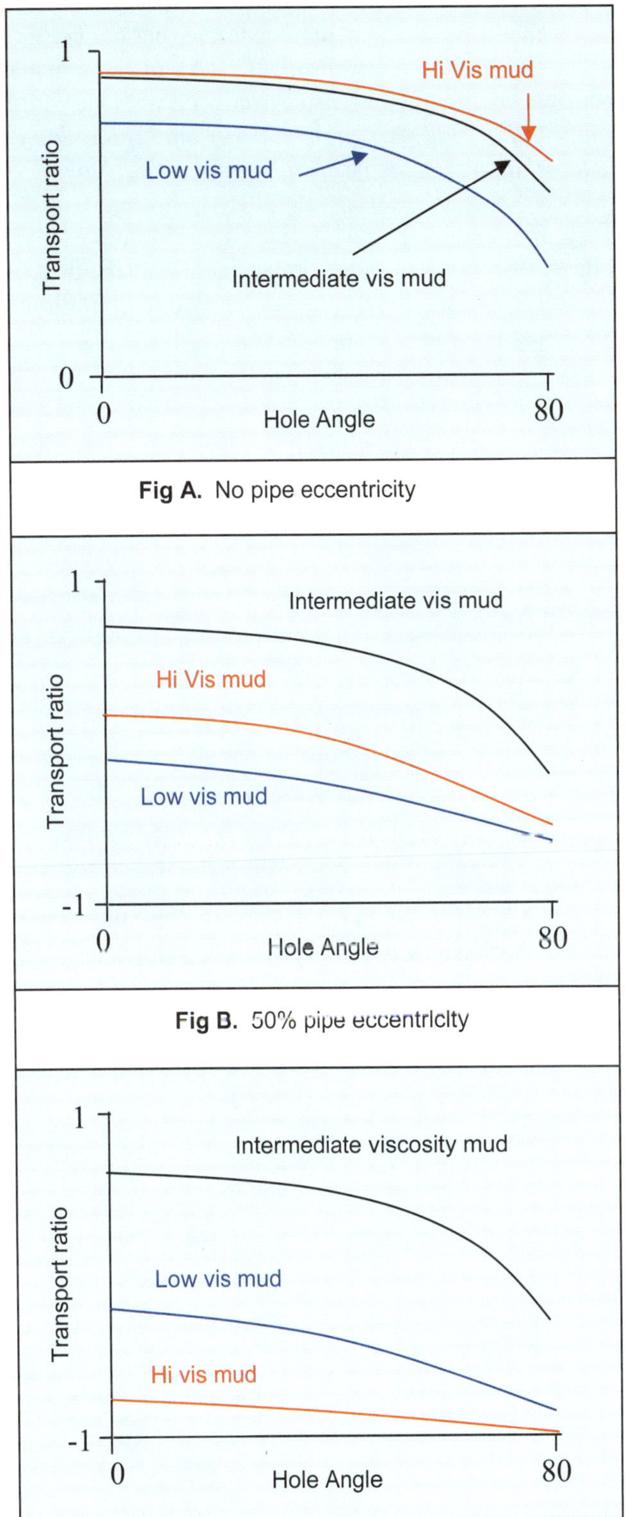

Fig A. No pipe eccentricity

Fig B. 50% pipe eccentricity

Fig C. 50% pipe eccentricity with reduced flow

Calculated transport ratios under the eccentric drill pipe at various eccentricities and flows.

Fig 7-27 Effect of viscosity on cuttings bed height

Sifferman's study[11] also demonstrated that *the height of cuttings beds decreases significantly as viscosity decreases at all angles above 45°*. (All of Sifferman's studies were done at angles above 45°) However, the cuttings bed height can also be reduced with very high viscosity muds, especially at lower angles. *This would suggest that both very high viscosity mud (plug flow) and very low viscosity mud (turbulent flow) would clean better than moderate viscosity mud (laminar flow).*

Turbulent and plug flow regimes are seldom possible in high angle holes, so we are limited to laminar flow. Fig 7-30 shows that, when in laminar flow, the best hole cleaning occurs with moderate viscosities. Very high or low viscous sweeps can be used to compliment the hole cleaning obtained with the best laminar flow possible.

Increasing the viscosity after drilling has stopped is only effective in low angle holes. The cuttings beds will not be disturbed in the high angle sections. The vertical section of the hole will be cleaned and the shakers will come clean, but the cuttings beds will still exist in the higher angle sections.

Here are the conclusions of a number of studies concerning viscosity in directional wells:

- Water in turbulent flow provides the best possible hole cleaning at inclinations above 65°.
- In the absence of pipe rotation, cuttings beds are always present in laminar flow no matter how high the flow rate is.[11]
- Cuttings beds do not exist in turbulent flow.
- A compromise must be struck between maximizing velocity behind the eccentric drill pipe and reducing the slip velocity at the low rates of shear that exist between the drill pipe and cuttings bed.[6]
- A change in rheology has less effect when the pipe is rotating[11]. This is because pipe rotation adequately disturbs cuttings beds.
- Pipe rotation is required more with high viscosity muds than with low viscosity muds.
- The viscosity effect is more pronounced with WBM than OBM.
- Cuttings beds tend to slide more with oil base muds than with water base muds.[11]

Shear Thinning

Another mud property to consider is **shear thinning**.

A shear thinning mud will have a lower viscosity near the wall, where the shear rate is highest, and a higher viscosity in the main body of flow, where the shear rate is lowest. (Fig 7-28) The more shear thinning muds also tend to have a higher shear rate near the wall and a flatter velocity profile in the center of the stream. Thus, the more shear thinning a mud is, the more likely it is to be in plug flow.

The more shear thinning a mud is, the better it will clean beneath the eccentric drill pipe.

The shear rate is very high through the nozzle jets, so viscosity is lowered as mud squirts through the jets. Recent PWD data suggests that it takes time for the viscosity in some muds to build back up again.[11]

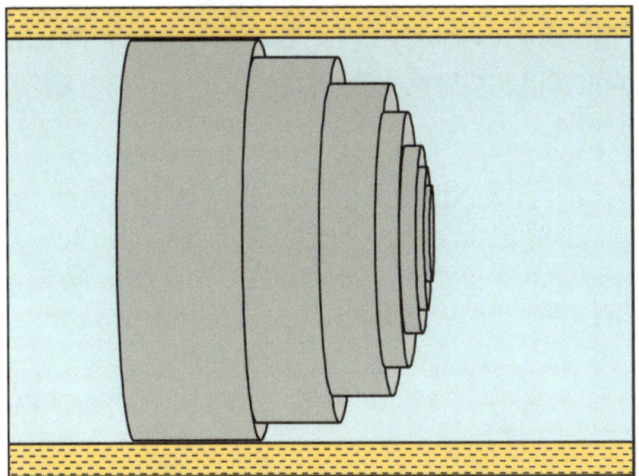

The speed at which layers of mud slip or "shear" past each other is fastest at the wall and slowest near the center of the flow area.

Fig 7-28 Shear rate

This means that the viscosity of these muds will be lowest in the annulus near the bit and become progressively higher farther up the well. This results in excellent scouring of the cuttings beds near the bit and weaker performance further away from the bit. Thus, when pulling out of the hole, we may not notice much drag until the collars are pulled up into the cuttings beds left by the more viscous mud.

The shear thinning properties of a mud decrease as the solids content increases. Therefore, the higher the YP/PV ratio, the more shear thinning the mud will be. For laminar flow in both high angle and vertical wells, an increase in the YP/PV ratio results in better hole cleaning.

Visco-Elastic Coupling

The final mud property I want to talk about is visco-elastic coupling. **Visco-elastic coupling** refers to a muds ability to stick to the drill pipe such that it imparts a spiral flow in the annulus when the pipe is rotating. The more visco-elastic the mud is the thicker the layer of mud will be that is spiraling around the pipe as it rotates. The mud will have a higher angular velocity around the pipe so that it can scour cuttings off the wall and out of cuttings beds.

Paint is an excellent example of visco elasticity. By design we want paint to stick to a surface. Think of a 5 gallon bucket of paint. If you stick a one inch diameter pipe on the end of a variable speed drill into the paint you will notice the entire bucket of paint spiraling around the pipe. The flow is almost the same speed as the pipe where it is closest to the pipe, and it gradually loses velocity as the distance from the pipe increases.

The more visco-elastic our mud is the better hole cleaning we will get as we are rotating.

Flow Rate (Hole Cleaning Factors in Directional Wells)

Annular velocity is generally regarded as the most influential factor affecting hole cleaning in a directional well. As previously noted however, the momentum transferred from the flowing mud to the cuttings is dependent on mud weight. *At lower mud weights, more annular velocity is required for adequate hole cleaning.*

To truly understand hole cleaning in a directional well, we must understand the difference between *flow rate* and *annular velocity*. The two terms are often used synonymously, but there is one important difference: *Annular velocity is a function of flow rate divided by cross sectional area.* The actual velocity depends on the size of the cross sectional area and the proximity of the mud to the wall.

$$v_a = \text{annular velocity} = \text{flow rate/cross sectional area of the annulus} \qquad \text{eq. 7.10}$$

In a vertical well, the cross sectional area of the annulus remains relatively constant with respect to flow rate. In high angle sections however, *the cross sectional area fluctuates with flow rate!* This distinction causes the flow rate to have an interesting effect on annular velocity throughout the well bore.

In a vertical well, the average annular velocity is dependent on flow rate. In a directional well, the average annular velocity above the cuttings beds is generally constant regardless of flow rate! This is because cuttings beds will be deposited until equilibrium is reached between deposition and erosion of the bed.

Cuttings will settle to the low side of the hole and form cuttings beds unless a critical "**threshold**" annular velocity is provided. This threshold velocity is the velocity that is just high enough to prevent cuttings deposition. An annular velocity high enough to prevent any cuttings deposition whatsoever is often unattainable, due to pressure or volume limitations. However, as cuttings settle and form cuttings beds, the annular space is reduced so the local annular velocity is increased. Eventually, the local velocity reaches the threshold limit and equilibrium is reached between bed deposition and bed erosion. (Fig 7-29) *In the absence of pipe rotation, an equilibrium bed height will be established, regardless of the amount of cuttings or flow rate*!

Adjusting the flow rate will not alter the local threshold velocity. The bed height will readjust to provide an annular clearance just large enough to provide this threshold velocity.[7,8]

A critical threshold velocity exists for each mud type that is just high enough to prevent further deposition of cuttings beds. As the cross sectional area of the annulus is reduced by cuttings bed deposition, the velocity must increase. Once the threshold velocity is reached, no further bed deposition will occur. If the flow rate is increased, the beds will erode until the threshold velocity is re-established. The threshold velocity varies with inclination angle, mud properties, and type of cuttings generated.

If any cutting bed exists, the local annular velocity is constant. If the flow rate increases, the bed will erode, causing the flow area to increase. The increase in cross sectional area causes the velocity to decrease back to the equilibrium state.

Velocity = flow rate/cross sectional area

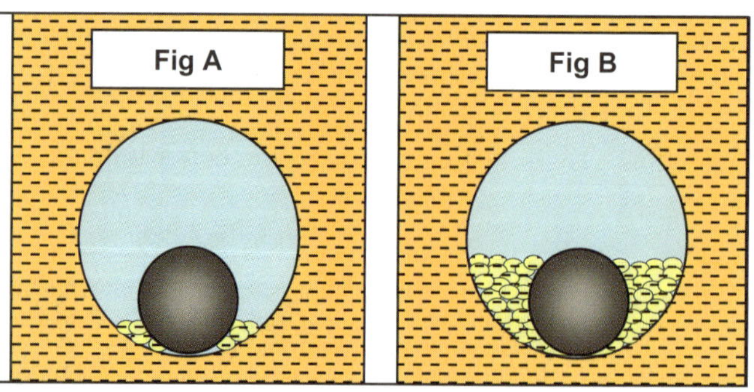

Fig 7-29 Velocity threshold

Obviously, with higher flow rates the threshold velocity will be attained with smaller cuttings beds. So, cuttings bed height decreases linearly with an increase in flow rate. (Fig 7-30A) The hole cleaning ratio[8] H/H_{crit} also increases linearly with flow rate. (Fig 7-30B)

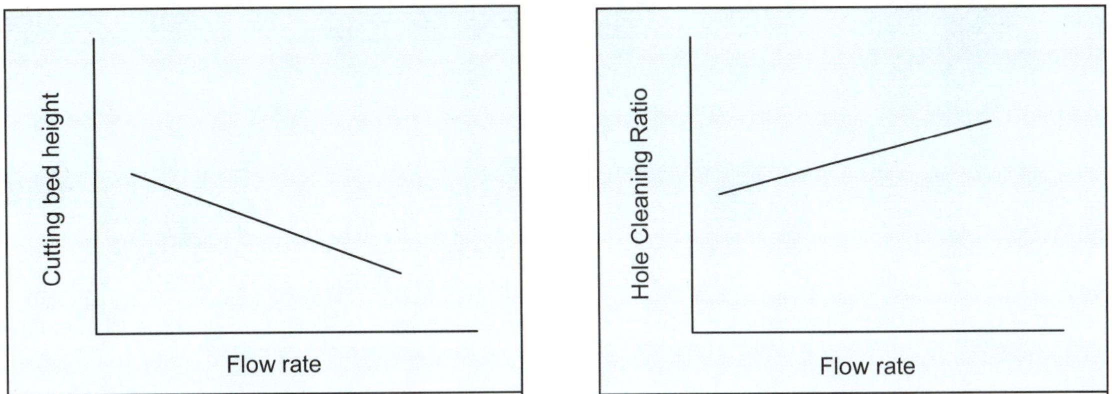

The cuttings bed height decreases linearly with an increase in flow rate

Fig A

The Hole Cleaning Ratio increases linearly with an increase in flow rate.

Fig B

Fig 7-30 The effect of velocity on cuttings bed height

The threshold velocity to prevent bed deposition is the Minimum Transport Velocity[10] necessary to initiate cuttings suspension. As previously mentioned, the MTV is defined as the velocity required to initiate cuttings transport. Or, in terms of threshold velocity, the MTV for suspension is the velocity that begins to erode the cuttings bed.

Here are the conclusions from Ford, Peden, Oyeneyin, Gao, and Zarrough's work[10] regarding the MTV for cuttings suspension:

- The MTV for suspension increases as cutting size increases and decreases as mud weight increases.
- Pipe rotation decreases the MTV for viscous fluids, but has no effect with water. (This is because there are no cuttings beds with water in turbulent flow.)
- The MTV to initiate cuttings suspension is less dependent on rheology than for cuttings rolling.
- The MTV is highest for medium viscosity and lowest for high viscosity; water is in between.
- The MTV is reduced for both transport mechanisms as viscosity is increased. However, it is lower for both mechanisms with water.

Here are some conclusions regarding annular velocity from Ford and Sifferman:

- AV and MW have significant effect on bed height; rheology and eccentricity have a small effect.
- Cuttings slide down the bed at low AVs in angles under 45 degrees.
- An increase in AV is less effective in laminar flow than in turbulent flow because shear thinning leads to faster settling, but is offset by more rapid axial transport.

Cuttings and Cuttings Beds (Hole Cleaning Factors in Directional Wells)

Cuttings beds are formed during periods of low or no pipe rotation, which is often the case in directional drilling. (Fig 7-31)

In the absence of pipe rotation a cuttings bed will be deposited. An equilibrium bed height will adjust itself to establish the threshold velocity that limits bed growth. The beds will then grow in length like a river delta.

Fig 7-31 Equilibrium bed height

When the BHA is pulled through these cuttings beds, they are deformed to form a "hill" of cuttings that piles up in front of the bit and stabilizers. When the bed height is small, this "hill of cuttings" reaches a constant height and a steady overpull is experienced as the BHA passes through it. (Fig 7-32). When the bed height is too high, the hill of cuttings grows into a plug that leads to sharply increasing overpulls and packoff.

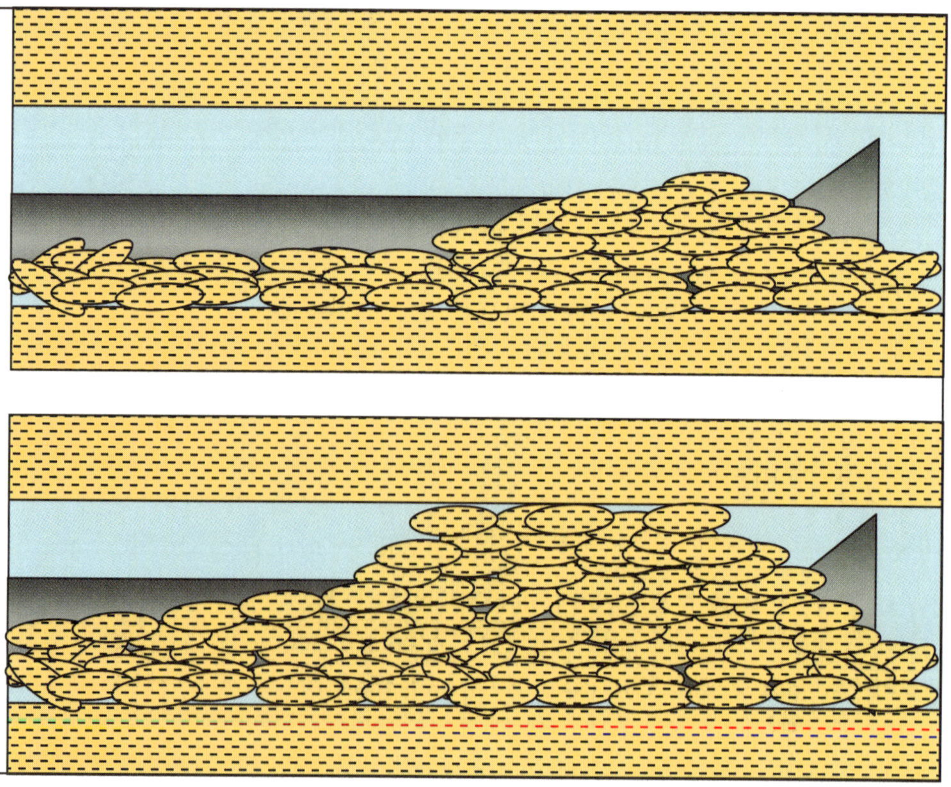

As the bit "ploughs" through the cutting bed the bed deforms to form a hill.

When the bed is small cuttings can squeeze through the open area around the bit.

When the cutting bed is too thick not all of the cuttings can squeeze past the bit.

The "hill" of cuttings grows into a plug that leads to sharply increasing overpulls and a packoff.

Critical Bed Height and the Hole Cleaning Ratio

The **Hole Cleaning Ratio (HCR)** proposed by Marco Rasi[8] is a method of evaluating hole cleaning efficiency using cuttings bed heights. Rasi argues that there is a maximum allowable cuttings bed height that a BHA can be moved axially through without getting stuck. This bed height is known as the **critical bed height (H_{crit}).** The height of annular space above the bed is called *free region height* (**H**). (Fig 7-33) This height provides the annular clearance that produces the equilibrium-producing threshold velocity. The free region height varies with fluid properties and flow rate. The H_{crit} depends only on the cross sectional area of the largest component in the BHA, usually the bit. (Fig 7-33)

The ratio of the height of annular space above the cutting bed to critical cutting bed height is called the **H**ole **C**leaning **R**atio. HCR = H/Hcrit If this value is greater than 1 we can pull the BHA without circulating . If not we can expect trouble.

Fig 7-33 Critical cuttings bed height

There must be enough annular clearance around the BHA and bit to allow the cuttings beds to pass by it as the bit is drug along the well bore. If the cross-sectional area of the bit is greater than the area of the open space above the cuttings bed, then the string cannot be moved axially. *The ratio A_{bit}/A_{open} must be less than 1 or the volume of cuttings in the bed cannot pass by it.* (Fig 7-34)

The cross sectional area of the open space above the cutting bed must be greater than the area of the bit or the bit can not be moved axially through the bed. In other words the cross sectional area of bit has to be able to fit within the open area above the cutting bed. Another way to look at it is to say the area of the cuttings bed must be less than the open area around the bit.

Fig 7-34 Critical cuttings bed height

The **Hole Cleaning Ratio** is the ratio of the height of annular space above the cuttings bed to the critical height of the cuttings bed.

$$HCR = H/H_{crit} \qquad \text{eq. 7.11}$$

Marco Rasi proposes that if the height of the free region above the cuttings bed is greater than the critical bed height, we will be able to pull through the cuttings bed without circulating. If $H/H_{crit} > 1$, we won't have problems. If $H/H_{crit} < 1$, we can expect problems.

From a study of 50 large diameter directional wells in the North Sea,[8] Marco observed that when the HCR was greater than 1.1, no stuck pipe incidents occurred. When the HCR was less than 0.5, stuck pipe always occurred.

As the HCR decreases, the tendency to become stuck increases.

As bed height increases, the annular space above the bed decreases. The larger the BHA, the smaller the cuttings bed must be to pull through it. In general, overpulls tend to increase as the BHA diameter increases or drill pipe diameter decreases. Our drill string, bit, and stabilizer selection should take these factors into account.

Estimating Bed Height

The bed height can be estimated mathematically through an iterative, trial-and-error procedure suggested by Rasi. The accuracy of this approach is questionable, however.

The bed height can be more accurately estimated by measuring the total amount of cuttings removed from the well. It can also be estimated by monitoring the total mud volume, assuming no lost circulation.[9]

The amount of hole being created must be filled with mud. The buildup of cuttings beds causes the surface volume to decrease less than expected. When cuttings beds are removed by intermittent pipe rotation, the surface volume is usually observed to reduce accordingly (Fig 7-35).

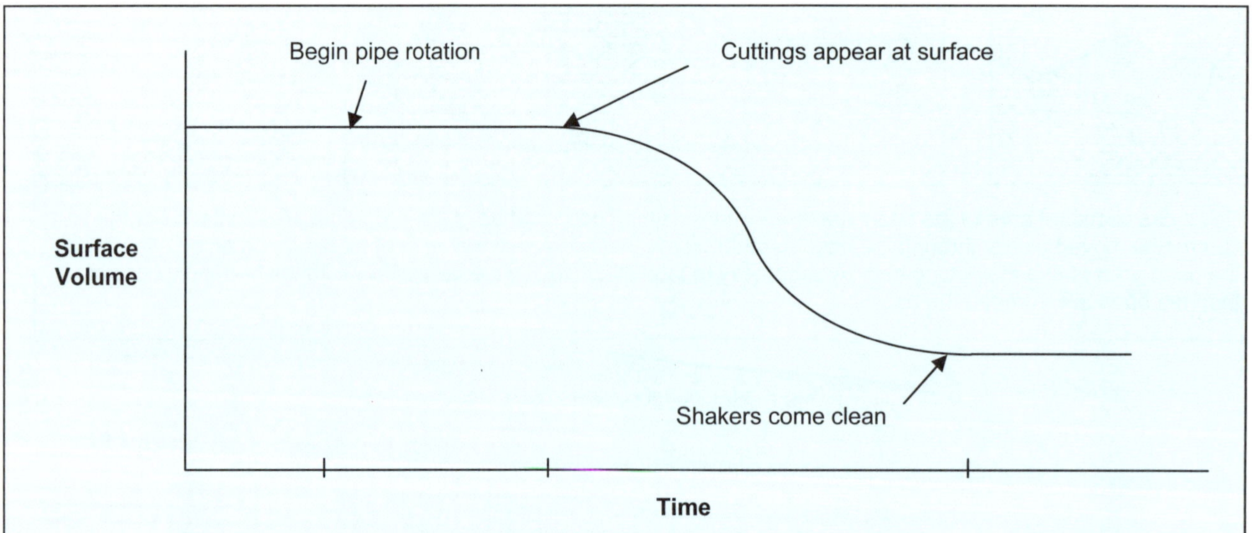

When cuttings beds are disturbed by pipe rotation the active surface volume will reduce as these cuttings are removed from the well.

Fig 7-35 Cuttings bed height vs. surface volume

Rig hands will also monitor drag trends while tripping to estimate the bed height. If overpulls continue to increase, or increase sharply, it is an indication that the beds are too thick. Torque trends while drilling also give an indication of bed height.

The three regions of cuttings bed formation

There are three distinct regions of cuttings bed formation. (Fig 7-36). **At very low angles**, cuttings are recycling on the low side of the well. The cuttings concentration is very high on the low side, but the cuttings are in heterogeneous suspension and not forming beds.

At moderate angles, the cuttings are forming beds on the low side but they are highly fluidized and easily disturbed. These beds will slide down the well if circulation is stopped. In fact, the beds may continuously slide and avalanche down the well, even while circulating, if the flow rate is to low.

At higher angles of inclination, the cuttings beds are well packed and tend to be stationary.

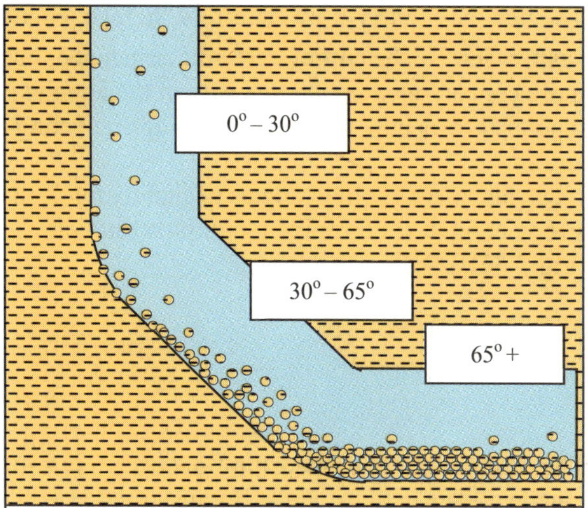

The three distinct regions of cuttings bed formation

Fig 7-36 Three distinct regions of cuttings beds

Cutting beds begin to form somewhere between 30° and 35°. Beds become stationary with pumps off somewhere between 55° and 65°.

Cuttings beds will form after rotation has stopped. They will reach an equilibrium height and then grow in length just like a river delta.

Ford et al.[10] identified seven distinct patterns of cuttings transport in high angles, which are described below. Each of the seven patterns falls into one of two general transportation mechanisms, *suspension* and *bed movement*. In suspension, the cuttings are suspended in the fluid as they move along the well bore. In bed movement, the cuttings are in contact with the low side of the well bore. The cuttings move along the well bore at a much faster pace in suspension flow, so this is the desired mechanism, but it can be difficult to obtain at higher angles of inclination.

Suspension Transport

The best mode of cuttings transportation is "**homogeneous suspension,**" where the cuttings suspension is uniformly distributed across the annular space. This is the type of transport expected with small cuttings in the vertical section. This can also be achieved with turbulent flow and pipe rotation at high angles of inclination. (Fig 7-37A)

The next best slurry flow pattern is "**heterogeneous suspension**". The cuttings are still suspended, but are more concentrated on the low side of the annulus. This type of flow occurs with heavy cuttings recycling at low angles.
It can also occur at high angles of inclination if the flow is strong enough and/or mixed with pipe rotation.
(Fig 7-37B)

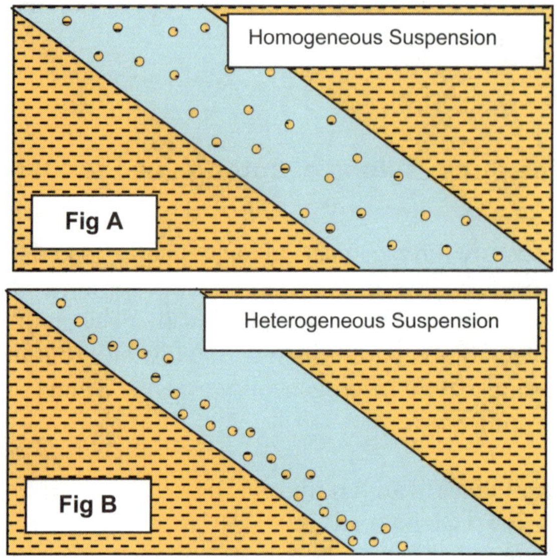

Fig 7-37 Cuttings transportation

The next pattern of suspension flow occurs in **sand clusters.** The cuttings are still transported in suspension, but in clusters, with all the cuttings in the cluster moving at roughly the same velocity. (Fig 7-38A)

The last pattern of suspension flow is referred to as **saltation**. Here, the cuttings are bouncing along the low side of the well bore as they are partially suspended in the flow stream. This pattern grades from predominately suspension to predominately bed contact. (Fig 7-38B)

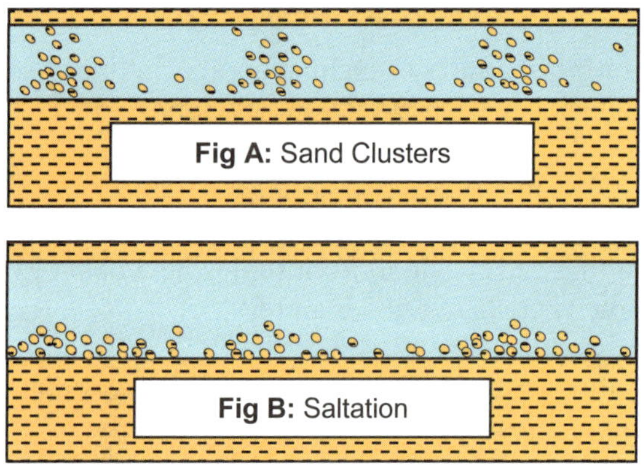

Fig 7-38 Cuttings transportation

All of the suspension forms of cuttings transport are more efficient than bed transport! In suspension flow, the cuttings move along at some percentage of mud velocity. In bed transport patterns, the cuttings roll or slide along the bottom of the well at a much slower velocity than the suspended cuttings. Without continuous pipe rotation, the cuttings cannot be transported along the well as fast as they are generated.

When pipe rotation is stopped, a cuttings bed will grow in size until a threshold velocity that supports suspension flow exists. Once this threshold velocity is established, additional cuttings will be transferred primarily in suspension, regardless of cuttings bed height. The initial cutting bed is formed a short distance away from the bit. *After the critical bed height is reached the bed cannot grow any higher but it does grow longer, just like a river delta.* It is not just the height but the length of the cutting bed that gives us trouble.

Bed Transport

There are three primary patterns of bed transport (Fig 7-39):

- **Dune transport:** Sand on the top of the bed rolls over stationary sand below it in a "leap frog" fashion. The general appearance is that the entire dune is moving forward.
- **Continuous moving bed:** A thin bed rolls or slides along the low side. All the cuttings move forward, but at different velocities.
- **Stationary bed:** A thicker bed is formed where the cuttings near the top of the bed roll forward, but cuttings inside the bed remain stationary. This is the least efficient pattern of cuttings transportation.

Stationary beds tend to form at angles above 65°. The dune and continuous moving beds are more common in the moderate angle sections. At these moderate angles, the cuttings beds must be supported with enough flow to prevent sliding and avalanching.

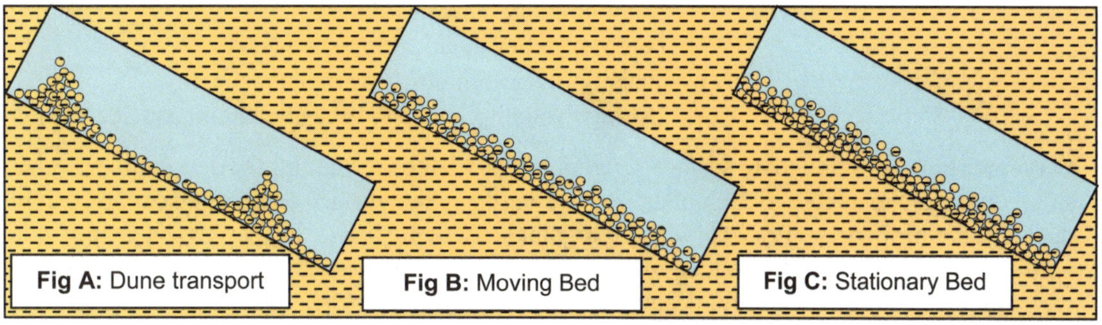

Fig 7-39 Bed transportation

Note that once an equilibrium bed height is established cuttings will be transported in suspension above the cuttings bed. If more cuttings are injected, they will be carried along the well bore. If drilling is stopped, the cuttings bed will not be reduced in height. The shakers may eventually come clean, but the cuttings beds will remain in place. It is only with a combination of flow rate and pipe rotation with visco-elastic mud that we can adequately clean the well.

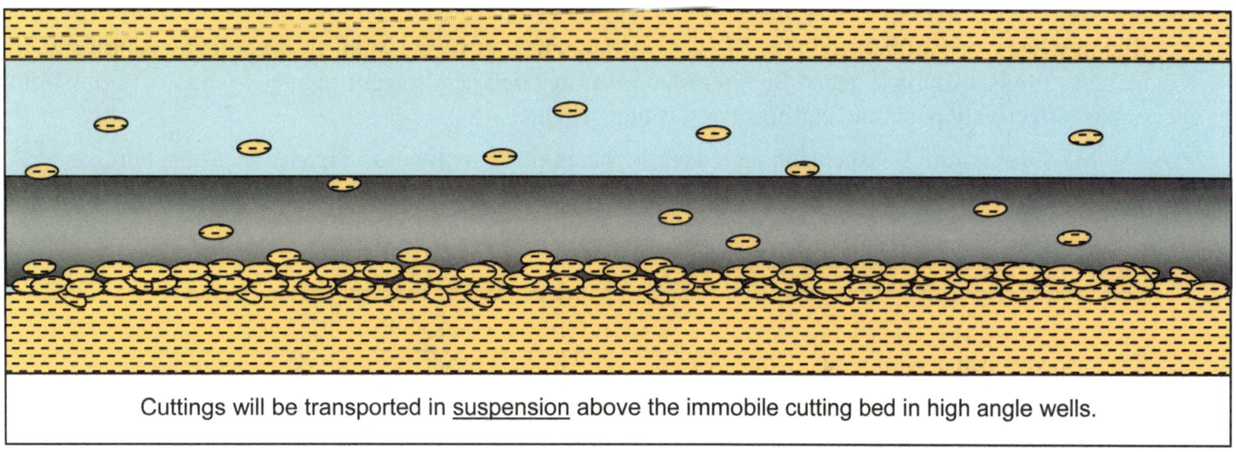

Fig 7-40 Typical cuttings transportation in high angle wells

In a vertical hole, the rate of penetration can be slowed down or stopped until the well bore cleans up. In the inclined sections of the well, the cuttings bed will not be reduced appreciably when drilling stops. This does not imply that drilling should continue if hole cleaning problems are evident. Reducing the rate of penetration may still clean the vertical portion of the well, but better hole cleaning methods may be called for in high-angle sections.

As previously mentioned, there is a Minimum Transport Velocity required to initiate cuttings transport in each mechanism. There is one MTV for initiating the movement of cuttings beds and another for initiating cuttings suspension. *The threshold velocity that prevents cuttings bed growth is the MTV for cuttings suspension.* Studies regarding the MTV to initiate transport have generated the following conclusions regarding cuttings beds.[7,3,8,10,11]

- Cuttings density and size affects the MTV and thus the depth of the cuttings bed. A minor reduction in cutting density can substantially reduce bed thickness.
- Large cuttings are more easily torn away from the cuttings bed than very small cuttings. This is probably because large cuttings feel a higher shear stress from the mud. Shear stress is a function of the fanning friction factor, which is determined by surface roughness and rheological properties.

  (Shear stress $\tau = f\rho v^2/2$)[8] eq. 7.12

  where: τ = shear stress in Newtons/meter^2
  f = fanning friction factor (dimensionless)
  ρ = density of fluid in kilgrams/meter^3
  v = average velocity above cuttings bed in meters/second

- Larger particles provide more surface roughness, thus a higher fanning friction number. Also, very small cuttings can adhere to each other very strongly. Their surface area to mass ratio is so high that the adhesive properties of fluid glue them together. Like the barite that settles out in the bottom of a mud pit, these silt size cuttings tend to form a pliable rock-like sediment that resists erosion. The barite on the bottom of a pit often cannot be jetted out, it must be shoveled out. The barite remains bonded together in large chunks as it is tossed from the pit. These silt size particles can bond together tightly enough that a 6" or larger opening is needed at the bottom of the tank to prevent the opening from being plugged by particles only microns in diameter!
- In a vertical well, large cuttings migrate to the wall and recycle, making them harder to remove. In a directional well, large cuttings are easier to tear off the cuttings beds, making them easier to remove. Pipe rotation is required to disturb cuttings beds, especially ones with small cuttings. Pipe rotation has less impact when the cuttings are larger. With water in turbulent flow, pipe rotation may not be required at all.
- Rheolgy has more influence on the MTV to initiate cuttings rolling than cuttings suspension. An increase in viscosity will generally not reduce cuttings bed height significantly because viscous fluids do not effectively penetrate cuttings beds at high angles.
- The MTV for cuttings suspension increases as the angle of inclination increases. Thus, annular velocity must increase to limit bed formation as angle increases. Bed thickness always increases as angle increases, so the local annular velocity at high angles will be higher than the velocity at lower angles with the same flow rate. (Fig 7-41)

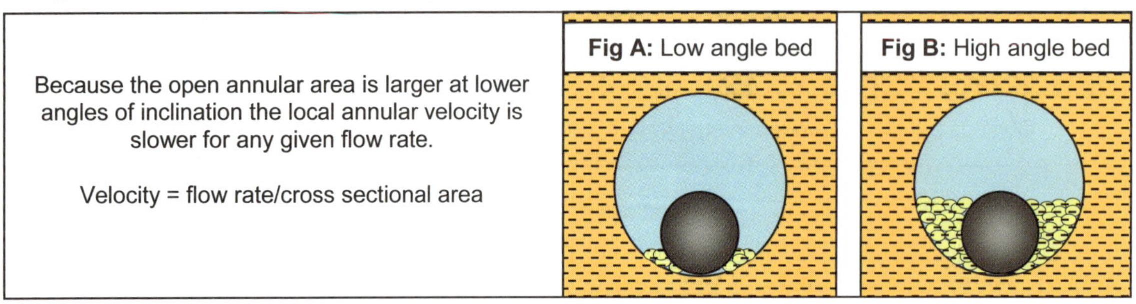

Fig 7-41 Annular velocity vs. hole angle

An increase in bed thickness is normally detected by an increase in torque. This is not always the case. When the bed slides around angles of 45°, torque tends to decrease. In some cases, cuttings beds have been found to lubricate the pipe, causing torque to decrease.[3]

If it becomes necessary to stop and circulate the well during a trip out, the circulation rate should be at least as high as it was to drill the well. Cuttings disturbed from stationary beds above 65° will be transported into the moderate angles where they can avalanche down the well and pack off the string.

Cuttings beds are the cause of most drilling problems in directional wells. They lead to stuck pipe and lost circulation due to packoffs. This is why they have been the focus of many hole-cleaning studies concerning directional wells. Here are some generalities gathered from some of these studies:

- Without pipe rotation, cuttings beds are almost certain to exist in high angle holes.
- Under normal conditions, as much as half the annular area can be filled with a cuttings bed.
- Cuttings beds are more likely to form in large diameter holes. This is due to slower velocities, including especially low velocities under the eccentric drill pipe.
- Cuttings beds do not usually cause problems while rotating. It is when the pipe is moved axially that it may become stuck.
- Large cross-sectional components in the bottom-hole assembly require very thin cuttings beds if they are to be moved axially without circulating.
- ROP has less effect on cuttings bed size than other factors, such as mud weight, pipe rotation, and flow rate.
- Cuttings beds become thicker and more packed as angle increases.
- Cuttings beds become thicker and more packed as mud weight decreases.
- Cuttings beds become thicker and more packed as flow rate decreases.
- When the hole is not being cleaned, the first noticeable warning sign is an increase in torque. Another trend that can be monitored is total active volume (Fig 7-35); surface volume will decrease as cuttings are removed from the well.

Rate of Penetration (Hole Cleaning Factors in Directional Wells)

The rate of penetration influences the size and amount of cuttings. However, the amount of cuttings has no affect on bed height. The cuttings bed reaches a steady state height regardless of the rate of penetration. The threshold velocity that limits bed growth does not change as the rate of cuttings production changes, so the equilibrium height of the bed cannot change. As more cuttings are generated, they are transported in suspension above the beds. The rate of penetration may have an effect on hole cleaning in the lower angle and vertical sections of the well, however.

If drilling is stopped, the vertical section of the well will be cleaned, but cuttings bed height in the high angle sections will remain unchanged. Monitoring the shale shakers for hole cleaning is misleading in this case. The shakers will come clean once all of the suspended cuttings are out of the well, but the cuttings beds remain intact. Pipe rotation or other measures must be taken to disturb the cuttings beds so they can be flushed into the vertical section and out of the well.

Eccentricity (Hole Cleaning Factors in Directional Wells)

Pipe eccentricity has very little effect in a vertical well. However, it has a significant effect in high angle wells. This is because of the effect it has on the flow profile. (Fig 7-42)

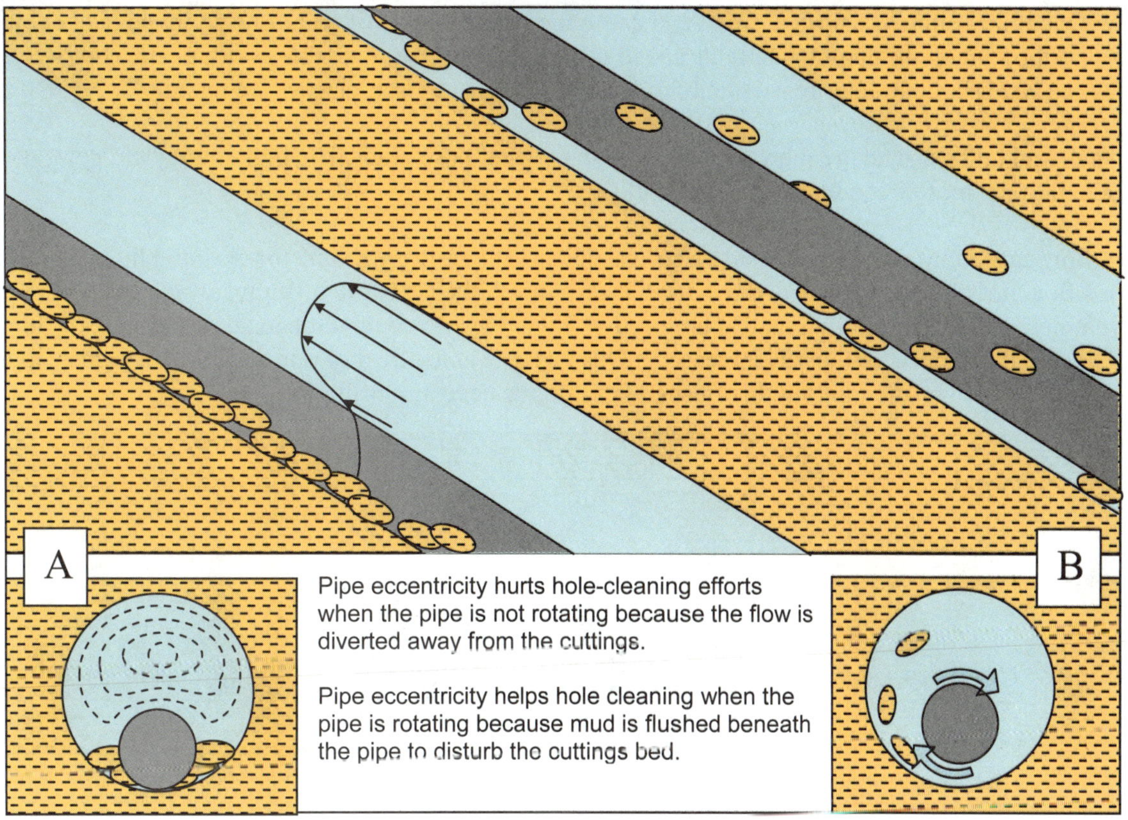

Fig 7-42 Effect of pipe eccentricity on cuttings beds

In inclined wellbores the pipe is laying on the low side of the hole. If the pipe is not rotating the flow profile is shifted away from the pipe and the cuttings bed. (Fig 7-42a) These effects are more pronounced with laminar flow than turbulent flow.

If the pipe is rotating visco-elastic fluid is dragged beneath it to disturb the cuttings bed. The flow becomes helical and the profile is flatter. The pipe eccentricity, combined with pipe rotation and visco-elastic coupling creates a much higher fluid velocity across the cuttings bed. (Fig 7-42b)

Pipe eccentricity hurts us when we are not rotating but helps us when we are rotating.

Eccentricity is also more effective with a more visco-elastic mud.

Visco-elastic coupling (Hole Cleaning Factors in Directional Wells)

Pipe rotation works in conjunction with *visco-elastic coupling* of the mud to the pipe. The more visco-elastic the mud is the more it tends to stick to the pipe and whirl around with the rotating pipe. This creates a spiral flow of mud in the well. (Fig 7-43) A good example is paint in a 5 gallon bucket. Paint is visco-elastic by design. We want it to stick to a surface. You can cause the whole bucket of paint to spin in the bucket by inserting a rotating bit on the end of a variable speed drill. If you do the same thing with a bucket of water the bit spins in the water but the water remains motionless. Water is not visco-elastic like paint.

Polymer mud and oil base mud are more visco-elastic than water base mud. The higher the yield point of water based mud the more visco-elastic it becomes.

The spiral flow caused by pipe rotation and visco-elasticity is the primary means for cleaning high angle wells. This flow not only washes away the cutting beds beneath the eccentric drill pipe, it increases the actual annular velocity. Since the spiraling mud is now following a longer path, and the axial component of flow has not changed, the annular velocity is increased. This has the same effect of pumping with a higher flow rate. The flow profile is flatter and fluid is flushed beneath the eccentric drill pipe.

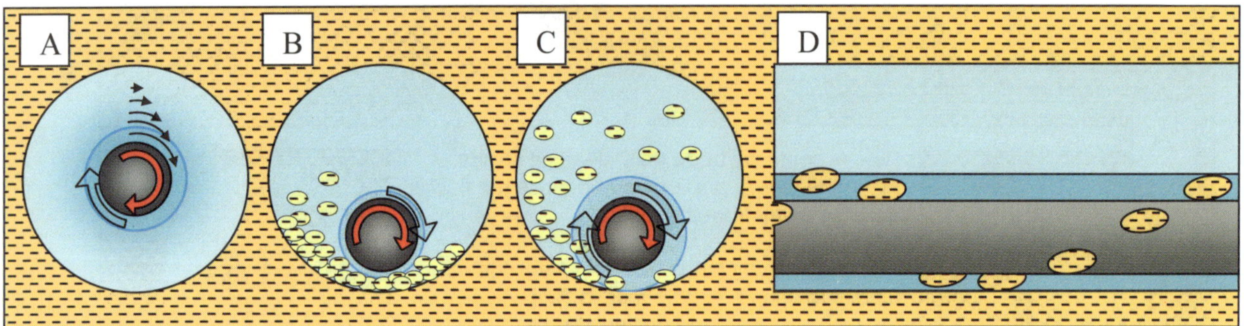

The spiral component of annular velocity increases with increasing viso-elasticity.

Fig 7-43 Visco-Elasticity

There is a greater spiral component of flow around the rotating pipe with a more visco-elastic mud. This is demonstrated in the above figure. There is more efficient hole cleaning in figure 7-43c than in 7-43b because the mud in fig 7-43c is more visco-elastic. The layer of fluid sticking to the pipe and rotating with it is thicker in the more visco-elastic mud.

Pipe Rotation (Hole Cleaning Factors in Directional Wells)

Pipe rotation has some effect in vertical holes by helping to drag cuttings away from the wall. It has a significant effect at high angles of inclinations by tearing up cuttings beds. In fact, pipe rotation may well be the single most important factor in cleaning a high angle hole. It is basically the combination of flow rate, visco-elastic mud, and pipe rotation that cleans high angle wells.

In order for pipe rotation to be effective at disturbing the cuttings beds, it must first reach a "threshold" rotational speed. At high angles and low rotary speeds, the drill pipe rolls up the side of the well and slides back down [3]. As rotary speed increases, but remains below the threshold rpm, the pipe climbs farther up the wall before sliding back down. Cuttings beds will dampen or eliminate this "rolling" motion of the drill pipe at low rotary speeds because the pipe is held in place by a cradle of cuttings. The pipe must reach a "critical" rpm to disturb the cuttings bed. (Fig 7-44)

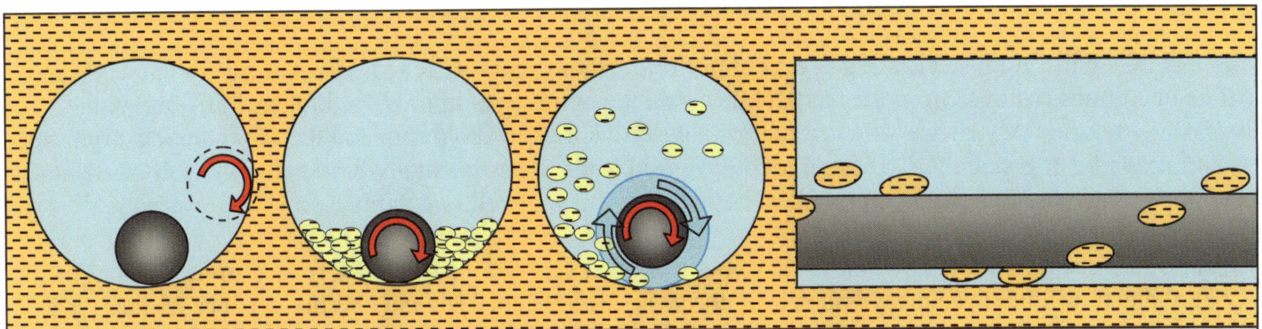

At low rpms, the drill pipe rolls up wall and slides back down. A cuttings bed acts as a cradle to dampen this motion. Once a "threshold" rpm is reached, the cuttings bed is disturbed and can be flushed from the wellbore.

It is a combination of pipe rotation and visco-elastic coupling of the mud to the pipe that disturbs the cuttings bed.

Fig 7-44 Threshold RPM

A lot of research has been done to determine the "critical" rpm required to disturb a cutting bed. If this critical rpm is not reached then the cutting bed is not disturbed and there will be little or no hole cleaning at all. Once the critical rpm is reached there is a step change in hole cleaning. (Fig 7-45)

There is some contention as to how fast the pipe must rotate to achieve this critical rpm. Various best practices recommendations suggest the critical rpm occurs between 100 and 120 rpm. However, if the mud is sufficiently visco-elastic, and well bore tortuosity is minimum, the critical rotational speed will be much lower. It is a combination of wellbore geometry and visco-elastic properties of the mud that determine the minimum rotational speeds. (Fig 7-46)

Typical recommendations for rotary speeds in extended reach wells:

- 75-80 rpm for 8 ½" diameter
- 100 to 120 rpm for 12 ¼" holes or larger
- 150 to 180 rpm for 12 ¼" and 17 ½" wells.

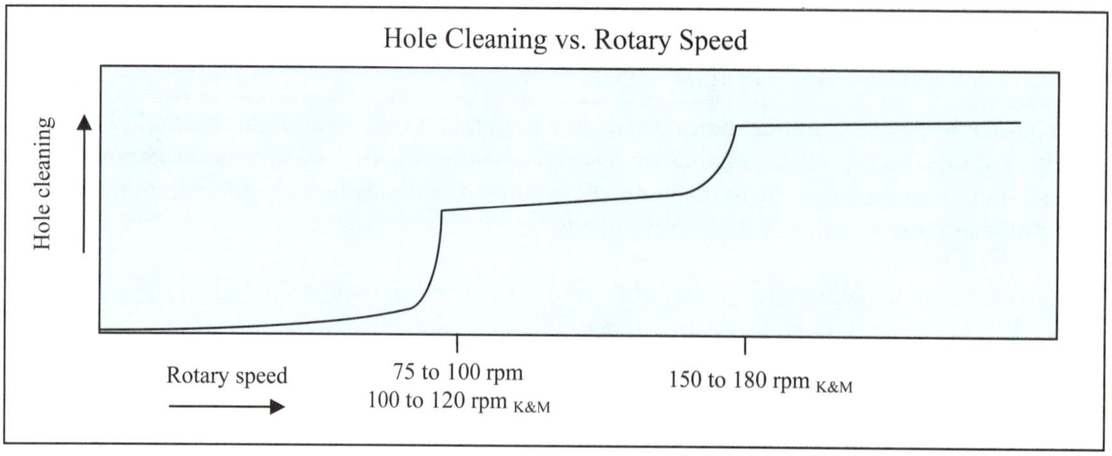

Fig 7-45 Threshold RPM

Note that no significant hole cleaning occurs until the threshold rpm is reached. Then there is a sudden step change in cuttings removed from the well. Hole cleaning continues to improve at higher rpm's but with diminishing returns. My advice is to get comfortably above the threshold rpm and then find the ideal rpm for the optimum rate of penetration. Excessive rotary speed leads to detrimental vibrations. This will cause a reduction in the rate of penetration as well as both drill string and wellbore damage.

A second step change in cuttings removal is sometimes seen between 150 and 180 rpm in some wells. This is probably due to wellbore geometry issues from tortuosity or hole enlargement. When this occurs there may be locations where the drill pipe does not lay on the low side of the hole. (Fig 7-46) If there also is insufficient visco-elastic coupling of the mud to the drill pipe, the pipe rotation may not be able to disturb all of the cuttings beds. In this case it may be necessary to increase rotational speed to the point that drill string vibration begins to "rattle" the pipe. While this may help the hole cleaning efforts it can also be destructive to the wellbore, the drill string, and the rate of penetration.

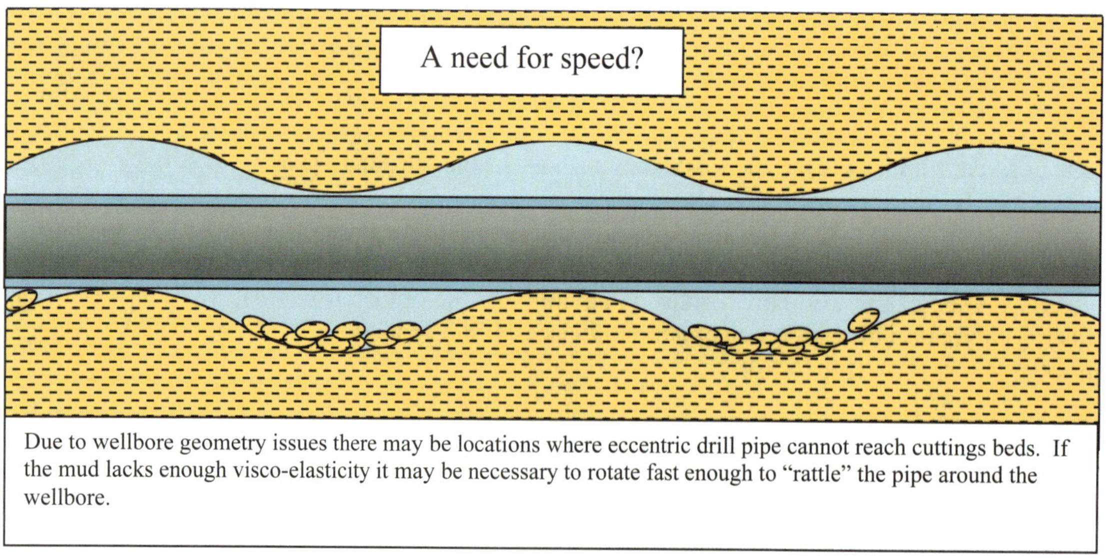

Fig 7-46 Threshold RPM and Tortuosity

It is also suspected that the flow transitions to turbulent flow beneath the pipe around 180 rpm. If this is the case it should be evident from a step change in bottom hole pressure.

Pressure While Drilling tools support my claim that the first step in hole cleaning occurs at or before 75 rpm.[12, 14] Field studies show that a typical threshold rpm for 5" drill pipe, in 12¼" and 8½" wells, occurs between 50 and 75 rpm. (Fig 7-47) Bottom hole pressure is seen to decrease or remain constant while sliding. As soon as the pipe is rotated fast enough to disturb the cuttings bed a small jump in bottom hole pressure is seen. This small increase is on the order of 25 psi. This is too small a pressure to see on a stand pipe gauge but it is easily spotted with PWD tools.

This effect has been demonstrated with both PWD tools and hydraulic models.[12, 14] and is more pronounced as pipe and hole size converge. Smaller well bores experience much higher pressure losses from pipe rotation than do large wellbores. In one study, the ECD from rotation in a 12 ¼" hole was approximately 100 psi once a threshold rpm of 50 to 75 rpm was established. The ECD in an 8-½" hole was as much as 200 psi once a threshold rpm was reached.

PWD tools can also be used to monitor hole cleaning trends. As cuttings beds are disturbed and move into the vertical portion of the well, bottom-hole pressure increases due to the annular cuttings load. As the cuttings are circulated out of the well the cuttings load is reduced and pressure drops. A characteristic pressure trend is produced for each slide-rotating sequence. Any divergence from this trend indicates a change in the hole cleaning efforts.

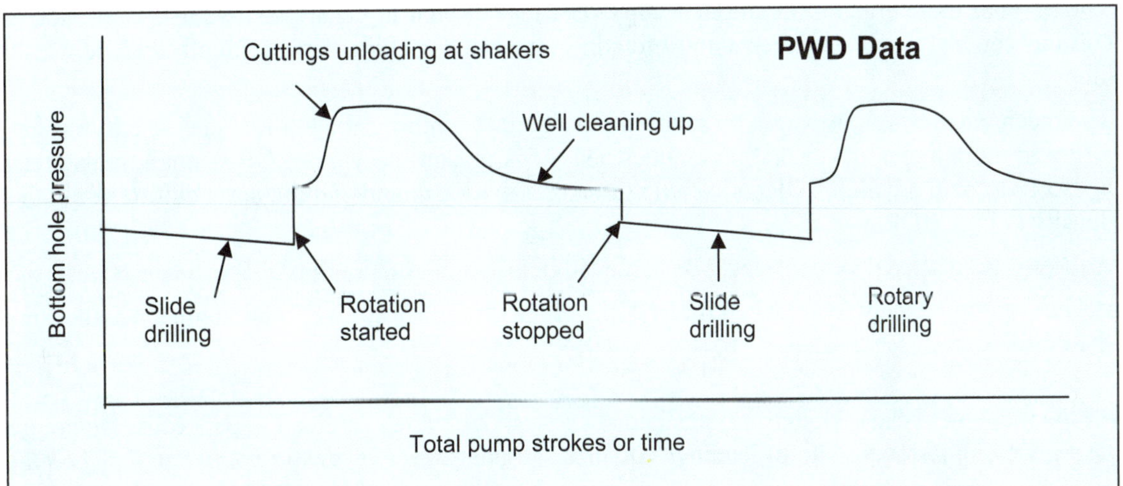

One method of monitoring the progress of pipe rotation on hole cleaning is to monitor bottom hole pressure. As cuttings are disturbed and moved into the vertical section of the well, the BHP increases. As the cuttings are cleaned from the well, the pressure returns to normal.

Fig 7-47 PWD data and hole cleaning

The effects of pipe rotation with other variables, such as fluid properties and cutting size, have been discussed in other sections of this chapter but are summarized again below:

- The influence of pipe rotation increases as the inclination angle increases. At moderate angles, the cuttings beds are loose and fluidized. They are easily disturbed with very little rotation. At higher angles the beds are packed and a threshold rpm is required to disturb the beds. Beds at moderate angles can be removed with improvements in rheology and flow rate, but beds at high angles rely on pipe rotation to disturb them. Thus, the effect of pipe rotation on cuttings bed height is more pronounced at higher angles.
- The effect of pipe rotation becomes less significant as flow rate increases, because higher flow rates reduce the height of the cuttings beds. (Note: Pipe rotation is still needed however)
- The effects of pipe rotation are more pronounced with high yield points. Viscous fluids are not as effective at penetrating the cuttings beds, so we must rely more on rotation.
- Pipe rotation is more effective when the mud is visco-elastic. Polymer mud and oil base mud are more visco-elastic than water base mud.
- Rotation has more effect on smaller cuttings. Because of their high surface area to weight ratio, small cuttings can be "glued" together with fluid adhesion forces, making them difficult to erode with fluid alone.
- The mechanical action imparted by rotating pipe against a cuttings bed diminishes as cuttings get larger or more numerous. Rotation reduces beds of large cuttings when ROP is small; it also reduces beds when there are high ROP and small cuttings. Rotation cannot handle large cuttings of high density at high ROP.
- A threshold rotational velocity must be achieved before pipe rotation has a significant effect.
- Higher rotational velocity may be required in wells with significant wellbore geometry issues or poor rheology.

Caution: Pipe movement can dislodge cuttings beds in a washout, which can lead to packoffs. Do not get too aggressive with the slide-rotation hole cleaning sequence. *In-gauge hole is always important for good hole cleaning.*

One final note about pipe rotation; Clearly the faster we rotate the better hole cleaning we get. However, we reach a point where vibrations become a liability. Vibration can limit the available weight on bit and slow down the rate of penetration. The pipe vibration is also detrimental to wellbore stability and drill pipe longevity.

Get comfortably above the threshold rpm to disturb the cuttings bed, and then find the optimum rpm to maximize the rate of penetration. Do not maximize pipe rotation simply because you can.

Slide Drilling (Pipe Rotation)

The drill pipe is often not rotated while drilling with mud motors. This results in cuttings bed formation while sliding. Intermittent pipe rotation will be required to disturb and remove these cuttings beds.

When the beds are disturbed, the cuttings will be transported along the well in heterogeneous suspension, but in a long slug of high concentration. When this slug reaches the vertical section of the well, it will affect bottom-hole pressure.

Abnormally high cuttings loading can occur when rotation is started after a period of sliding. A clear indication of cuttings overload can be seen by PWD when first picking up off bottom. These pressure surges of 100 to 300 psi. are often masked by the mud motor.

Better sliding results from rotation prior to sliding. This gets the cuttings beds away from the BHA. Cuttings beds cause excessive torque and can frustrate steering efforts.

Coiled Tubing (Pipe Rotation)

With coiled tubing drilling, pipe rotation is not possible. Therefore, the formation of cuttings beds at high angles is inevitable.[13]

Frequent wiper trips with full circulation and bit rotation must be used to clean the hole. Circulating with axial movement of the tubing is not enough to disturb the beds. The beds can only be disturbed by the rotating bit. The cuttings will quickly resettle to form beds again, so the tubing must be pulled occasionally to remove the beds from the hole. The cuttings beds are basically chased up the hole by the bit as it is pulled from the hole. (Fig 7-48) Note that the bit will be oriented downward to dig into the bed as it is pulled.

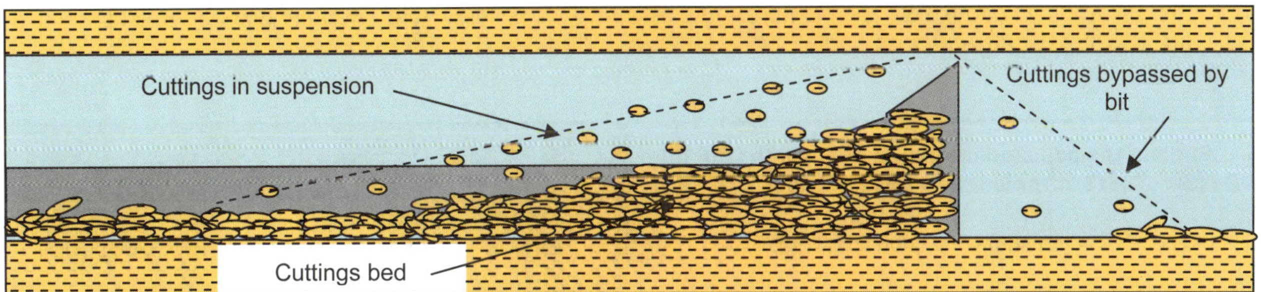

In coiled tubing drilling, cuttings are "chased" up the hole by the rotating bit. The cuttings disturbed by the bit quickly resettle. The bit must periodically be pulled to the vertical section of the well to chase the cuttings all the way into the vertical section.

Fig 7-48 Coiled tubing drilling

Although it is relatively easy to perform a wiper trip to the vertical section of the well during coiled tubing drilling, it is desirable to minimize the need for such trips. The mud properties and flow rates are generally optimized to extend the time between these trips.

Time (Hole Cleaning Factors in Directional Wells)

It takes more time to transport cuttings along an inclined well bore than it does in a vertical well bore. *The time to effectively clean the well bore increases as the angle increases.*

Most mud logging companies can estimate the time it takes to transport cuttings from bit to surface using their own in house slip velocity software. If these facilities are not available to the rig crew, they should establish their own empirical estimates for hole cleaning time. The Training to Reduce Unscheduled Events manual[15] presents one empirical method for predicting hole cleaning time for directional wells.

This method assigns a circulating stroke factor to each section of the well based on how many "bottoms up" would have to be circulated to completely clean the well. These **C**irculating **S**troke **F**actors are determined empirically for the types of formations drilled and muds in use. They are rough guides only. A typical chart of CSFs is presented in the following table.[15]

Circulating Stroke Factors (CSF) or Number of Bottoms Up to Clean Hole[15]

	27 1/2"	**17 ½"**	**12 ¼"**	**8 ½"**
0° – 30°	2.25	1.75	1.5	1.25
30° – 65°	2.75	2.5	1.75	1.5
65° +	3+	3	2	1.75

The numbers that go into this chart must be arrived at empirically from field experience with the various hole angles, hole sizes, flow rates, and muds in use. The goal with this chart is to find the total number of strokes necessary to clean the well. To do this, the well is first divided into sections, depending on hole size and angle. The measured depth of each hole section is then multiplied by the appropriate CSF and added together to get a total adjusted depth. We then calculate the total strokes necessary to circulate one bottoms up from the total adjusted depth. (Fig 7-49)

Finding the adjusted measured depth with CSF for a 12 ¼" hole drilled to 12,000 feet.[15]

Adjusted MD = the sum of (section length x CSF)

= (4000 x 1.5) + (2000 x 1.75) + (6000 x 2)

= (6000) + (3500) + (12000)

= 21,500 Total adjusted MD

To clean this well, one bottoms up should be circulated as though it was a 21,500' vertical well.

Another way to look at it is to circulate 1.8 bottoms up.

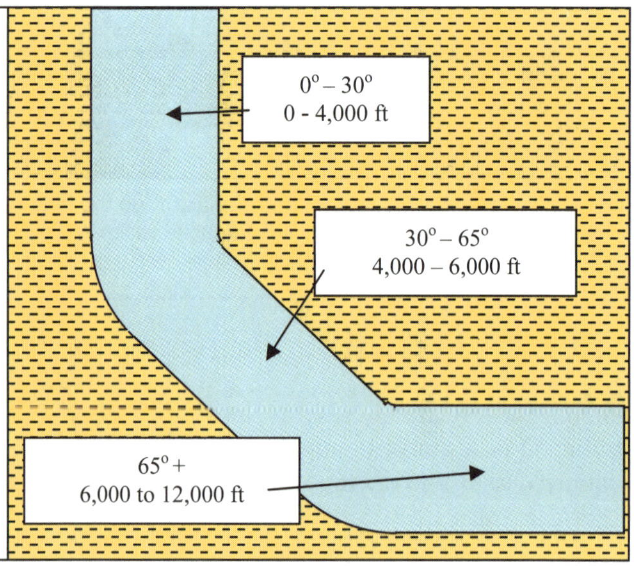

Fig 7-49 Circulating time

Viscous Sweeps (Hole Cleaning Factors in Directional Wells)

Viscous sweeps are generally ineffective in high angle wells when the pipe is not being rotated. Viscous mud does not penetrate the cuttings bed as well as thinner mud. The flow profile of the more viscous mud is also shifted away from the cuttings bed. (Figure 7-50)

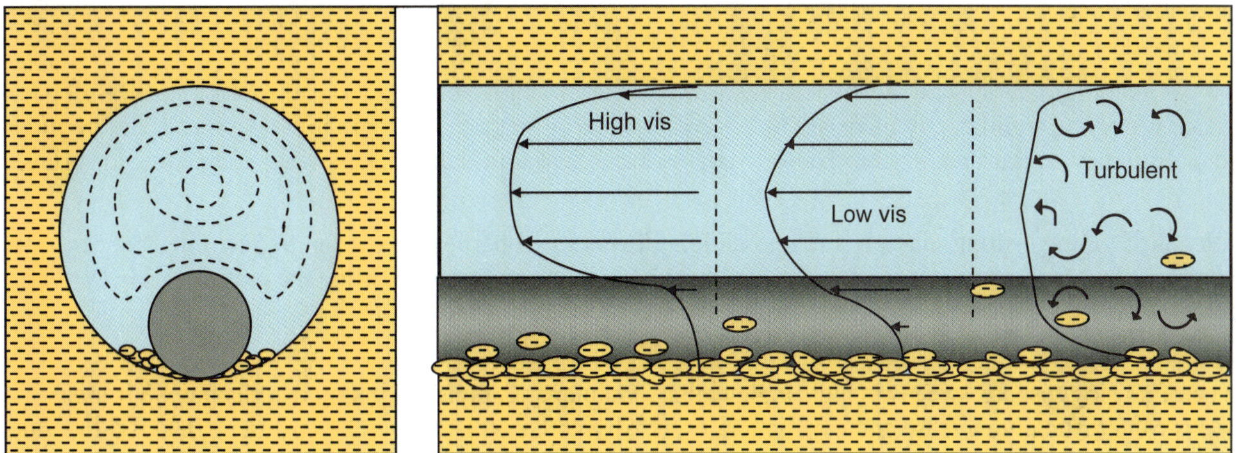

The velocity profile is shifted up away from the drill pipe and cutting beds laying on the low side of the well. This causes the velocity on the low side to be too low to erode the cutting bed. This problem is less pronounced with lower viscosity muds.

Fig 7-50 Viscous Sweeps

With adequate flow rate and pipe rotation the cuttings beds will be disturbed and circulated from the well without the help of viscous sweeps.

There are times however when viscous sweeps may provide some benefit:

- When pipe rotation is not possible such as when building the curve.
- With large hole size and mud that is not very visco-elastic
- Drilling with coiled tubing

When building a curve with a motor and bent sub pipe rotation may not be possible. (For example, one operator does not allow rotating a motor with a bend higher than 1.83 deg.) Ideally we would slide 20 feet and rotate 10 feet of each joint. If rotation is not allowed at all in the build section we have no choice but to pump viscous sweeps. Viscous sweeps are very effective in the vertical hole but are largely ineffective in the horizontal hole. The higher the inclination angle, and the more measured depth the sweep must travel before reaching the low angle section, the less effective the sweep will be.

A low vis sweep followed by a high vis sweep is generally more effective here. A turbulent low vis sweep penetrates and disturbs the cutting bed. The viscous sweep then carries the cuttings some distance along the well bore before the cuttings can settle again. Several combo sweeps are required to move the cuttings any distance along the well bore. The sweeps become less effective with increasing hole angle. Though not the ideal way to clean the well it may allow us to build a curve with high bends on a mud motor. A wiper trip with rotation may be required to clean the well prior to running casing.

Sweeps are acceptable here because there is very little measured depth beneath the end of the vertical section.

Pipe rotation is less effective in a large hole size especially if the mud is not very visco-elastic. A high vis sweep with visco-elastic properties, used in conjunction with pipe rotation, is often effective in this case. They are only effective in combination with pipe rotation. Several long sweeps may be required because once the sweep passes through the cuttings they will tend to settle into beds again. Visco-elastic sweeps will tend to "wet" the wire in the shaker screens so it is likely that the mud will "walk" off the shakers when the sweep comes to surface.

Lost Circulation Material, (LCM), sweeps are also popular. In large concentrations the LCM absorbs all of the free water and significantly increases the viscosity. The aggregate nature of the sweep resists deformation and tends to have a flat flow profie. These sweeps are also lost across the shakers when they come to surface.

With coiled tubing drilling rotation is not possible. Ideal mud properties are chosen to maximize the carrying capacity of the mud, while minimizing the thickness of the mud so it can penetrate the cuttings bed. Different operators have different philosophies regarding the use of sweeps in this case.

Be careful using viscous sweeps in high angle wells. Annular friction is a function of viscosity. Viscous sweeps can drastically increase annular friction, especially with small annular clearances. ECD is a function of both annular friction and the ratio between measured depth and true vertical depth, MD/TVD. As the MD/TVD ratio increases, or the annular clearance decreases, the ECD increases. Viscous sweeps can lead to lost circulation.

If you get too aggressive with your sweeps you may pack off due to boycott settling at the bottom of the build section.

Remember, viscous sweeps are generally ineffective in high angle wells. They lead to higher ECD's in slim holes or extended reach wells. They should only be used in conjunction with pipe rotation or when rotation is not permissible such as when using a motor with a high bend to build the curve.

Viscous sweeps are very effective in vertical holes when the combination of flow rate and rheology is insufficient to clean the hole, such as large diameter surface hole. They rapidly lose effectiveness as we build hole angle. They are generally ineffective if we are not also rotating the pipe. They will raise annular friction losses and can lead to lost circulation and wellbore instability. Therefore they are generally only used when needed, and only if they are demonstrated to be effective. Typically they are only used in vertical, large diameter wells or in the build section when rotation is prohibited by downhole tools.

Air and Foam Drilling

Air drilling presents unique hole cleaning challenges. The weight of air is almost negligible, so much higher annular velocities must replace the missing effect of buoyancy. Air is also compressible. As the pressure along the annular path from bit to surface changes, so does the annular velocity and thus hole cleaning efficiency.

Compressibility

A fixed quantity of air behaves according to Boyle's principle, which states that pressure times volume is a constant. This is expressed mathematically as:

$$P_1V_1 = P_2V_2 = \text{constant} \qquad \text{eq. 7.13}$$

If the pressure is changed, then the volume must also change. In the following example, a full bottle of compressed air and an empty bottle are joined together (Fig 7-51). One bottle has 10 gallons at 100 psi, other has 10 gallons at zero psi. When we open the valve and let gas from the full bottle bleed into the empty bottle, the total quantity of air does not change. The pressure is reduced by half and the volume doubles.

The quantity of gas in the bottle on the left does not change as it bleeds into the empty bottle on the right. The volume doubles but the pressure is reduced by half.

Fig 7-51 Boyles law

The same sort of thing happens down hole. As pressure increases from annular friction losses, the volume decreases. The reduction in volume results in a reduction of annular velocity. If the well bore has a constant diameter, the annular velocity will be lowest at the bottom of the well, where the pressure is highest. As the pressure reduces further up the well, the volume expands and annular velocity increases. The maximum velocity will be at the top of the well. (Fig 7-52)

The largest cuttings with the highest slip velocities will be near the bottom of the well. As they are transported up the well, they are broken into smaller pieces with less slip velocity.

The combination of low velocity and large cuttings make the lowest part of the well the hardest to clean. As the well is deepened, more air must be injected into the well to maintain adequate bottom-hole velocity.

Annular velocity = air volume ÷ cross sectional area

Since the volume of air is proportional to pressure so is the local annular velocity.

Pressure is greatest near the bit so the air volume and thus velocity is smallest near the bit.

As the air moves up the well the pressure is reduced, its volume expands and annular velocity increases.

Wellbore pressure is partly due to the hydrostatic effect of the cuttings concentration, but mostly to annular friction losses.

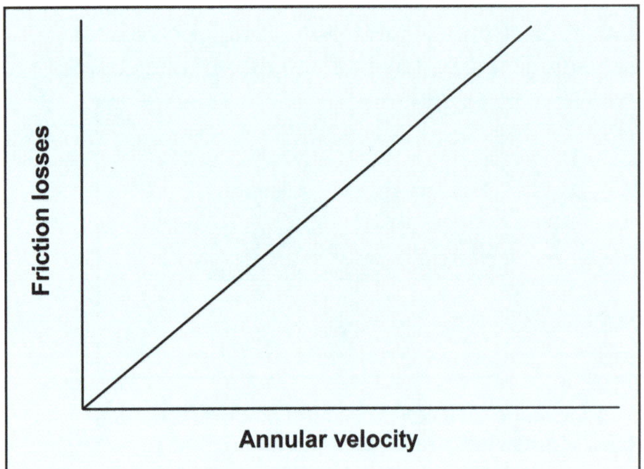

Annular friction loss is linearly proportional to annular velocity. Therefore, wellbore pressure increases linearly with depth, and annular velocity decreases linearly with depth due to friction losses.

B

Cuttings are broken into smaller sizes as they are circulated up the well.

The worst hole cleaning problems occur near the bit.

This is because the cuttings are larger with higher slip velocities and the annular velocity is at its lowest.

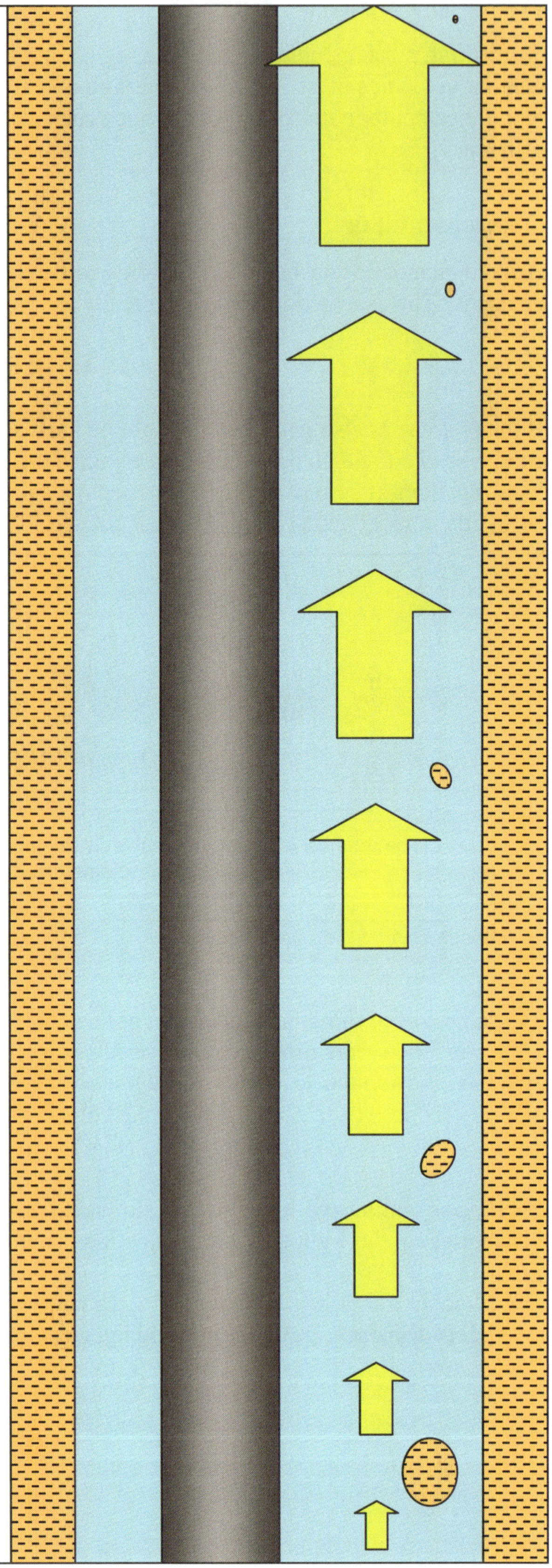

Fig 7-52 Effect of compressibility on annular velocity

Bottom-Hole Pressure

We are obviously very concerned with the annular velocity, and thus bottom-hole pressure, just above the bit and collars. It is the combination of annular friction and hydrostatic head that controls this pressure. When the flow rate is high and the amount of liquid and cuttings in the air stream is very small, annular friction predominates. When the cuttings and liquid concentration is high and flow rate is low, hydrostatic pressure may predominate (Fig 7-53).

The weight of the cuttings increases the effective density in the annulus, which further increases bottom-hole pressure. Care must be taken not to generate more cuttings than the airflow can handle. As the cuttings load becomes greater, the annular pressure increases and the volume (and thus velocity) decreases. The decrease in velocity causes the cuttings concentration to increase, which in turn causes a further increase in pressure and reduction in velocity.

The pneumatic transport industry recognizes a point at which the velocity cannot be reduced any further without setting off an irreversible collapse of cuttings. This velocity is known as the "**choking velocity.**" If the cuttings concentration becomes high enough to "choke" the flow, the drill string will pack off with rapidly settling cuttings, even though strong airflow can still be seen at the blooie line.

To avoid choking the flow, the air injection rate must be high enough to handle the penetration rate. As the flow rate increases, annular friction losses (and thus bottom-hole pressure) increase. Air compressor energy consumption climbs rapidly as pressure increases, so excessive airflows are avoided. There exists an optimum airflow that will minimize annular friction losses while providing enough annular velocity to minimize the cuttings concentration (Fig 7-53).[16]

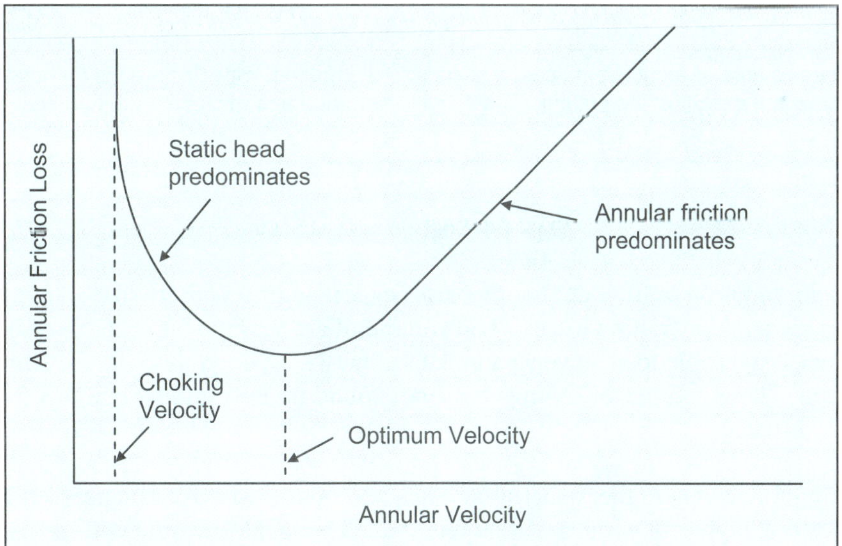

There is an optimum flow rate that minimizes compressor energy consumption while providing adequate hole cleaning.
Strive to stay <u>above</u> this optimum flow rate!

Fig 7-53 Optimum annular velocity

The optimum annular velocity increases as angle increases, because cuttings recycling and cuttings beds increase the cuttings concentration. This increases both the static head and the frictional losses in the annulus. More air is required in higher angle holes.

Hole Cleaning Efficiency in Air Drilling

Frictional drag contributes more to hole cleaning than buoyancy in air drilling. It is the specific surface area of the particle that controls the amount of frictional drag it feels. The specific area is equal to the particle's surface area divided by its weight. Therefore, the larger a cutting's surface area is, and the smaller its volume is, the more lift it will feel.

In Fig 7-54, a cube with 1" sides is broken in half on each axis to make 8 cubes with ½" sides. The surface area of each cube is now ¼ as much as the 1" cube, but the volume of each cube is only $^1/_8$ as much. Thus, the specific area of the cube doubles each time the cube is broken down. The lift felt by the cutting, and thus the cuttings transport ratio, increases exponentially as the cutting size decreases.

Large spherical cuttings will transport very slowly (if at all) while small flat particles will travel at nearly the velocity of the air. When circulation is stopped, the cuttings fall rapidly.

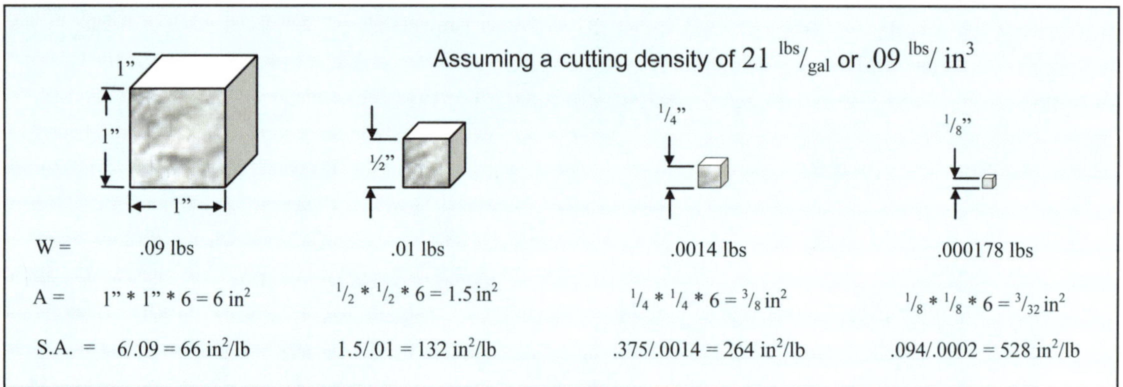

The specific area of a cutting increases as its size decreases. Specific area is the total surface area divided by the weight of the cutting. The specific area doubles each time the diameter of the cutting is halved.

Fig 7-54 Specific area of cuttings

The annular velocity required to lift a cutting increases exponentially with its size. An increase in annular velocity increases the annular friction losses. The volume of air near the bit is compressed into a smaller volume. It may not be possible to achieve an annular velocity high enough to lift the largest cuttings generated at the bit. These cuttings or cavings must be ground up into smaller cuttings by the rotating pipe before they can be lifted.

Larger cuttings are produced at higher penetration rates and/or at lower overbalances. If higher rates of penetration are desired, higher flow rates will be required. The ideal flow rate will be one that balances the benefits of penetration rate against air compression cost. Experience shows that a flow rate that gives 3,000 ft/min annular velocities is adequate for most air drilling applications.

If hole cleaning efforts are ineffective, bottom-hole pressure will climb, annular velocity will decrease, and hole cleaning will further deteriorate. Obviously, we would like to monitor hole cleaning by monitoring bottom-hole pressure. While drilling with mud, this can be accomplished by monitoring standpipe pressure and downhole sensors. We will try to do the same with air drilling, but this is complicated by the compressibility of air.

Pressure waves propagate at the speed of sound. The speed of sound propagation is much slower in air than it is in mud because air is much more compressible than mud. It only takes a few seconds for a 200psi annular restriction to be noticed on the standpipe while drilling with mud. It will be several seconds to several minutes for a restriction to be noticed while drilling with air. While drilling with mud, the full 200psi restriction will be reflected on the standpipe gauge. While drilling with air, only a percentage of this restriction will be seen on the standpipe gauge, if any restriction is seen at all!

The pressure drop across the bit causes much of the problem in detecting bottom-hole pressure. Even with no jets (which is the norm in air drilling), the bit provides a major pressure restriction in the drill string. As the air rushes through the small cross-sectional flow area of the bit, its velocity increases. If the velocity of air reaches the speed of sound, then the flow is said to be "sonic". Sound and pressure waves can no longer propagate from the annulus into the drill string. It is analogous to a fish trying to swim up stream against a current that is flowing faster than the fish can swim. The greater the pressure drop across the bit, the higher the velocity through the bit.

From basic fluid mechanics, the pressure drop that results in sonic flow can be calculated for any type of gas. A rough guide is that when the pressure just upstream of the bit is twice the pressure immediately downstream of the bit, the airflow through the bit will be in sonic flow.

The onset of sonic flow occurs when $P_{inside} \div P_{outside} \approx 2$ eq. 7.14

If the flow is sub-sonic, bottom-hole pressure can be calculated from standpipe pressure with basic fluid mechanics principles.[17] It is important to note that *a small change in standpipe pressure will reflect a much larger change in bottom-hole pressure.* The larger the pressure drop across the bit, the smaller the change in standpipe pressure for any change in bottom-hole pressure.

When the flow is supersonic through the bit no change in standpipe pressure will be detected as the annulus begins to pack off! (Fig 7-55)

The graph in Fig 7-55 depicts the pressure changes reflected on the standpipe gauge through a range of bit pressure drops. As flow rate increases so will the pressure drop across the bit and the harder it will be to detect pack offs. *We must pay very close attention to small changes in standpipe pressure and respond quickly!*

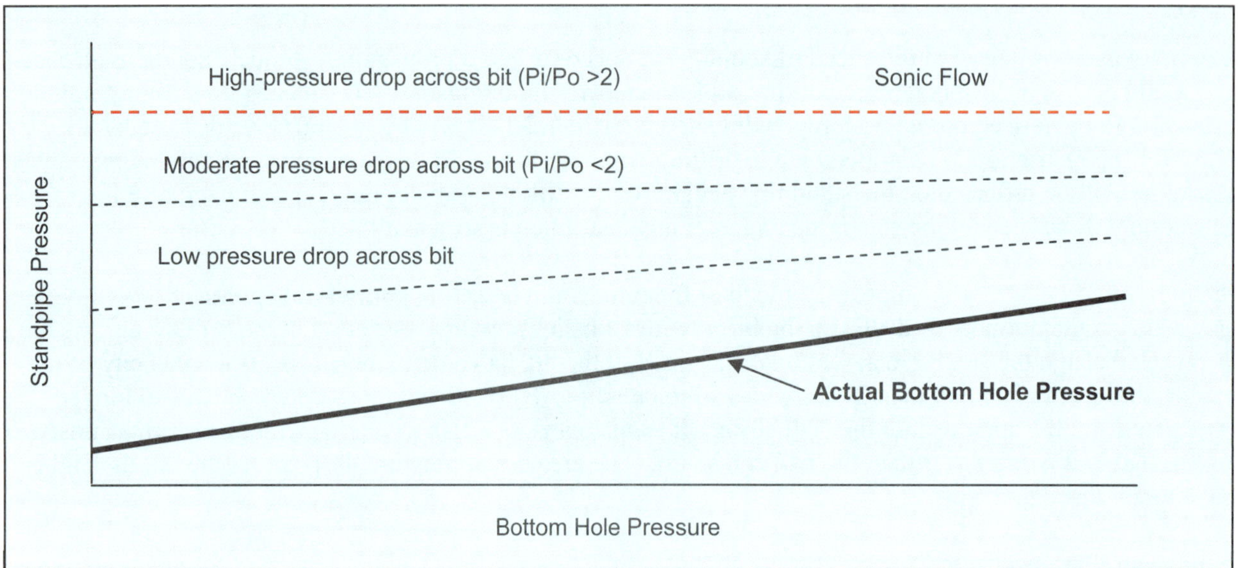

Standpipe pressure is less effective for monitoring increases in bottom hole pressure with air drilling than with mud drilling. Thus annulus loading and pack offs are much harder to detect.

Fig 7-55 Standpipe pressure vs. bottom hole pressure

Mud Rings

Air does not provide an impermeable filter cake, but some filter cake does exist. By definition, a filter cake refers to solids filtered out of the drilling fluid as it flows into a permeable formation. When a permeable formation is exposed, air will flow into it if the formation pressure is less than well bore pressure. Solids will bridge across the pore openings, but may not substantially reduce the flow into the formation. The bridge will grow in size until equilibrium between deposition and erosion is reached. (See filter cakes in the differential sticking section.) If there is no moisture in the air, the bridge may be very thin. If moisture is present, the bridge will pack and become thick and hard. The thick, hard filter cake that forms across permeable formations when moisture is present is referred to as a "**mud ring**."

If moisture is present, mud rings will also form just above the drill collars or in washouts where the annular velocity is very low. Mud rings restrict annular flow and thus increase bottom-hole pressure. The increase in pressure reduces flow rate and increases temperature. If gas is present, the increase in temperature and pressure may be enough to start a down hole fire that may go undetected for some time. Down hole fires can cause the loss of the drill string and will often result in sidetracks.

Even if the mud ring can be broken up, it will form large chunks that will be too heavy to circulate up the well. If the mud ring was caused by moisture, then these chunks will be "sticky" and will often stick together again to form new mud rings. Eventually, the pipe may become stuck in a packoff.

Mud rings can be detected by:

- A loss of air volume at the blooie line due to partial or total lost circulation.
- "Dusting" stops. The dust particles are sticking together and to the walls.
- A small increase in standpipe pressure.

When mud rings are suspected, action must be taken quickly to avoid packoffs.

Misting

Once moisture is encountered, the mud rings must be washed away with an excess of water. Water and surfactants are added to the air stream in what is called "**misting**." A small amount of moisture serves both to cement the grains in the filter cake together and to pack them tighter. This increases the strength of the bridge. (See rock mechanics in the well bore instability section) Additional water saturates the bridge and turns it into soup. The saturating water coats the grains to lubricate them and force them apart. This weakens the bridge and allows it to be easily eroded. The surfactants also help to coat and separate the dust particles.

Unfortunately, the droplets of water add to the total load the air must carry up the well. Adding water to the well has the same effect as adding additional cuttings to the well—more air must be added to maintain adequate bottom-hole annular velocity. Experience has shown that 30% to 40% more flow rate is required once misting begins.

If more air is not available, then the penetration rate must be reduced.

If too much water is added, or if formation water is flowing into the well, the flow may begin to "slug." **Slugging** refers to the separation of air and water into slugs of water and air. This causes severe pressure surges as the slugs of water are blown out of the well. Bottom-hole annular velocity is greatly reduced and well bore instability becomes a concern. (Fig. 7-56)

Slugging occurs when the water and air separate from a homogeneous mixture and coalesce into pockets of air and slugs of water. This can lead to pressure surges and well bore instability.

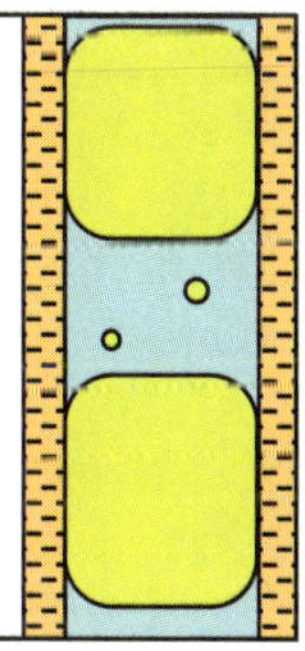

Fig 7-56 Slugging

Stable Foam

If too much water is flowing into the well, or if cuttings or cavings are too large to effectively clean from the well, then the next step is to convert to stable foam.

With stable foam, there is enough liquid to trap and surround all of the air in the form of little bubbles. Liquid forms a continuous phase in which the liquid is in continuous contact with itself, while the air bubbles in the discontinuous phase are isolated from each other by the liquid. (Fig 7-57)

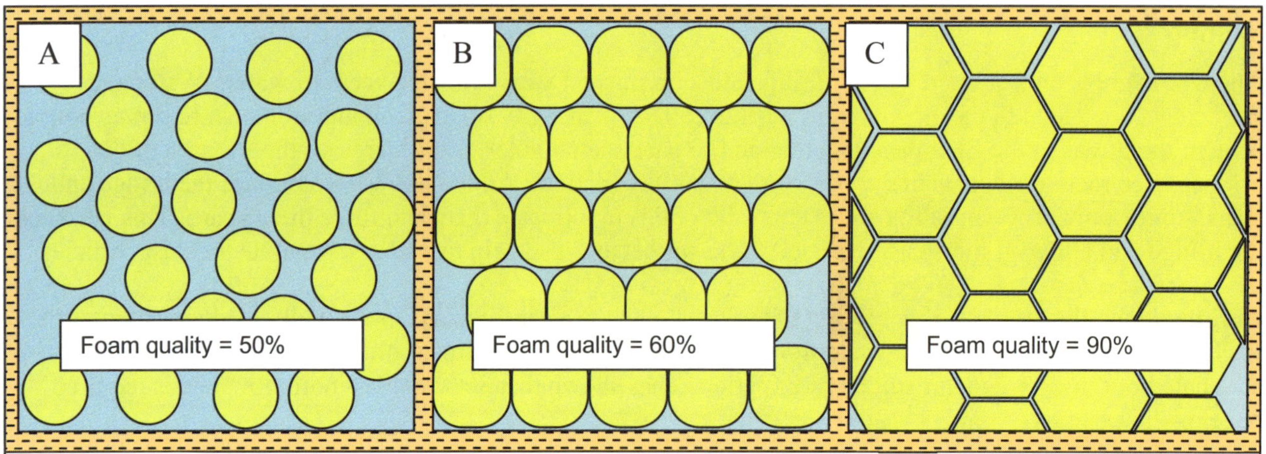

Foam is a mixture of liquid and gas. The continuous liquid forms a cellular structure encapsulating tiny bubbles of gas. Foam quality refers to the volume fraction of air in the mixture. The higher the ratio of air to liquid the higher the foam quality.

Fig 7-57 Foam quality

Foam is described by its "quality." **Foam quality** refers to the ratio of air volume to total volume in the mixture.

$$\text{Foam quality} = \text{Volume of air} \div (\text{air volume} + \text{liquid volume}) \qquad \text{eq. 7.15}$$

If air occupies less than 55% of the total volume, the bubbles will form perfect spheres that do not contact each other. At this point, the foam is really just an aerated liquid. Once the volume of air in the mixture reaches 55%, the bubbles contact each other and begin to deform. The bubbles form small flat spots where they are in contact with each other.

As the quality increases further, the bubbles began to form polyhedrals as they are crammed closer and closer together. The bubbles must deform to occupy more of the space between the bubbles. The bubbles do not want to be deformed. By their nature they want to be perfect spheres. This is where their potential energy is the lowest. (A sphere has the highest volume to surface area ratio of any shape.) In order to deform into a polyhedral, the bubbles skin must stretch and its internal pressure must increase.

Foam Viscosity

Bubbles are analogous to balloons, which try to form the shape of a sphere. If a room is partially filled with balloons, we will be able to walk through it by pushing the balloons out of our way. A cutting could push its way through a foam with less than 55% quality in the same manner. (Fig 7-57A)

If we try to cram additional balloons into the room the balloons will begin to deform. At this point we would find it difficult to walk through the room. (Fig 7-57B) The balloons must deform even more as we try to move them out of our way. Eventually we could cram enough balloons into the room that we could not move through it at all. (Fig 7-57C)

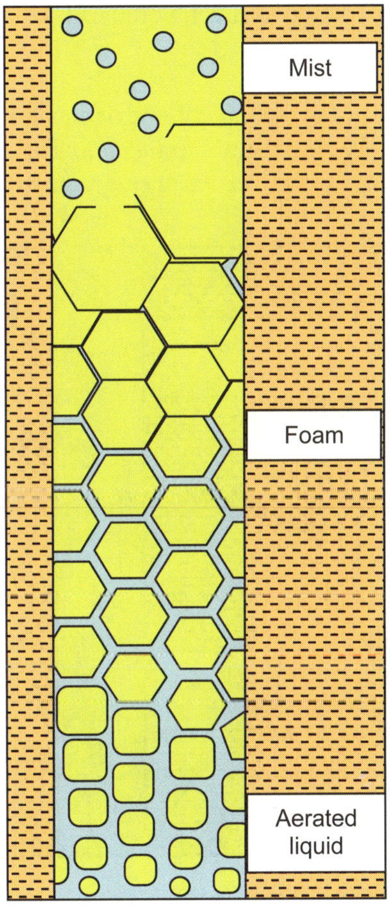

The viscosity of a foam increases with its quality. (Fig 7-58) At the bottom of the well, where the pressure is highest, the quality will be lowest. As the foam moves up the well, the pressure is reduced and the volume of air increases. The volume of liquid remains constant, so foam quality increases as pressure is reduced. If the air expands too much, it may break down into a mist. In this case, air is the continuous phase surrounding droplets of water. (Fig 7-58)

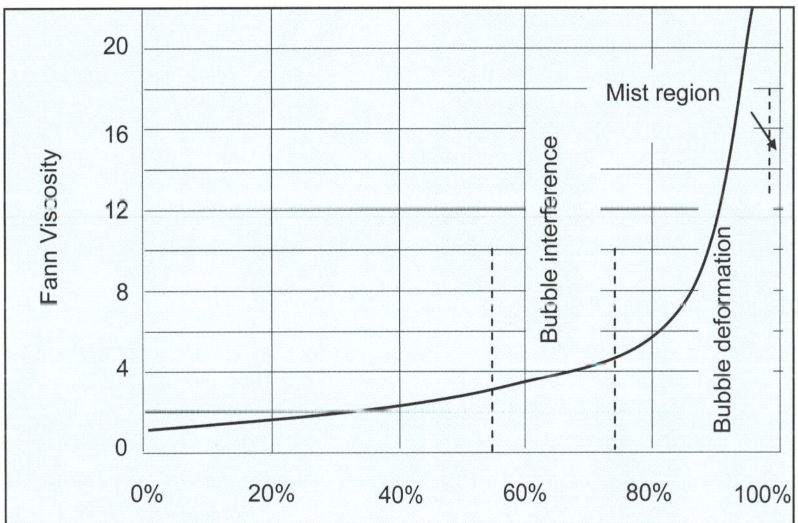

Viscosity increases with a foams quality (from Mitchell 1971[18]).

Fig 7-58 Foam viscosity

The viscosity of foam can be as much as ten times the viscosity achievable with mud. It is possible to lift fist size rocks out of a 30" diameter hole with stable foams with relatively low annular velocities. Stable foams can also lift large quantities of liquid produced by drilled formations.

If the foam breaks down into a mist before it reaches the surface, it loses its viscosity and cuttings carrying capacity. One means of preventing this is to inject more liquid into the foam. However, if the quality of the foam at the bit is less than 55%, will not be able to lift the cuttings at the bit. *Our goal is to manage the quality of the foam such that the quality is at least 60% at the bit and no more than 98% at the bell nipple.*

This is achieved by limiting the expansion of the bubbles through the application of back pressure at the surface.

The problem is that excessive annular frictional and hydro static pressures can compress 98% quality foam to less than 55%. When a back pressure is applied this compresses the bubbles near the bottom even more. More gas must be injected to overcome the compression of the gas due to friction and back pressure. *With a combination of controlling air and liquid injections rates, and applying annular back pressure at the surface, a suitable foam quality can be achieved throughout the well.*

The effect of backpressure on air drilling can be seen in Fig 7-59. In Fig 7-59A, the bottom-hole pressure is 500 psi and top-hole pressure is 20 psi. A unit quantity of air will occupy 25 times as much volume at the surface as it did on the bottom. Thus, the annular velocity will be 25 times higher at the surface than at the bottom. In Fig 7-59B, 500 psi of backpressure has been applied at the surface. Bottom-hole pressure is now 1,000 psi and top-hole pressure is 500 psi. A unit quantity of air occupies only twice as much volume at the surface as it did near the bit, so annular velocity is only doubled.

By increasing the backpressure to the point that bottom-hole pressure is doubled, twice the volume of air must be injected just to achieve the same bottom-hole velocity that we had with no backpressure. *Enough air must be injected to ensure we have sufficient velocity to clean the bottom of the hole, after it has been compressed by the combination of friction and back pressure.*

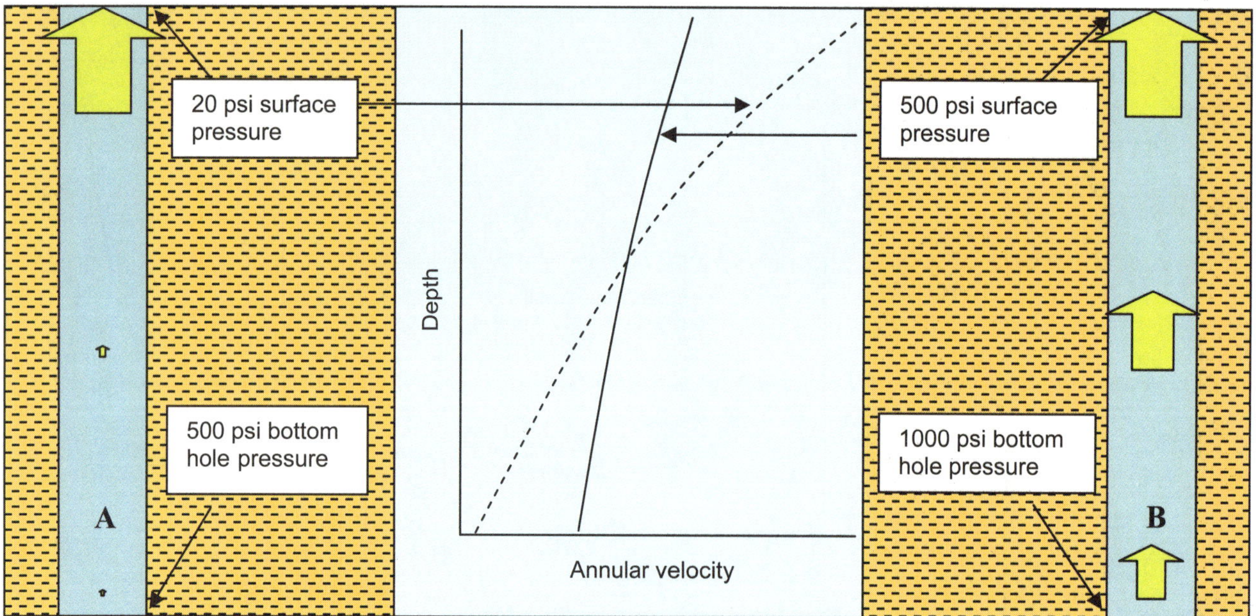

The ratio of top to bottom hole annular velocity is proportional to the ratio of top and bottom hole pressure.

Fig 7-59 Annular velocity vs. depth

This same reasoning also applies to foam, but with one important difference—foam has more density than air. When drilling with foam, the density of the liquid must be taken into account. A much higher percentage of bottom-hole pressure will be due to the hydrostatic effect of the liquid in the foam. The annular velocity will also be much lower with foam, so the percentage of bottom-hole pressure due to annular friction losses is less. Well bore pressure will not decrease as linearly with foam as it did with air as foam progresses from the bottom of the hole to surface. (Fig 7-60)

Annular velocity with foam is much less than that of air or even mud. Typical annular velocities at the bit for air will be on the order of 3000 ft/min, mud will be about 300 ft/min, while foam may be as low as 100 ft/min. As previously mentioned, foam quality and annular velocity at the bit must be managed with a combination of back pressure and air and liquid injection rates to maintain adequate hole cleaning.

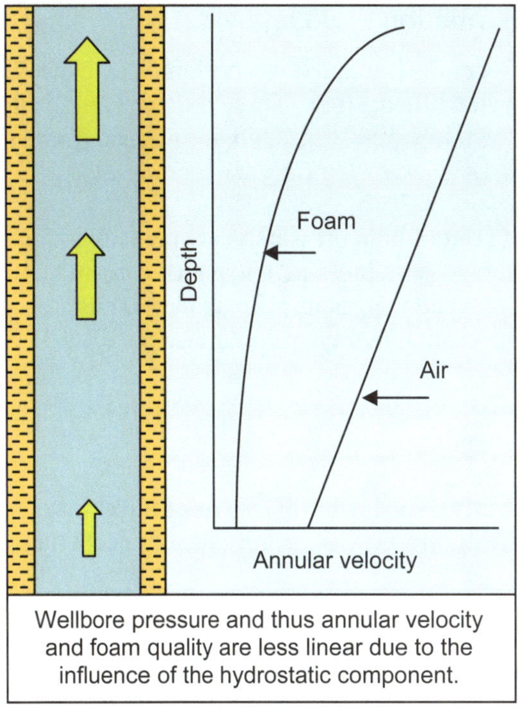

Fig 7-60 Air vs. foam

Complex computer programs are currently used to calculate bottom-hole pressure, velocity, and foam quality from surface measurements.[19,20] Companies that provide air and foam drilling services usually provide these programs.

As a general rule, more air and liquid must be injected as depth or penetration rate increases. More back pressure must also be applied.

Because of the hydrostatic effect, there is a greater pressure drop with foam than with air. Pressures throughout the well bore will be higher with foam. Even though air pressure is higher, the air volume requirements are much lower with foam. Compressor fuel consumption is substantially lower with foam than with air. To optimize fuel consumption, backpressure is usually minimized while maintaining 65% foam quality at the bit and 95% quality at the surface.

The higher pressures from foam drilling contribute slightly to well bore stability. The water component of foam and mist may reduce well bore stability slightly, so the liquid component of mist and foam drilling may have to be treated to minimize well bore instability. Also, the liquid components must be compatible with drilled contaminants.

Formation Fluid Inflows

Generally, air and foam drilling is an under-balanced operation. Formations fluids are likely to flow into the well while drilling. These inflows complicate things. Gas inflows can increase quality, possibly breaking it down into a mist. Liquid inflows can decrease quality and increase bottom-hole pressure. *Gas inflows can either increase or decrease bottom-hole pressure.*

If bottom-hole pressure is dominated by the hydrostatic component of the foam, a gas inflow will reduce bottom-hole pressure and increase foam quality near the bit. If the bottom-hole pressure is friction dominated, a gas inflow can increase bottom-hole pressure and decrease foam quality near the bit, while increasing quality at the surface.

Water or gas inflows can be detected at the surface by monitoring the blooie line for foam quality and slugging.

If there is a gas influx, we must inject more liquid. We may also have to inject more air to maintain quality at the bit. If an oil or water influx is encountered, we will have to inject more air and surfactant.

Stiff Foam

By adding bentonite and or polymers to the soap solution, we will make stiffer foam. It will have a higher viscosity and move in plug flow up the annulus. Stiff foams are most effective at low foam qualities, such as just above the bit.

Aerated Muds

Aerated muds are used when a reduction in well bore pressure is desired. Either some air is pumped into the circulating fluid or air is introduced somewhere into the annulus between casing strings. As with foam and air drilling, the compressibility of the air will cause the annular velocity to fluctuate. Unlike air and foam drilling, the cuttings transport mechanism is essentially the same as it is with mud drilling—at least until the volume of air exceeds 55% by volume.

At some point in the annulus, the aerated fluid may convert to foam. Since the carrying capacity with aerated liquid was sufficient to carry the cuttings, the foam will as well. However, there remains a risk of the mixture breaking down into a mist near the surface. If this happens, there will be slugging and insufficient carrying capacity near surface.

Summary

When to Expect Hole Cleaning Problems

In vertical wells, hole cleaning problems generally take the form of cuttings recycling. This is likely to occur with low density, low viscosity muds, and at low annular velocities. These circumstances usually occur in the large diameter surface hole. Problems may also occur at high drilling rates or when the well bore is unstable.

In directional wells, hole cleaning problems can always be anticipated. Hole cleaning problems will be worst at lower flow rates, with low or no pipe rotation, and with poor mud rheology.

Stuck pipe and tight hole problems generally show up while pulling out of the hole. (Fig 7-61) When the problem is more severe, symptoms will show up after a connection when the pumps are shut off. If very large pieces of debris or large quantities of recycling cuttings exist, excessive drag may be noticed when first picking up for a connection.

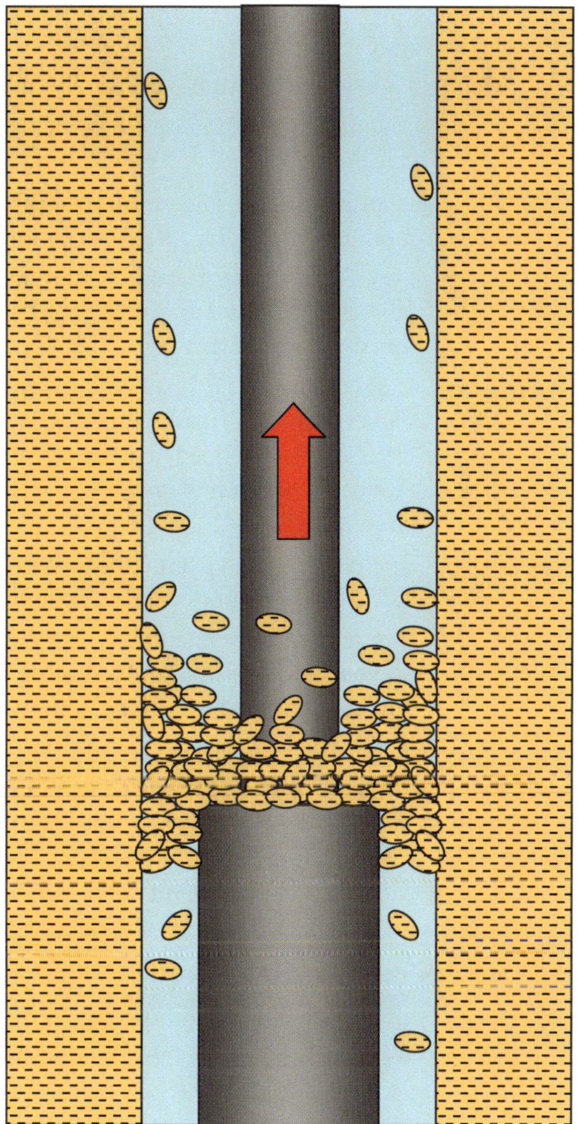

Upward motion of drill string into settled cuttings wedges them tightly together to form a packoff.

Fig 7-61 Packoff

Preventive Measures

To prevent stuck pipe due to hole cleaning problems, the causes of poor hole cleaning must be avoided. In general:

- Maintain adequate flow rate—especially in directional wells.
    - A simple rule of thumb for vertical wells: The annular velocity should be twice the cuttings settling rate.
    - Another general rule of thumb: 1000 gpm or more are required in 17½" hole, 750 gpm or more in 12¼" hole, 500 gpm or more in 8½" hole.
    - Lou, Bern, Champter, and Kellingray of BP Amoco have produced a set of simple charts to help choose flow rates in deviated wells.[5] These charts assume full pipe rotation and are based on typical drilling in the North Sea. (Included in Appendix A)
    - If circulating out of the hole in high angle holes, it is important to circulate at the same rate or higher than we drilled with. Cuttings beds will slide at low circulation rates and when circulation is stopped at angles between 30° to 65°. We must not circulate out of hole at reduced rates on high angle holes. Rotation speeds of 80 to 180 rpm are required while circulating.
    - There is no hole cleaning benefit in high angle holes if the drill pipe is not rotated at a sufficiently high rate while circulating.
    - It is always advisable to circulate the last three stands to bottom. This is done not only to prevent a plugged bit, but to ensure we don't bury the bit in cuttings and pack off when circulation is started. In high angle holes, circulate at full rate to avoid cuttings bed sliding at moderate angles.
- Rotate the pipe in high angle holes. Determine what the minimum rotating speed is to disturb the cuttings beds and maintain rpm above this rate. This is especially important while circulating the well clean before a trip. Rotating speeds as high as 120 to 180 rpm may be required in some cases.
- It is the combination of flow rate, pipe rotation, and visco-elastic mud that cleans the high angle hole.
- Control the rate of penetration. Hole cleaning becomes more difficult as more cuttings load the annulus, due to the shifting of the flow profile away from the cuttings beds.
- Stop drilling when hole conditions dictate. If hole cleaning becomes a problem, adding more cuttings to the annulus will only make the problem worse.
- Plan wiper trips. This will dislodge cuttings stuck to the wall, disturb cuttings beds, and tell us how successful our hole cleaning efforts are.
- Circulate the hole clean before POOH and circulate cuttings away from BHA prior to making a connection. Maximize string motion when circulating the hole clean. Reciprocate with pipe rotation in deviated wells to disturb the cuttings beds. Move the string enough for every part of the well to be passed by a tool joint during the reciprocation.
- Maintain adequate mud properties. A high YP/PV ratio is desired in vertical holes.. Lower YP and PV are desired in high angle holes but high visco-elasticity is also desired.
- Use high viscosity sweeps for vertical wells, and a combination of low viscosity then high viscosity and high density sweeps for the build section of directional wells when pipe rotation is not possible. The high viscosity sweeps will place the flow regime in plug flow. This is effective in vertical wells but high viscosity sweeps by themselves are not effective in directional wells. (This is because the high viscosity fluid may cause the flow to divert away from the cuttings beds, especially if the sweep has time to mix with adjacent mud such that the flow regime is not in plug flow.)
- Use visco-elastic sweeps in conjunction with rotation when large annular clearances are present and a non visco-elastic mud is in use.

- Do not get too aggressive with size or time between sweeps. It is possible to pack off while pumping a sweep, especially at moderate angles of inclination.
- Minimize connection time through better planning and organization.
- Establish over-pull limits. We don't want to pull so hard into a packoff that the pipe cannot be freed with downward movement. The harder we pull, the tighter the packoff will become. It is better to "walk" through cuttings beds in small increments of overpull. When an obstacle is encountered, pull into it a short way, ensuring that downward motion is free, then pull a little harder and back off again to ensure downward motion is free. This process is repeated until the obstacle is passed or the overpull limit is reached. An overpull limit is established so that a decision will be made by the foreman instead of the driller on how hard to pull into a packoff before giving up and trying something else.
- Monitor the hole cleaning trends. Record free rotating weight, pick up weight, slack off weight, off bottom torque, on bottom torque, and circulating pressures. The best way to stay out of trouble is to watch for the early warning signs and take corrective action before the problem gets out of hand.
- Evaluate torque and drag trends, and the rate of cuttings returned to monitor hole cleaning. These trends should be recorded by the driller prior to each connection. A well site engineer can prepare a spreadsheet to monitor the volume reduction of active system. Convergence and divergence of actual vs. expected trends gives indications of hole cleaning. For example, a reduction of off bottom rotating torque may indicate the hole is loading up with cuttings.
- Use backreaming only as a last resort. If back reaming is required while tripping out be sure to run back in the hole several stands below the first sign of trouble.
- Record any tight spots during trips or connections.
- Understanding hole cleaning leads to fewer trips, less back reaming, optimized circulation time, and maximum penetration rates. The first wiper trip plus diverging/converging trends indicate when next trip is required.

The driller must always be aware of where his BHA is with respect to well bore geometry and known trouble spots. The drill collars are more likely to pack off in a dogleg or in full gauge hole just above a large washed out section. As the drill collars pass through a dogleg the collars plow cuttings and cavings in front of them. This debris is packed together where the bit and drill collars are forced against the well bore wall. (Fig 7-62)

A lithographic chart with drilling trends and caliper logs should be available on the rig floor prior to tripping. The driller can drag a paper model of the BHA along the chart as the bit is pulled to anticipate potential trouble. (See Fig 13-1 and the accompanying discussion on trend monitoring)

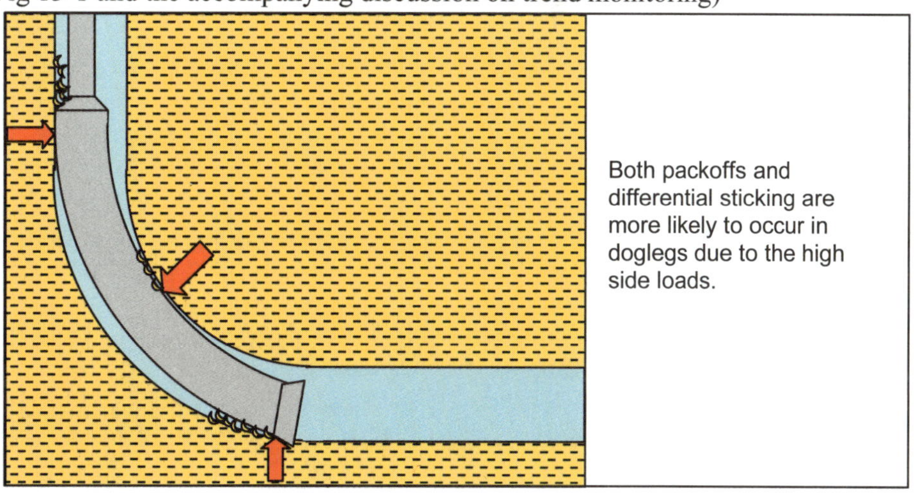

Fig 7-62 Packoffs in doglegs

Warning Signs

The rig crew must be alert to the warning signs of poor hole cleaning. Most stuck pipe due to insufficient hole cleaning shows up in drilling and tripping trends long before the pipe becomes stuck. The obvious warning signs are:

- Insufficient cuttings return for the rate of penetration.
- Erratic cuttings return.
- More solids being removed downstream from the shakers than normal. (The longer it takes a cutting to be transported up the well, the more it breaks down into smaller pieces)
- An increase in plastic viscosity, mud weight out, sand content, or low gravity solids. (As cuttings break into smaller and smaller pieces, the solids control equipment can no longer remove them.)

The poor hole cleaning trends that can be observed on the Geolograph are (Fig 7-63):

- Connection trends:
  - An increase in overpull off of slips and a pressure surge to start circulation. Cuttings have settled around the BHA and as we pick up off the slips and/or start the pumps, we start to pack off.
  - An increase in pump pressure to break circulation. As the annulus loads with cuttings, the difference in hydrostatic pressure between the clean mud in the drill string and the cuttings-laden mud in the annulus increases.
- Drilling trends:
  - A linear increase in pump pressure. As more cuttings are added to the annulus than are removed, the bottom-hole pressure increases. This increase occurs at the same rate of the cuttings overload, which is normally steady. It is important not to confuse normal pressure trends due to increased depth with cuttings overload. The driller must become familiar with normal trends in order to spot abnormal trends.
  - Erratic pump pressure. Sliding cuttings beds or extreme cuttings recycling is attempting to pack the string off at this point.
  - An increase in both torque and drag. Torque and drag also become more erratic. The build up of cuttings eventually begins to interfere with pipe movement.
  - Gradual decrease in ROP. The increase in bottom-hole pressure increases the apparent rock strength of the formation being drilled and thus lowers the rate of penetration. (See Rock Mechanics)

- Tripping Trends:
  - *Swabbing*: As the cuttings pack together around the bottom-hole assembly, they act as a piston. The well does not take the correct amount of fluid for drill string displacement as steel is removed from the well. Swabbing also increases, with an increase in plastic viscosity.
  - If the drill string is being pumped out, *string pistoning may be experienced.* This is when trapped pressure beneath the bit begins to pump or "piston" the drill string out of the hole. Some or all of the string weight will disappear. In extreme cases, the drill string moves upwards even when the drill line is slack. Thus, the overpull trend becomes masked.
  - *Excessive or erratic drag*: The cuttings are wadding up around the drill collars and bit and interfering with the pipes motion. The "tight spots" may be hard to pin down, meaning that the restrictions are not stationary. The conglomerations of cuttings or cuttings beds are shoved along the well bore and occasionally pushed out of the way. Tight spots due to doglegs and keyseats are stationary.

As hole cleaning problems develop the following trends are observed:
•Increasing overpull picking up off the slips after a connection.
•Increasing pump pressure to break circulation after a connection.
•Pressure surges or spikes.
•Increasing pump pressure, torque and drag while drilling.
•Torque, drag, and pump pressure becoming more erratic.
•Fill on bottom or difficulty in establishing weight on bit.
•Weight on bit decreasing (as bottom hole pressure attempts to pistons bit off bottom.)

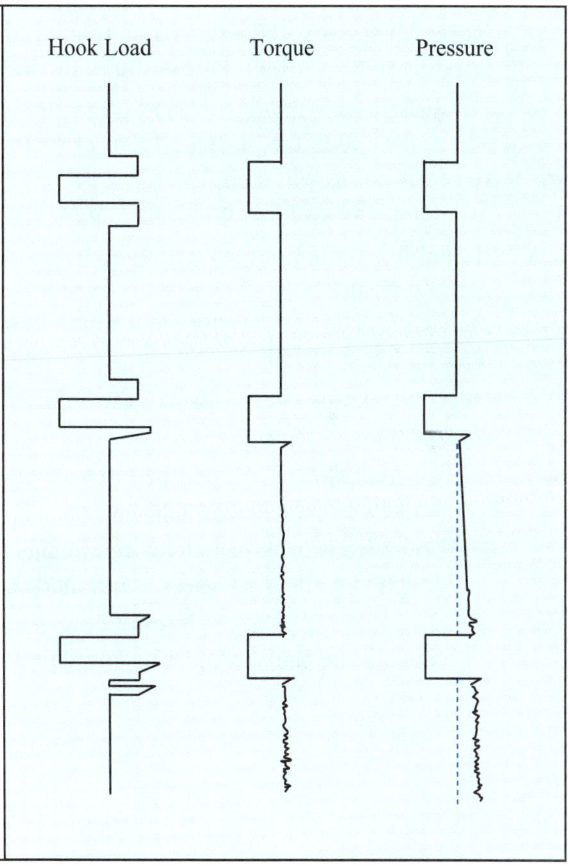

Fig 7-63 Geolograph trends for poor hole cleaning

Freeing Procedures

First Action

The first action to take for any type of packoff is to **bleed off any trapped pressure and apply 200 psi to 500 psi to try to re-establish circulation, then torque and slump the pipe. Jar down if jars are in the string.**

- Trapped pressure will pump the bit further into the packoff and make matters worse. Also, we will want to move downward and the pistoning effect will reduce the amount of downward force available. Low pressure is applied to re-establish circulation once any movement is established.
- Torque is applied to help establish pipe movement and circulation.
- The cuttings are moving downward and are wedged together when the pipe moves upward. Thus, the best direction for pipe movement is downward to reduce the wedging forces. If downward motion can be established, the packoff usually loosens, circulation can be re-established, and the packoff can be broken up with circulation and pipe movement.
  Note: Most packoffs occur while moving the pipe upward. In high angle holes however, it may be possible to packoff while running in the hole. In this case, move up not down.
- Jar with the maximum trip load if jars are in the string. Torque must be used carefully and in accordance with the jar manufactures recommendations. Torsional and tensile stresses are additive, so never jar up while applying torque. (It is okay to jar down while applying maximum torque. Torsional stress and compressive stress are not additive.)
- If the bit is on bottom, as it may be when making connections with a top drive, apply torque and low pump pressure only. Try to work pipe upward gradually. If warnings indicate the risk of packoff is increasing, consider laying out a single before making a connection to keep the bit off bottom while making the connection.
- Once circulation is established, the hole must be cleaned before further drilling or tripping can commence.
  - Viscous sweeps are pumped in vertical wells and a combination of low and high viscosity/high density sweeps in deviated wells.
  - Low viscosity sweeps with surfactants and lubricants may be required in both vertical and deviated wells if progress is not made soon after circulation is reestablished.
  - Note that the problem has not gone away once we are freed from the packoff. The cuttings causing the packoff must be circulated out of the well or we are likely to become stuck again.

Secondary Freeing Procedures

If the first action is unsuccessful, there are a number of other techniques that have proven successful.

Low Frequency Resonance Tools

A technique that is gaining popularity is the use of **Low Frequency Resonance** tools[21]. These tools impart resonant standing wave energy via wire line to the stuck points in the string. These vibrations break down and "fluidize" the rock and debris near the drill string. The drill string also dilates and contracts, which further reduces the friction forces.

Resonant pipe vibration can impart substantially more energy to the stuck point than any conventional mechanical means, such as jarring. Large chunks of debris or ledges are broken into small grains that are then "fluidized". When granular particles are exited by vibrational energy, they are transformed into a fluid-like material, which allows objects to pass through them as they would through a liquid. This can be very effective in packoffs due to cuttings beds or settled cutting. When the rock grains are fluidized, they will move out of the way of tool joints rather than become wedged between the drill string and formation. (It is a little like shaking a cup of ice as you push a straw into it)

Pulling and Pumping

Other techniques (used in desperation) include pulling until something breaks and pumping until something breaks. Sometimes the drill string can be pulled through a packoff. This should not be our first choice because pulling into the packoff tends to pack the debris tighter and only sticks us harder. If all else fails however, it may be possible to pull the pipe through the packoff if the drill string is strong enough. The driller should be aware of how much tension his string can withstand and not exceed it.

At times, the string has been pulled successfully with tensions beyond the advertised strength of the string. This should only be done if one is willing to discard some damaged string. Be aware that tensile failures are more likely to occur in the upper portion of the string which may result in a sizeable fish and more violent movement on surface when the pipe parts. Remember also that tensile stress and torsion stress are additive. Be certain that no torsion is applied to the drill string while pulling near its tensile capacity. (See combined loading formula in appendix B)

Sometimes it is possible to "pump" the pipe out of the hole with the mud pumps. This is not our first choice because pressure below the bit tends to piston the drill string into the packoff, making it worse. If the formations below the packoff are competent enough to withstand the pressure, it may be possible to "piston" the debris up the well. This has the same effect as moving down with the pipe. This technique is sometimes used when the bit is on bottom.

Fishing Technique

One other successful means of freeing pipe from a packoff is to **back off** above the stuck point and **wash over** the stuck pipe with wash pipe. This fishing technique will only be successful if the conditions leading to poor hole cleaning are addressed prior to washing over the stuck pipe. The length of wash pipe must be carefully selected with regards to stuck pipe from other mechanisms, such as differential sticking and well bore geometry.

A **fishing jar** can also be installed just above the stuck point after a back off. This is often employed if no drilling jars were in use or if they failed to fire.

Nomenclature

AV = Annular Velocity
ECD = Equivalent circulating density
d_c = Diameter of cutting
f = Fanning friction factor
g = Gravitational constant
HCR = Hole Cleaning Ratio
H_{crit} = Critical bed height
H = Free region height above cuttings bed
K = Plastic viscosity in power law equation
M = Momentum
MTV = Minimum Transport Velocity
MW = Mud Weight
n = Flow index
P_i = Pressure inside the drill string just above the bit
P_o = Bottom hole pressure outside the drill string
PV = Plastic viscosity
v = Average velocity above cuttings bed
V_a = Annular velocity
V_c = Velocity of cutting
V_s = Slip velocity
YP = Yield point
γ = Specific shear rate
θ = Shear stress measured in degrees of rotation
ρ_c = Density of cutting
ρ_f = Density of fluid
τ = Shear stress
τ_0 = Yield stress a zero shear rate

Test your understanding of Hole Cleaning

1. What is the volumetric cuttings concentration? (Eq. 7.1, pg. 46)
2. What is the transport ratio? What two approaches help increase the transport ratio? (Eq. 7.2, pg. 46)
3. List 7 factors influencing hole cleaning in a vertical well. (Pg. 47)
4. Which factor has the most influence on the slip velocity of the cutting? (Pg 48)
5. How does the flow profile affect hole cleaning? (Fig 7- 3, 7- 4 pg. 49)
6. What do we mean by the phrases "flow regime" and flow profile? (Fig 7- 6, 7- 7, pg. 51)
7. What is yield point? (Fig 7- 8, pg. 52)
8. What are the units of measurement of yield point? (Pg. 53)
9. What is plastic viscosity? (Fig 7- 12, pg. 55)
10. How do yield point and plastic viscosity affect hole cleaning? (Fig 7- 7, pg. 51)
11. Why was the "power law" model for shear stress developed? (Fig 7- 10, eq. 7.8 pg. 56)
12. What is shear thinning? What does it do for us? (Pg. 57 & 71)
13. How does the rate of penetration affect the flow profile and our hole cleaning effort in a vertical well? (Fig 7- 15, pg. 59)
14. How does pipe rotation and eccentricity affect the flow profile and our hole cleaning effort? (Fig 7-16, 7- 17, pg. 60)
15. Give two reasons hole cleaning is more difficult in directional wells. (Pg. 61- 70)
16. List 7 factors influencing hole cleaning in a directional well. (Pg. 62)
17. At what angle do cutting beds begin to form? (Fig 7- 19, 7- 20, pg 62)
18. What affect does hole angle have on cutting bed height? (Fig, 7-20, Pg 63)
19. What is Boycott settling? (Fig, 7-21, Pg. 64)
20. Why would we want laminar flow in the vertical section of the well but turbulent flow in the horizontal section? (Fig 7- 26, pg. 68)
21. Why will our annular velocity decrease in the horizontal section when we increase our mud weight? (Fig 7- 25, 7- 29, pg. 72)
22. Why does an increase in yield point improve hole cleaning in a vertical well but hurt hole cleaning in a directional well? (Fig 7- 26, 7- 27, pg. 68, 69)
23. Why do we want a high shear thinning mud in the high angle section of the well? (Fig 7-28, pg. 71)
24. Is the height of our cutting beds influenced by the rate of penetration? (Fig 7-29, pg. 72- 73) (It affects cutting size, which will affect the MTV, but otherwise the cuttings bed won't change)
25. If we increase our flow rate while drilling with a down hole motor why won't annular velocity increase in the horizontal section? (Pg. 72- 73)
26. What is the relationship between cutting bed height and annular clearance around the bit and BHA? (Fig 7- 34, pg. 75)
27. List the cutting transport mechanisms. (Fig 7- 37 to 7-39, pg. 78)
28. Which transport mechanism is the most efficient? (Pg. 78)
29. Are we likely to have a suspension type of cutting transport in the horizontal section? (Fig 7-40, pg. 79)
30. How can the monitoring of surface volume warn us of the existence of large cutting beds? (Fig 7- 35, pg. 76)
31. When the shakers come clean after circulating at TD is it safe to assume the high angle section of the well is clean? (Fig 7- 40, pg. 79)

32. What happens to the annular velocity when the pipe is rotated? (Fig 7- 42, pg. 84)
33. What happens to the flow profile when the drill pipe is rotated? (Fig 7- 42, pg. 84)
34. What is visco-elastic coupling and how does it influence hole cleaning? (Fig 7- 43, pg. 85)
35. Why is pipe rotation so important in high angle holes? (Fig 7- 44, pg. 85)
36. Why is a threshold rotational velocity required to disturb cutting beds with the drill pipe? (Fig 7- 44, pg. 85)
37. Can pipe rotation lead to turbulent flow in the annulus? (Fig 7- 45, pg. 86)
38. Under what conditions rotation of 150 rpm to 180 rpm would be needed? (Fig 7- 46, pg. 86)
39. How can PWD data be useful to determining how our hole cleaning efforts are going? (Fig 7- 47, pg. 87)
40. Why is drilling with a compressible medium, such as air, so different from drilling with a non-compressible medium, such as mud? (Fig 7- 48, pg. 95)
41. When might viscous sweeps be effective in a high angle hole? (Fig 7- 50, pg. 91)
42. What is the problem with viscous sweeps in high angle hole? (Fig 7- 5, pg. 91)
43. What does foam quality refer to and how does it affe52ct hole cleaning? (Fig 7- 57, 7- 58, pg. 100)
44. Why do we need to manage back pressure while drilling with air or foam? (Fig 7-59, pg. 102)
45. Under what conditions can we expect to see hole cleaning problems? (Pg. 105)
46. What can we do to prevent hole cleaning problems? (Pg. 106)
47. Why are we more likely to pack off in a dogleg? (Fig 7- 62, pg. 107)
48. What are the warning signs we are likely to see if we are not adequately cleaning the hole? (Pg. 108, Fig 7- 63, pg. 109)
49. What first action will we take if we become stuck do to a hole cleaning problem? (Pg. 110)
50. What are some secondary freeing procedures we might try? (Pg. 111)

Bibliography

1) Sifferman, T.R., Myers, G.M., Haden, E.L, and Wall, H.A.: "Drill-Cutting Transport in Full Scale Vertical Annuli," J. Petrol Tech. (Nov 1974)
2) Williams, C.E., & Bruce, G.H.: "Carrying Capacity of Drilling Muds" Trans. AIME (1951)
3) Becker, Thomas E., & Azar, J.J.: "Mud-Weight and Hole-Geometry Effects on Cuttings Transport While Drilling Directionally" SPE 14711 (Aug 1985)
4) Gray, George R. & Darley, H. C. H.: "Composition and Properties of Oil Well Drilling Fluid's" fourth edition, Gulf Publishing Company (1980)
5) Yuejin Luo, Bern, P.A., Chambers, B.D., & Kellingray, D.S.: "Simple Charts to Determine Hole Cleaning Requirements in Deviated Wells" IADC/SPE 27486, 1994 IADC/SPE Drilling Conference in Dallas Texas (Feb 1994)
6) Patrick Kenny, Egil Sunde, & Terry Hemphill: "Hole Cleaning Modeling: What's 'n' Got To Do With It?" IADC/SPE 35099, 1996 SPE/IADC Drilling Conference in New Orleans LA (March 1996)
7) Okrjni, Slavomir S., Azar, J.J.: "The Effects of Mud Rheology on Annular Hole Cleaning in Directional Wells" SPE reprint series no. 30 "Directional Drilling" (1990)
8) Marco Rasi: "Hole Cleaning in Large, High-Angle Well bores" IADC/SPE 27464, IADC/SPE Drilling Conference, Dallas, Texas (Feb. 1994)
9) Guild, G.J., Tom Hill & Associates, Wallace, I.M., & Wassenborg, M.J.: "Hole Cleaning Program for Extended Reach Wells" SPE/IADC 29381, 1995 SPE/IADC Drilling Conference in Amsterdam (Feb 1995)
10) Ford, J.T., Peden, J.M., Oyeneyin, M.B., Erhu Gao, & Zarrough R.: "Experimental Investigation of Drilled Cuttings Transport in Inclined Boreholes" SPE 20421, 65th Ann. Tech. Conference of SPE in New Orleans (Sept. 1990)
11) Sifferman, T.R., & Becker, T.E.: "Hole Cleaning in Full-Scale Inclined Well bores" SPE 20422, 65th Ann. Tech. Conference of SPE in New Orleans (Sept. 1990)
12) Easton, M.D.J., Nichols, J., & Riley, G.J.: "Optimizing Hole Cleaning by Application of a Pressure While Drilling Tool" SPE 37612, 1997 SPE/IADC Drilling Conference in Amsterdam (March 1997)
13) Leising, L.J., & Walton, I.C.: "'Cutting Transport Problems and Solutions in Coiled Tubing Drilling" IADC/SPE 39300, 1998 IADC/SPE Drilling Conference in Dallas Texas (March 1998)
14) McCann, R.C., Quigley, M.S., Zamora, M., and Slater, K.S.: "Effects of High-Speed Pipe Rotation on Pressures in Narrow Annuli" SPE 26343 presented at the 1993 SPE Annual Technical Conference and Exhibition in Houston (Oct 1993)
15) "Training to Reduce Unscheduled Events" a pre-spud training course developed and owned by BP Amoco. (1996)
16) Supon, S.B and Adewumi, M.A.: "An Experimental Study of the Annulus Pressure Drop in a Simulated Air-Drilling Operation", SPE Drill. Eng. (March 1991)
17) Lyons, W.C.: "Air and Gas Drilling Manual", Gulf Publishing Co. (1984)
18) Mitchell, B.J.: "Test Data Fill Theory Gap on Using Foam as a Drilling Fluid", Oil and Gas Journal (Sept 1971)
19) Krug, J.A., and Mitchell, B.J.: "Charts Help Find Volume, Pressure Needed for Foam Drilling," Oil and Gas Journal (Feb 1972)
20) Guo, B., Miska, S. and Hareland, G.: "A Simple Approach to Determination of Bottom-hole Pressure in Directional Foam Drilling," presented at the 1995 ASME Energy and Environmental Expo 95, Houston, TX (Jan 1995)
21) Buck Bernat, Henry Bernat, Vibration Technology LLC Shreveport: "Mechanical Oscillator Frees Stuck Pipe Strings Using Resonance Technology" Oil and Gas Journal (Nov 3, 1997)
22) Hopkins, C.J. and Leicksenring, R.A.: "Reducing the Risk of Stuck Pipe in the Netherlands." Paper SPE/IADC 29422 presented at the 1995 SPE/IADC Drilling Conference, Amsterdam (Feb 1995)

Chapter 8
Well Bore Instability

Introduction

Well bore instability is responsible for the most serious type of stuck pipe. When packoff is due to well bore collapse, we often lose our tools and have to sidetrack. As the name implies, well bore instability refers to an unstable well bore that tends to cave or collapse on us. Unconsolidated formations, fractured formations, and chemically or mechanically stressed shale are all unstable formations that can cave and cause a pack off.

The goal of this chapter is to bring an understanding of the mechanics of wellbore stability to the men who supervise the drilling of the well. The more one understands a problem the better armed he is to deal with it effectively.

The chapter begins with a brief discussion of rock strength and then unfolds into a deeper explanation of rock mechanics as it pertains to drilling. Some of this discussion involves engineering subjects such as "Mohr's Failure Envelopes". These are tools that help drilling engineers pick mud weight windows and well bore trajectories. They are offered in this chapter as a tool to help the drilling supervisor better understand the concepts of stability, and to reason more effectively with the well design team.

The factors affecting stability are discussed in detail. Then the different types of failure are addressed. If the rig crew can recognize the type of failure, and understands the factors attributing to the failure, they will be better prepared to take corrective action.

A summary of preventive measures, warning signs, and first actions to take in the event of trouble is presented next. With an understanding of the mechanics of well bore stability, the reasoning behind these warning signs and actions can be explained. This is where the effort in the first parts of the chapter pay off. *The better the rig crew understands the science behind their actions, the more confident, and successful, they will be in applying them.*

Shale Instability

Most of our instability problems occur in shale. This is because it is one of the most common formation types encountered and it's one of the weakest. Most of our discussion of well bore stability will involve shale, so a brief explanation of the different types of shale and how they are formed will be useful.

Shale Formation

Shale is formed in a marine environment. It is composed of very fine sediments that gradually settle to the sea floor and compact together. (Fig 8-1) As the shale is first deposited each grain is completely surrounded by water. The grains are barely touching each other and all of the water between the grains is continuous. In fact right at the seabed there may be so much water in between the grains that it may be difficult to determine where the sea ends and sediment begins.

As more and more sediment is deposited, the weight of this overburden compacts the grains closer together and squeezes out the water. As the grains pack closer together, porosity and permeability are reduced.

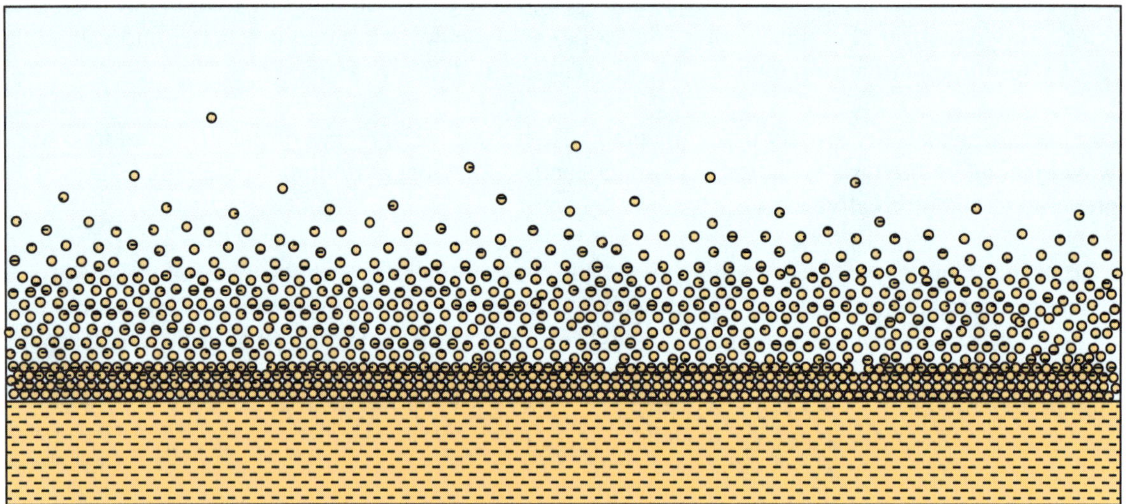

Shale is formed in a marine environment. Each micron size particle is completely surrounded by water as it is deposited. As the shale is compacted by overburden, much of the water is expelled and the shale becomes progressively firmer and more impermeable with depth.

Fig 8-1 Shale formation

Porosity is defined as the percent of pore or void space within the rock. **Permeability** is a measure of the ability of fluid to flow through the rock. Permeability is a result of the pores being connected with each other such that there is a passage for fluid to flow.

Shale is the most porous of all the sedimentary rocks. It also tends to be the least permeable. This is because it is composed of very small grains and the connections between the pore spaces become so small after compaction that water can barely flow through them. Eventually, the shale becomes compacted to such a degree that it is relatively impermeable and water can no longer escape from the pore spaces.

In areas of rapid deposition, a shale sequence can become so thick that even though it is still permeable, fluid cannot escape from its pores. This is because there is no place for the fluid to go. The amount and type of fluid trapped in the pore spaces has a significant effect on a shale strength and behavior. The type of sediment that comprises the shale, as well as the amount of compaction and cementation, also has a significant effect on its strength and behavior. Young shale near the surface tends to be soft and plastic, while older shale at depth tends to be hard and brittle.

A high percentage of the sediments in shale are clay particles. This is because the larger sand and silt particles settle out in the river deltas, leaving only the small colloidal- size clay particles to settle farther out in still waters. Clay particles are solids smaller than 2 microns and consist of a mixture of minerals that form a flat sheet-like crystalline structure similar to mica. As the clay is deposited and compacted, the crystalline structures grow to form larger flat sheets. These sheet-like structures orient themselves such that they stack together like a deck of cards. The sheets themselves are fairly strong, but these sheets are easily separated from each other. (Fig 8-2)

The properties of any specific clay depend on the minerals of which it is comprised. There are several types of clays that give rise to several types of shale. The most common clay mineral groups are Smectites, Illites, Kaolinites, and Chlorites.

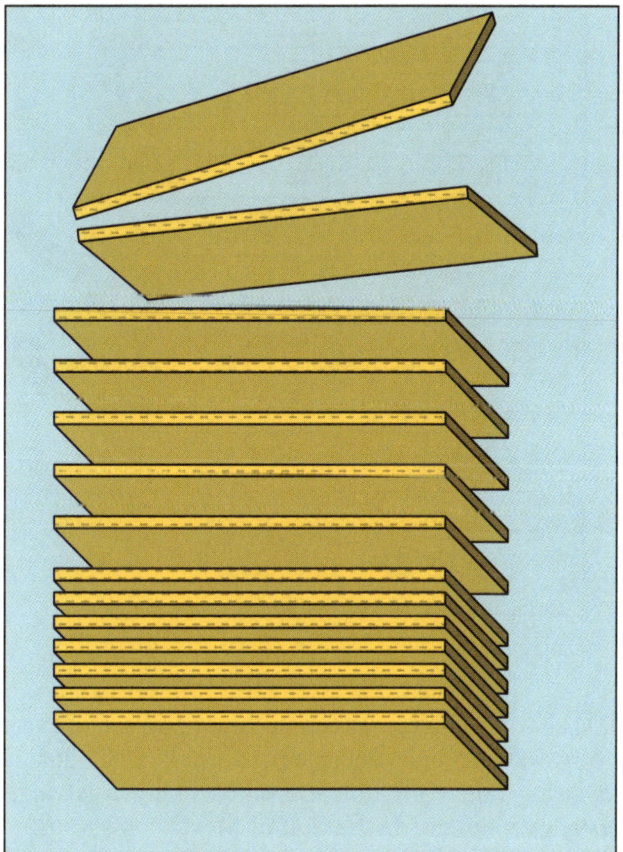

Shale is comprised of microscopic sheets of clay that stack together like a deck of cards.

Fig 8-2 Shale formation

Clay Mineral Groups

Smectites are clays that swell in the presence of water. The bonding between the layers of crystalline sheets is weaker than any other type of clay. Water can easily penetrate between these layers and force them apart. As the layers of clay expand the clay is said to swell. The individual platelets of clay can continue to expand to such a degree that they are no longer associated with each other. This degree of expansion is known as **dispersion**. (Fig 8-3)

Sodium Montmorillonite is a common and troublesome member of the Smectite group. It is more commonly known as **bentonite**. Bentonite can swell up to 20 times its size before it completely disperses.[1] Bentonite and other smectites are frequently found near the surface in young, recently deposited clay formations. Smectites tend to make soft, spongy shale that drills easily but squeezes in and reduces the well bore diameter. Hole enlargement also occurs in smectites because of dispersion and caving.

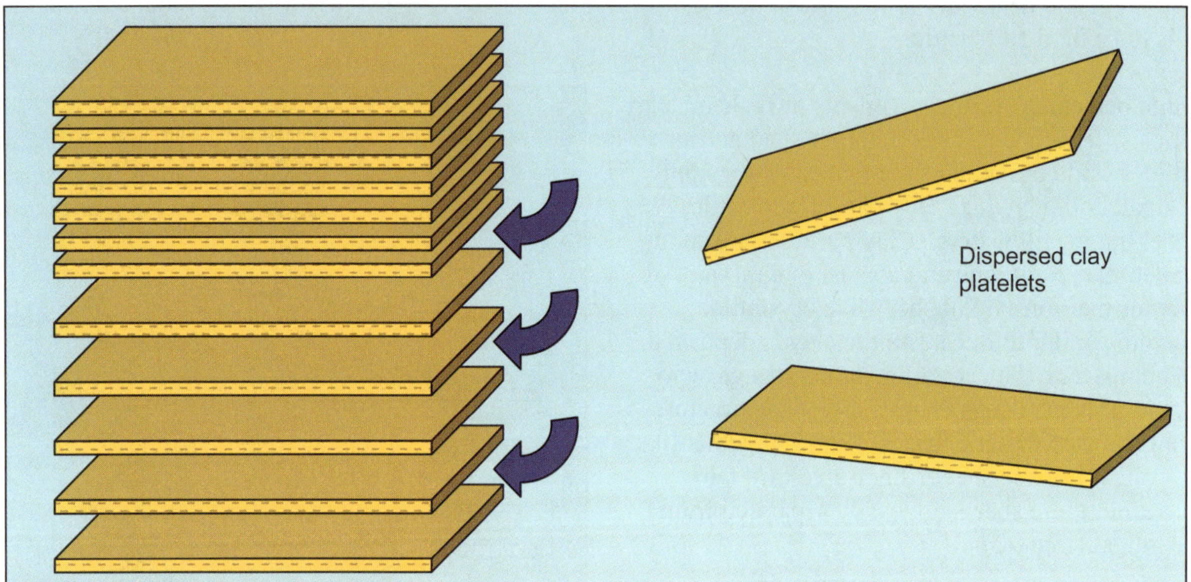

Water can easily penetrate between the layers of bentonite clays. This causes the clay layers to expand and eventually disperse.

Fig 8-3 Dispersion

Illites have a structure similar to Montmorillonite except that they do not have an expanding lattice. This means water cannot penetrate between the individual layers of clay. Montmorillonite will chemically and physically convert to Illite in a process known as *diagenesis* as it is subjected to the heat and pressures from heavy overburden. In the Gulf of Mexico, this transformation takes place between 10 and 14 thousand feet. Illites make harder, more brittle shale that tends to drill more slowly. *Shale instability with Illites usually results in hole enlargement.*

Kaolinites and Chlorites do not expand much in the presence of water either. Chlorite may expand more than Kaolinite or Illite, but not as much as the smectites.

We will examine the swelling mechanisms of clays in the section on chemically induced stress.

The clay mineralogy influences the physical and chemical properties of the shale. Other factors that influence these properties are:

- The age and amount of compaction
- The amount, orientation, and strength of bedding planes
- Permeability
- Interbedding and contamination with salts and sands
- Pore fluid and pore pressure
- The amount of tectonic and overburden stress to which the shale is subjected

These different properties give rise to many different types of shale. Some are soft and plastic, others hard and brittle, some are young and others old, some shale has weak bedding planes, others don't. These different shales have different strengths and modes of failure. But before we can understand the various modes and causes of well bore failure, we must first learn something about rock strength.

Rock Mechanics

To find out how strong a particular formation is, we must test a core sample in a lab, where it will be compressed in a hydraulic press (Fig 8-4). We will gradually apply compressive force until the sample breaks. The stress felt in the rock at the moment it fails is called the uniaxial **compressive rock strength**.

As the core is compressed, we take very careful measurements of its length and diameter with calipers or strain gauges. From our measurements, we find that the rock gets shorter as it is compressed. We also find that as it gets shorter it increases in diameter. This occurs because the core is trying to maintain its original volume. The change in vertical length under stress is called **axial strain**, and the change in diameter under stress is called **radial strain**.

The ratio of radial strain to axial strain is a function of Poisson's ratio and will be explained later.

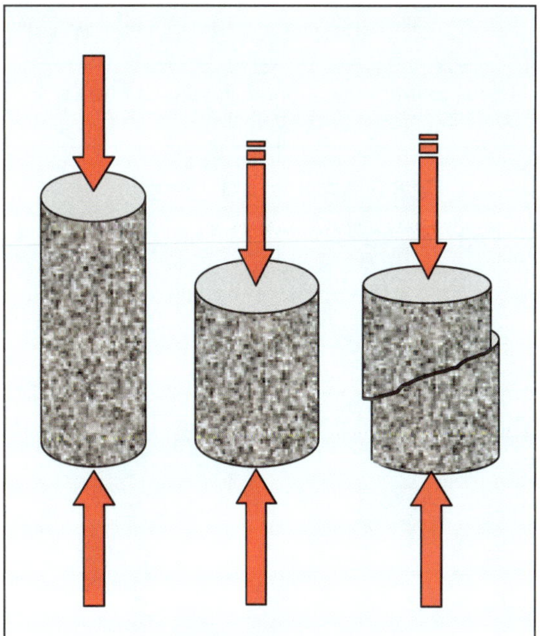

Rock strength is determined by compressing a core sample in a laboratory.

Fig 8-4 Rock strength

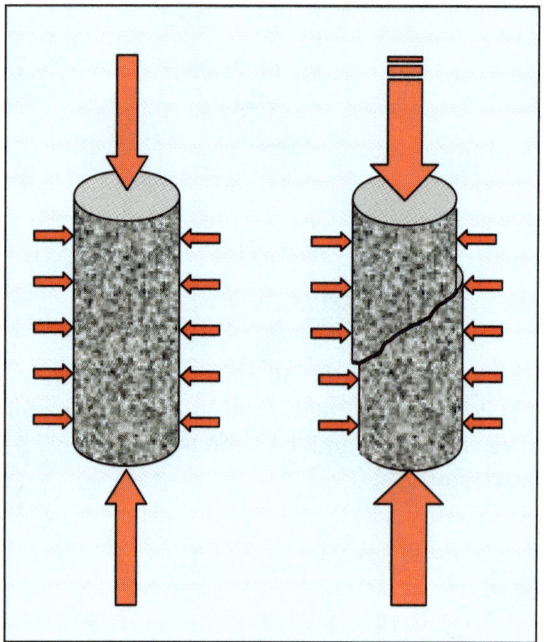

The apparent rock strength increases when a confining pressure is applied.

Fig 8-5 Apparent rock strength

If the core is placed in a cylinder, such that a constraining force can be applied to resist its growth in diameter, we find that it becomes harder to make it shorter. (Fig 8-4) We also find that a higher compressive force is now needed to break the rock. This is because the rock is stronger when it is subjected to a constraining force. The force required to break the rock when a constraining force is applied is called the **apparent rock strength**. Higher constraining forces give higher apparent rock strengths. The constraining force is often referred to as confining pressure. (Fig 8-5)

The rock also becomes less elastic and more plastic with increasing confining pressure. It becomes more likely to deform than break when it is subjected to very high confining pressure.

The rock becomes stronger and more plastic with increasing confining pressure.

Where does the rocks strength come from?
What physical properties resist deformation and failure?

To answer these questions, lets look at the rock under a microscope (Fig 8-6). In order for the rock to break or change shape, there must be movement between the individual grains of rock. One part of the rock moves one way and the other part moves in the opposite direction. *Friction and cementation between the individual grains prevents this movement and gives the rock its strength.*

Rock strength is determined by the amount of friction and cementation between the grains.

Overburden and the shape and orientation of the grains influence the friction.

The age of the rock affects both the cementation and the shape and orientation of the grains.

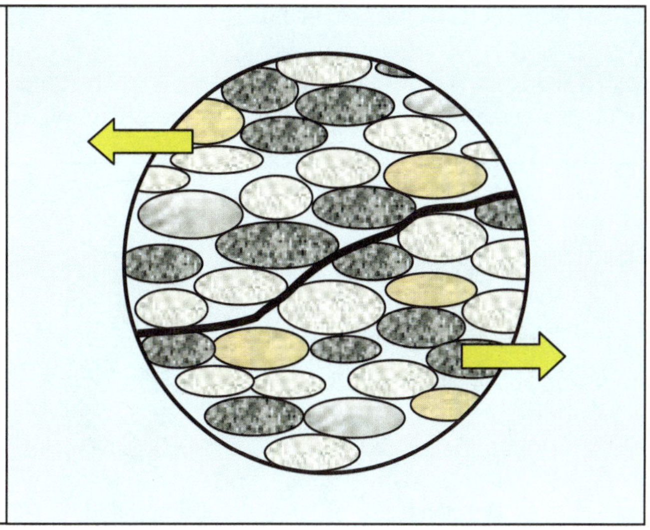

Fig 8-6 Rock strength

Most of the rock strength comes from **friction** between the grains. Several things influence this friction, including:

- The size and shape of the grains
- The orientation of the grains
- The compressive forces across the grains
- The amount of lubricating fluid in the pore spaces

Friction is defined by the following formula:

$$\text{Friction} = \mu N \qquad \text{eq. 8.1}$$

Where:
μ = coefficient of friction
N = Normal force

The coefficient of friction is a measure of surface roughness and the normal force is a measure of how hard the two surfaces are pressed together. Friction then is a function of how hard the grains are pressed against each other.

Cementation comes from minerals that precipitate out of the water passing through the formation. You may have noticed the rock-like material that forms on the bottom of an electric teapot. This is similar to the precipitates that come out of solution in a formation. These precipitates cement the grains together and increase the rock strength.

The concepts of friction and cementation can be illustrated by sand on a beach. (Fig 8-7) When the sand is dry, it can be piled in a conical shape, much like a volcano or pyramid. The reason the sand does not just run out flat like marbles would is because of the friction between the individual grains. The friction between the grains gives the sand enough strength to form a hill or hold what is known as an **angle of repose**.

If the sand is damp, it becomes strong enough to make sand castles. The walls will now stand vertically. The cohesive and adhesive properties of water act as a cement to bind the grains together. The water also helps the sand grains pack tighter together, which increases the friction between the grains. The damp sand makes a stronger building material than dry sand.

However, if the sand is thoroughly saturated with water it has no strength and will run out flat. This is because the water now acts as a lubricant. Without friction, the sand has no strength.

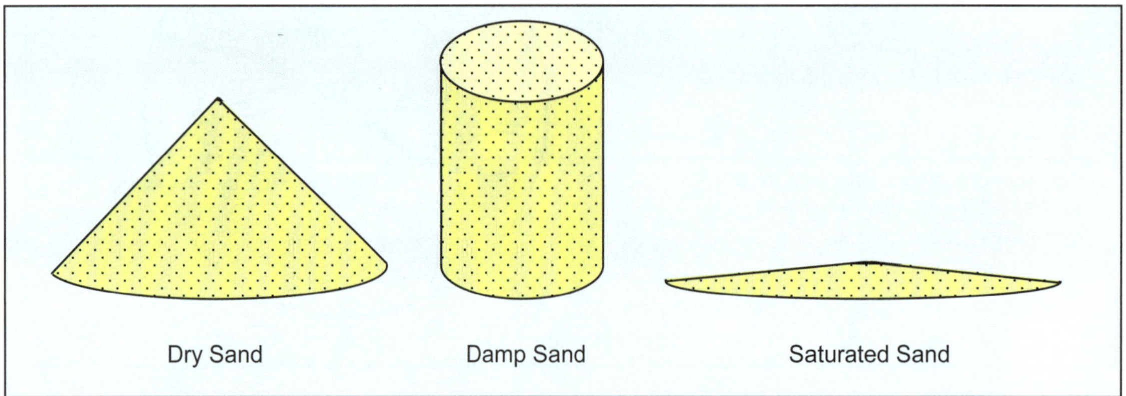

The relative strength of a pile of sand can be determined from the angle of its slope. There is enough friction between the grains of dry sand for it to form a cone. The grains in damp sand are drawn closer together, which increases the internal friction and makes it strong enough to form a vertical wall. Excessive water in saturated sand forces the grains apart and lubricates them. With very little friction between the grains, the sand flattens.

Fig 8-7 Rock strength analogy

Imagine digging a hole in the beach. The wall of the hole in the dry sand will have the same slope as a hill of dry sand. When damp sand is reached, the walls will be vertical. When the water table is reached, we can dig no deeper—the sand flows into the hole as it is dug.

Now lets return to our core sample in the laboratory. When a confining pressure was applied to our core sample, the internal friction between the grains was increased. This is why the rock was stronger when subjected to the confining pressure.

Rock Mechanics Terminology

We will talk a lot about stress and strain as we explain rock strength and well bore instability so before we go any farther lets define some of the various terms we will use.

Stress

Just as forces are transferred through liquids by pressure, forces are transferred through solids by stress. **Stress** is force divided by area and has the same units as pressure. It is represented by the Greek letter sigma, **σ**. Unlike pressure however, stress can be positive or negative. A solid can be subjected to the following stresses:

- Compressive stress
- Tensile stress
- Shear stress

Compressive stress occurs when material is in compression.
Tensile stress occurs when material is in tension, like a cable suspending a load.
Shear stress resists lateral movement within the material. (Fig 8-8)

A solid can be subjected to all three stresses simultaneously.

When a page is torn, it is subjected to shear stress. It is important to note that most well bore failure occurs from excessive shear stress. Shear stress increases as the difference between perpendicular stresses increases. The difference in perpendicular stresses causes an object to deform. In order for the object to deform, lateral movement must occur between the elements within the object. Shear stress causes this lateral movement and it is represented by the Greek letter tau, τ.

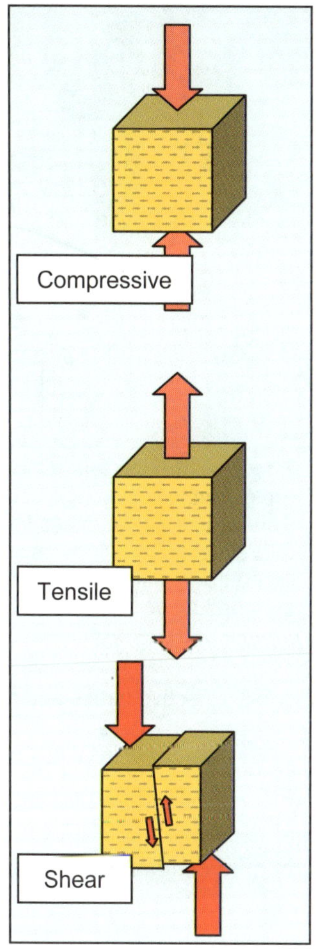

Fig 8-8 Stress states

Effective Stress

Not all of the stress in the formation is carried by the rock matrix. Some of this stress is carried by the fluid trapped in the pore spaces within the rock. (Fig 8-9)

As previously mentioned, when shale is first deposited, each grain is surrounded by water. As the shale is compacted, this water is squeezed out. However, the shale sequence is sometimes too thick or the permeability becomes so reduced that fluid can no longer be squeezed out of the shale as it is compacted. When this occurs, the fluid in the pore spaces begins to accept some of the load, similar to how air pressure in a tire supports the load of the car. The total stress felt in the formation is divided between the stress carried by the rock matrix and the stress carried by the pore fluid.

That part of the stress felt by the rock matrix is called **effective stress**. *Intergranular stress* and *matrix stress* are other names for effective stress.

The stress carried by fluid in the pore spaces is expressed as **pore pressure**. The combination of pore pressure and effective stress is the total stress.

$$\text{Total stress} = \text{pore pressure} + \text{effective stress} \qquad \text{eq. 8.2}$$

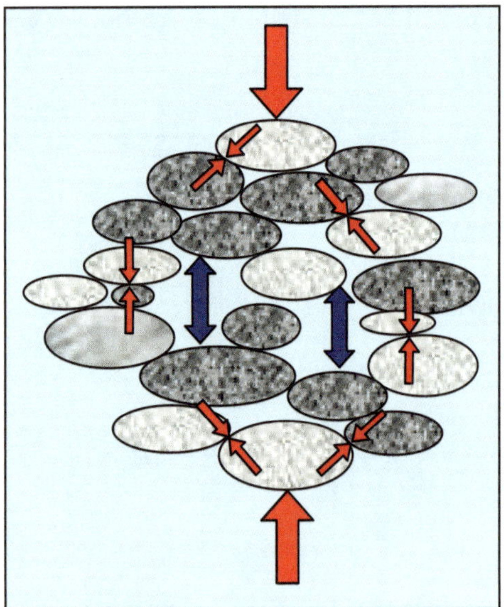

Effective stress is felt at the grain to grain contacts.
Pore pressure is a stress felt at the fluid to grain contacts. It helps support the overburden just as air pressure in a tire supports a car.

Fig 8-9 Effective stress and pore pressure

The deformation and strength of a rock specimen is dependent only on the effective stress. It is the stress felt between the grains that controls the movement of these grains relative to each other. Intergranular slippage and deformation is independent of pore pressure. Therefore, it is the effective stress we are concerned with when investigating rock strength.

Remember from fig 8-6 that rock strength is largely caused by friction between the grain to grain contacts. The harder the grains are pressed against each other the higher the friction. Effective stress is a measure of how hard the grains are pressed together.

As we will see later, **an increase in pore pressure reduces the effective stress, and thus weakens the rock**.

Strain

Strain is defined as a materials change in length or width under the influence of stress. It is represented by the Greek letter epsilon ε. The more stress the rock is subjected to, the more strain it experiences.

When the core sample was compressed in the laboratory, it was subjected it to compressional stress. From our measurements, we observed both axial and radial strain when the core was subjected to this stress.

The graph in Fig 8-10 illustrates a typical stress strain relationship for a core sample under compression. Note how this curve resembles the curve from a leak-off test. During a leak off test, the rock is subjected to tensile stress and the drilling fluid to compressional stress. Both the mud and the rock experience strain under these stresses. The mud is compressed while the well bore gets larger. The amount of additional mud required to fill the well is a measure of the combined strains. Thus, the LOT chart represents a stress strain relationship measured in bbls and psi.

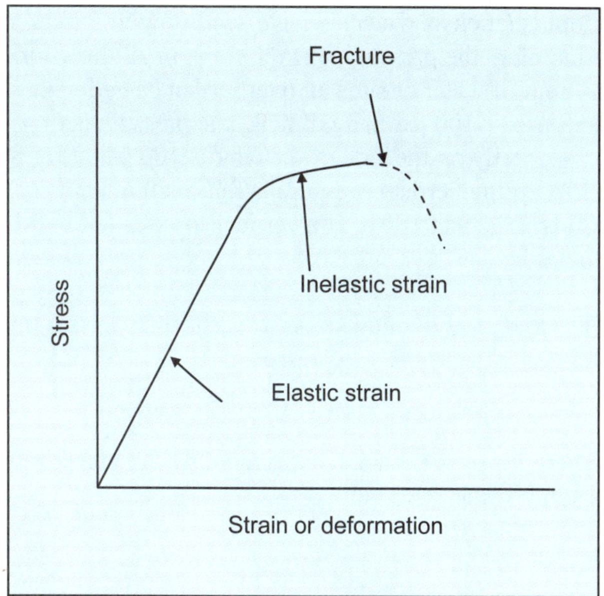

A leak off test is a stress vs. strain graph.

Fig 8-10 Stress-strain relationship

When the level of stress is low, the rock behaves **elastically**. The rock will return to its original dimensions when the stress is removed, just like a rubber band. The straight-line portion of the graph represents the region of **elastic strain**. Once a level of stress is attained, such that the elastic limit is exceeded, the material will be permanently deformed or fractured. This is referred to as **in-elastic** or **plastic strain**. With most rocks, some permanent deformation occurs before the ultimate strength is reached. Once the ultimate strength is exceeded, the rock breaks. Soft shale may experience a lot of permanent deformation before its ultimate strength is reached. Brittle limestone may shatter shortly after its elastic limit is exceeded.

Brittle vs. Ductile

Brittle rocks exhibit very little inelastic strain before failure. Ductile rocks exhibit substantial inelastic strain before failure. (Fig 8-11)

Brittle rocks fail when their ultimate stress is reached. They also reach their ultimate strain at this point. With brittle rocks, an increase in deformation results in a decrease in strength. Ductile rocks can still support a load after some deformation. Their failure is less catastrophic.

Both brittle and ductile rocks see an increase in ultimate strength and ductility with an increase in confining pressure. If the confining pressure becomes high enough, there will be a transition from brittle to ductile behavior in all rocks.

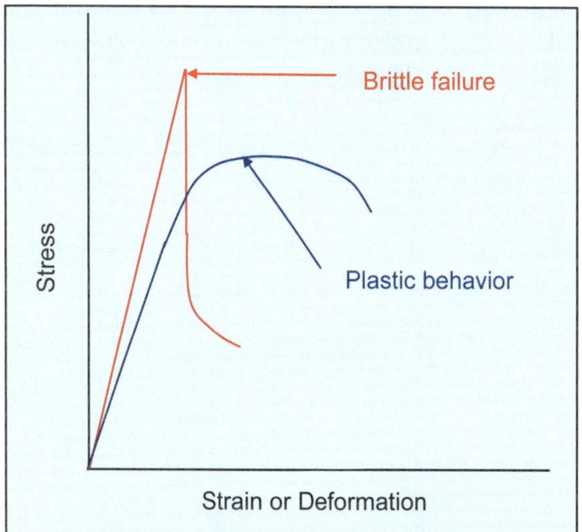

Brittle rocks loose their strength at failure while plastic rocks deform but still retain strength.

Fig 8-11 Stress-strain relationship

Poisson's Ratio

Liquids (such as water) behave according to Pascal's principle—the pressure acts the same in all directions. If a column of water creates an overburden or hydrostatic pressure of 100 psi in a tall tank, the pressure acting horizontally on the side of the tank is 100 psi. (Fig 8-12) 100% of the vertical pressure is felt in all other directions. In solids, less than 100% of a vertical force is felt in other directions.

The ratio of horizontal stress to vertical stress is a function of Poisson's ratio and is expressed as:

$$\sigma_h / \sigma_v = \nu/(1-\nu) \qquad \text{eq. 8.3}$$

Where Poisson's ratio, (represented by the greek letter Nu, ν) is some value less than 0.5.

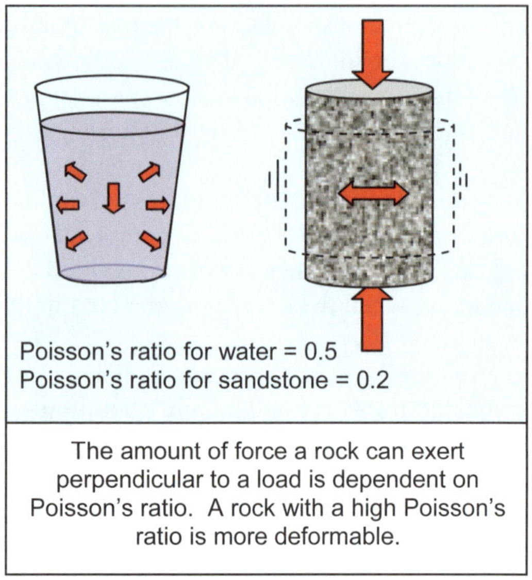

Fig 8-12 Poisson's ratio

The ratio of horizontal strain to vertical strain is also related to Poisson's ratio and is expressed as:

$$\varepsilon_h / \varepsilon_v = \nu \qquad \text{eq. 8.4}$$

Water has a Poisson's ratio of 0.5. 100% of the pressure or stress felt in the vertical direction is felt laterally. We could measure Poisson's ratio for the rock in the core sample by finding the ratio of radial strain to axial strain.

With soft, recently deposited clays and shale, Poisson's value is high, so the horizontal in-situ stresses are high. The strong, brittle rocks (like old dolomites) have a smaller Poisson's ratio, so less vertical force is felt horizontally. (Fig 8-13) In other words, some rocks are more deformable than others; they have a relatively high Poisson's ratio. Rocks that are more brittle have a lower Poisson's ratio; they experience less deformation under stress.

The Poisson's ratio for hard sandstones can be as low as .01. Limestones vary from 0.15 to 0.31.

The values for shale vary from 0.17 to 0.28, and clay can be as low as 0.17 or as high as 0.50 (if the clay is very wet).

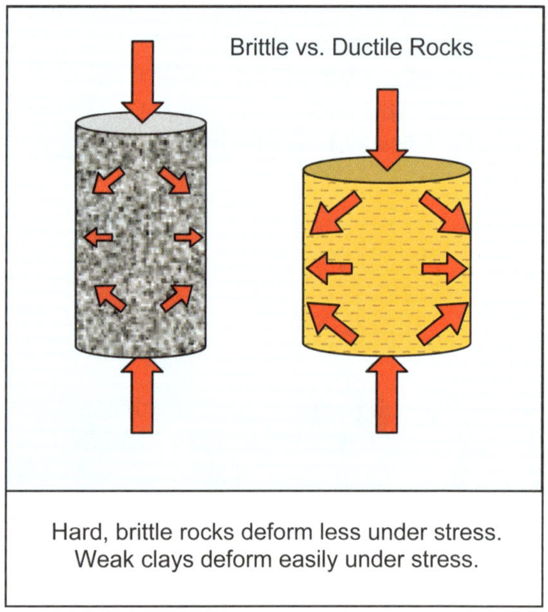

Fig 8-13 Brittle vs. ductile behavior

Why do we want to know Poisson's ratio for the material we drill through? Because plastic rocks tend to squeeze in on us due to the weight of the rock above it. A higher mud weight will be needed for rocks with higher Poisson's ratio to prevent the well from squeezing in or collapsing.

The values of Poisson's ratio, as determined in a lab test, assume that the confining pressure is equal from all directions. This is not the case in the earth's crust.

Tri-axial Stress State & Principal Stresses

Solids do not obey Pascal's principal. The stress felt in one direction is not necessarily equal to the stress felt in orthogonal (perpendicular) directions. To understand the stress strain relationship of a solid, we must represent the stress in a three dimensional or tri-axial state. The three dimensional stress state of an element is represented by three principal stresses. (Fig 8-14)

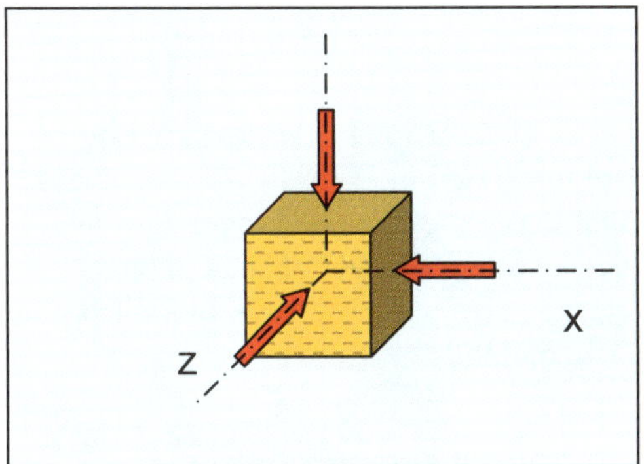

The tri-axial state of stress represents the principal stresses. The principal stresses are the maximum and minimum stresses and the stress orthogonal to both of them.

Fig 8-14 Three-dimensional stress state

The term **principal** means that one of these stresses is the maximum stress the element is subjected to and one of them is the minimum. The third or intermediate stress is orthogonal to both the maximum and minimum stress.

The vertical stress felt from the overlaying rock is referred to as **overburden stress** or $\boldsymbol{\sigma_v}$. The horizontal forces derived from horizontal strain are referred to as $\boldsymbol{\sigma_H}$ and $\boldsymbol{\sigma_h}$. σ_H is the larger of the two horizontal stresses and σ_h is the smaller of the two horizontal stresses. These stresses are often referred to as the major and minor horizontal stresses. All three of these principal stresses are orthogonal or perpendicular to each other. (Fig 8-14)

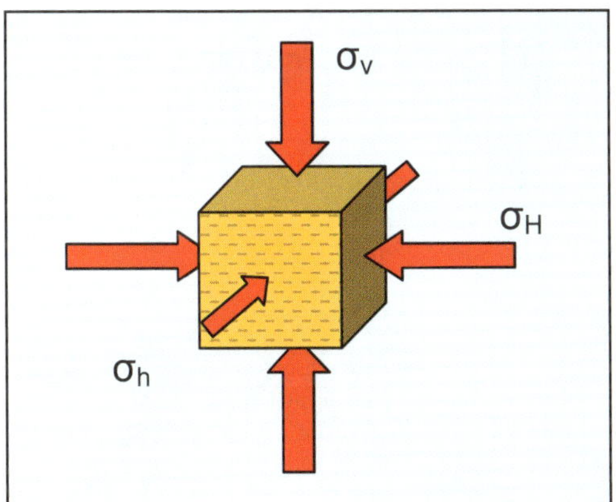

The principal stresses in a rock element are caused by the overburden and horizontal forces. The horizontal stresses are seldom equal and are referred to as the major and minor horizontal stresses, σ_H and σ_h.

Fig 8-15 Major and minor stresses

If the weight of the overlying rock were the only force a rock was subjected to, then σ_H and σ_h would be equal in magnitude. This is seldom the case with rocks in-situ.

Tectonic forces from the movement of the earth's crust tend to make the horizontal force in one direction larger than in another direction. (Fig 8-15)

Stress Components

The principal stresses in the earth's crust are vertical and horizontal in direction. Our wells are not always exactly vertical or horizontal, and *it is the stresses parallel to and perpendicular to the well path that we are most interested in.* (Fig 8-16) To find the stresses in our well, we must look at stress components.

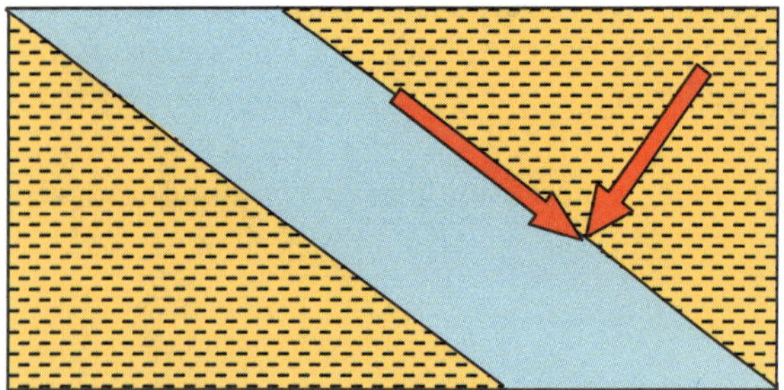

Well bore stresses are those stresses that act parallel and perpendicular to the direction of the well bore.

Fig 8-16 Well bore stresses

A stress component is that portion of the stress acting in the direction we are interested in. (Fig 8-17) Any stress acting at some angle to our well can be broken into two stress components, one along the well path, and one perpendicular to it. The size of the stress is found from simple trigonometric methods.

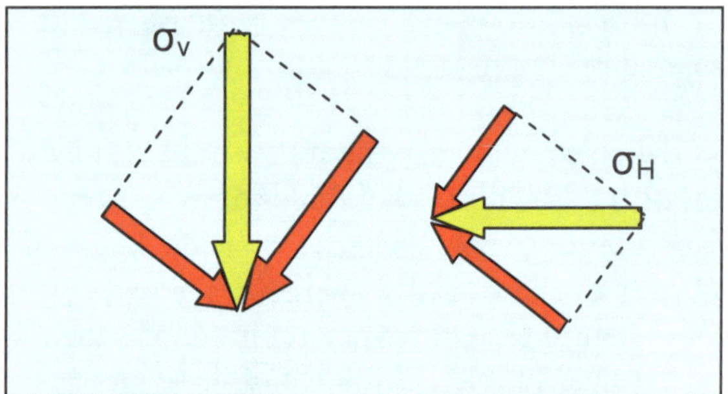

The vertical and horizontal stress components are combined to find the well bore stress components

Fig 8-17 Well bore stress components

Stress In-Situ

Now lets look at a little rock element in-situ. In-situ means the rock is in place and undisturbed in the formation. (Fig 8-18 When the rock is in place, it is in equilibrium. The weight of the overlying rock presses down on the rock, just as the hydraulic press did in the laboratory. This overburden stress will attempt to make the rock shorter and fatter, just like the core sample in the laboratory. All of the neighboring rock feels the same overburden and is also trying to get shorter and fatter. These neighboring rock elements push out in all directions as they try to expand, thus apply a confining pressure to our rock element. As the overburden increases with depth, so does the confining pressure. The apparent rock strength remains high enough to keep it from failing. Eventually, the overburden and confining pressure becomes so great that the yield strength of the rock is exceeded. However, this doesn't happen until depths of about 80,000 to 90,000 feet[1].

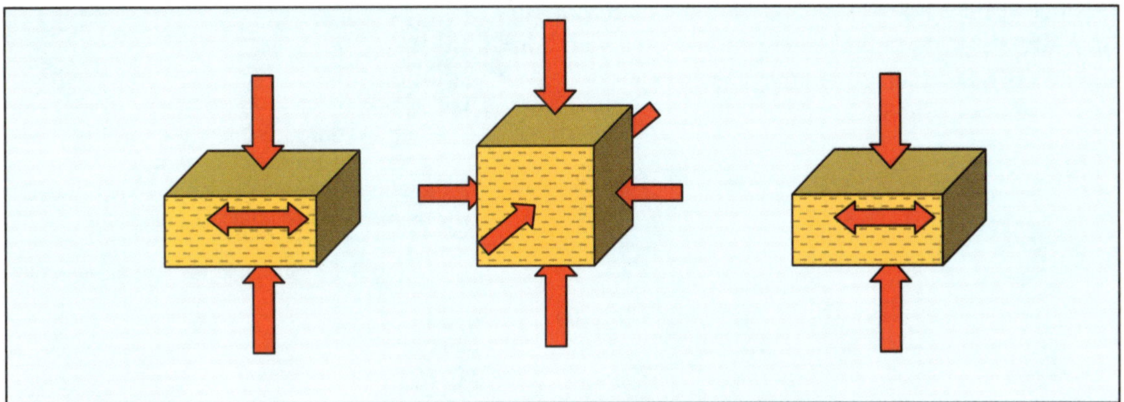

The rock element "in-situ" is in equilibrium. The overburden stress on all the surrounding rock tries to deform the rock. This produces a lateral or confining pressure that prevents the neighboring rock elements from deforming. Each rock element supports, and is supported by, its neighbors.

Fig 8-18 Stress In-Situ

Stresses Around the Well Bore

Our little rock element was happy when it was in place and surrounded by all his little buddies, as in Fig 8-18. When we drill a well however, we remove some of the rock elements that were applying a horizontal confining pressure to it. (Fig 8-19) The stress that existed in the removed material has to be replaced by fluid pressure in the well and the remaining rock elements along the well bore wall. If there is no fluid in the well, then 100% of the stress is transferred to the wall as hoop stress. Hoop stress is a stress that is tangential to the well bore wall. **Hoop stress σ_θ** is often called *tangential* or *circumferential* stress. (Fig 8-22)

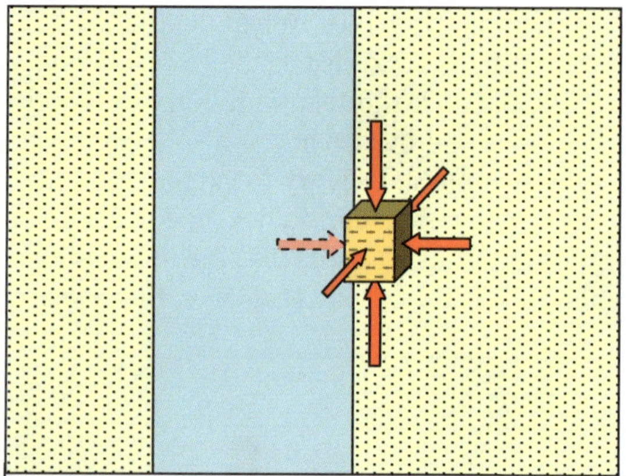

When we drill a well and remove the rock adjacent to our rock element, we also remove the supportive stress with it. The stress that was removed will have to be accounted for.

Fig 8-19 Stress in the well bore wall

Hoop Stress

We will talk so much about hoop stress that it's worth our time to explain it more thoroughly. We will use the analogy of a pressure vessel to explain hoop stress. Look at the vessel in Fig 8-20. If we were to split the vessel in half along the vertical plane, the two halves would try to blow apart. The internal pressure acting on the cross sectional area of each half of the vessel provides a force that acts to push the two halves apart. The force that holds the two halves together comes from the tensile stress in the wall of the pressure vessel. The stress times the cross sectional area of the wall provides a force that equals the force trying to separate the two halves. This "hoop" stress is equal all around the vessel, as long as the wall of the vessel is the same thickness all the way around.

Now imagine the same vessel pressurized from the outside, as if it were submerged deep in the ocean. The external pressure acting on the same cross sectional area is holding the two halves together. The hoop stress anywhere in the vessel is now in compression.

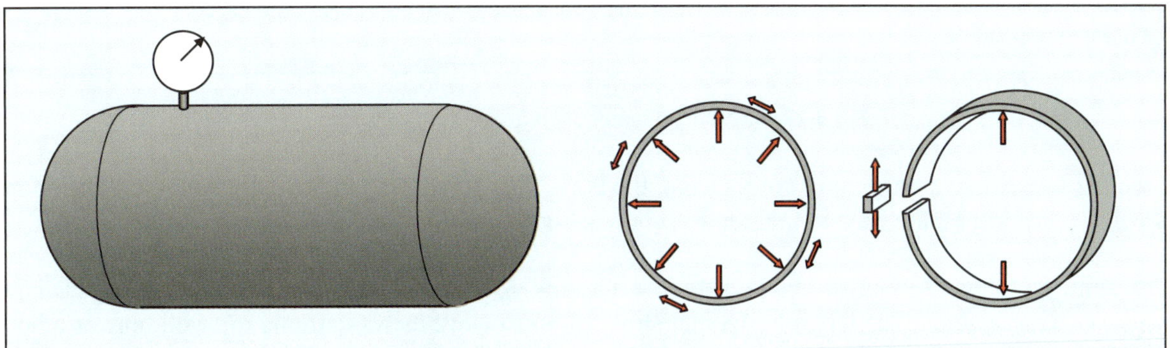

Internal pressure tries to split this pressure vessel apart. This pressure causes a tensile "hoop" stress in the wall of the vessel. If this vessel were submerged deep in the ocean, a compressive hoop stress would be felt in its wall.

Fig 8-20 Hoop stress

Now look at the vessel in Fig 8-21. With the 10,000-pound force acting along the y-axis, the hoop stress along the x-z plane is 1,000 psi. With a 5,000-pound force acting along the x-axis the hoop stress along the y-z plane is 500 psi. *Two forces of unequal magnitude produce hoop stresses of different magnitudes.*

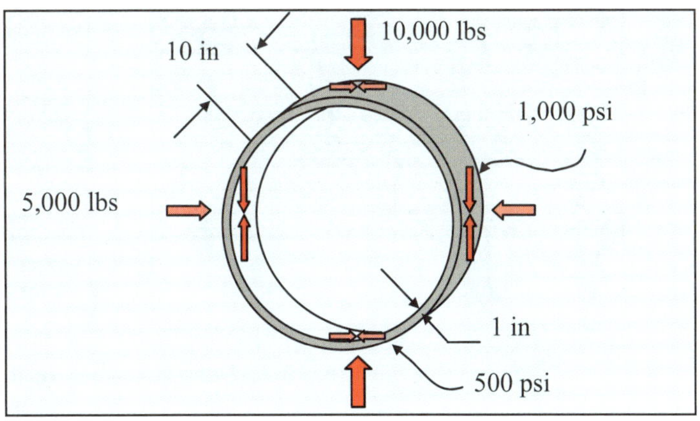

Fig 8-21 Hoop stress

Now lets return to our little rock in-situ. (Fig 8-19) When we drilled a well next to it and removed the supporting rock on one side, the missing stress is replaced with hoop stress. Another way of looking at it is that the stress fields must now work their way around the well bore, as in Fig 8-22.

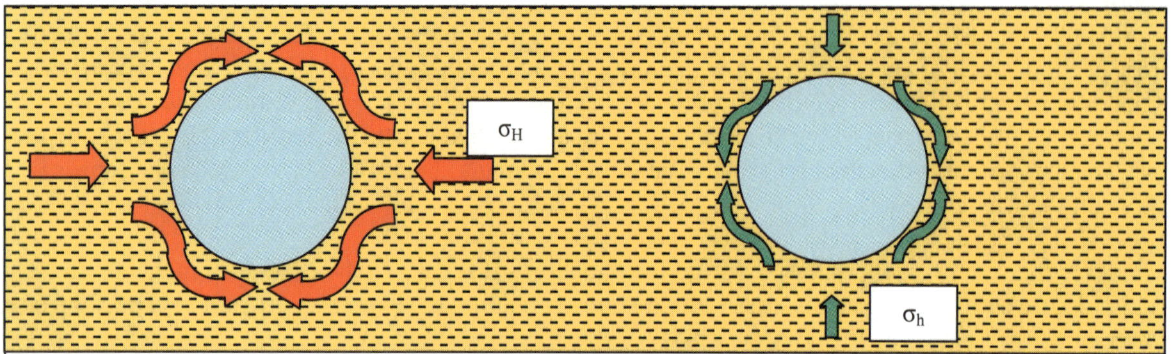

When material is removed by the drilling process, the stress fields must be re-distributed around the well bore as hoop stress.

Fig 8-22 Stress field-distribution around the well bore

If the well is vertical, and no tectonic forces exist, the horizontal forces (σ_H and σ_h,) are equal and the hoop stress will be uniform all the way around the well bore. (Fig 8-23A)

Because some tectonic stress exists everywhere, the horizontal force will be higher in one direction than the other in any stress regime. The larger horizontal stress σ_H must work its way around the well bore as a hoop stress. The same is true for the smaller horizontal stress σ_h. This produces a hoop stress that is greatest at 90° and 270° from the direction of the largest stress. (Fig 8-23B)

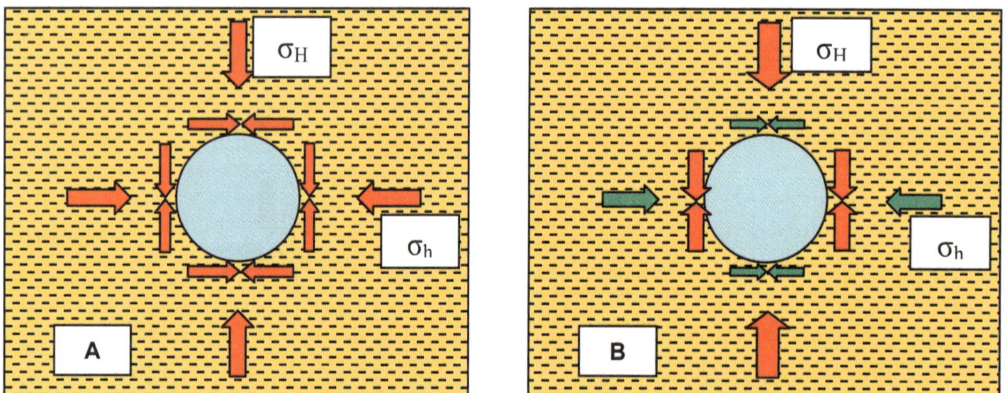

When σ_H and σ_h are equal, the hoop stress is uniform around the wellbore.

When σ_H and σ_h are unequal, the hoop stress is not uniform around the wellbore.

Fig 8-23 Hoop stress around the well bore

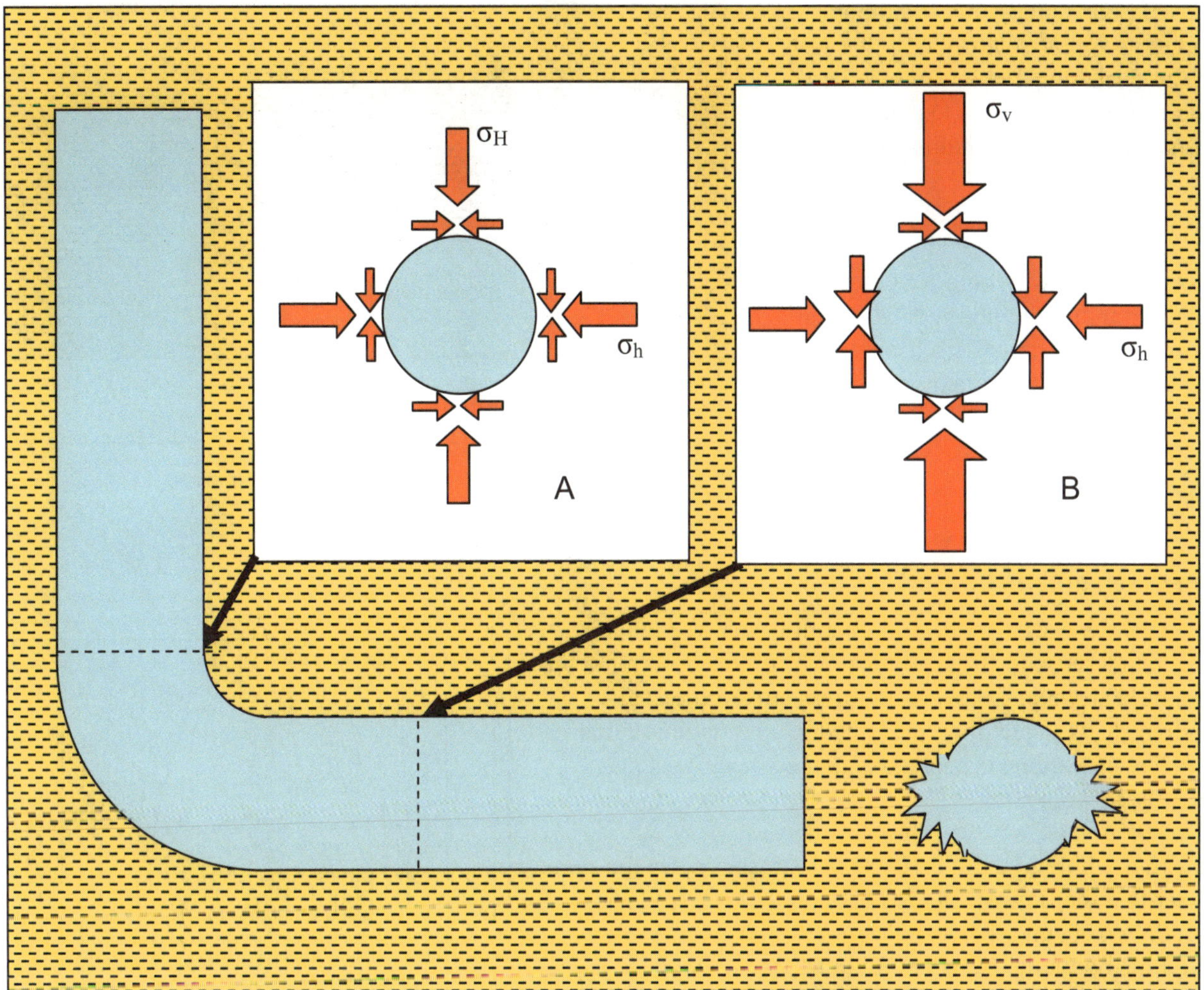

Overburden causes the hoop stress to be highest on the walls of a horizontal well, in a normal fault regime.

Fig 8-24 Anisotropic stress distribution

Look at the well in Fig 8-24. Assuming no tectonic stresses ($\sigma_H = \sigma_h$), the hoop stresses in the well bore wall will be uniform all the way around the well bore, as in Fig 8–23A. In the horizontal section however, the overburden produces more force in the vertical direction than we have in the horizontal direction. This produces a hoop stress that is highest in the wall and weakest on the ceiling and floor. (Fig 8-23B) If the mud weight is too low, the well will cave in at the wall.

The mining industry has been aware of this for centuries. That is why timbers in mine shafts are often thickest along the wall, rather than on the ceiling or floor. (Most mine shafts cave in at the walls, not from the ceilings, as portrayed by Hollywood movies.)

Large stress differentials are more likely at shallow depths. This is because rocks become more plastic at depth, due to the higher confining pressures. The plastic rocks will deform until the stress is equalized. This phenomena is referred to as "Heim's rule"[3]. In mountainous regions where tectonic stresses are high, we often see stress fields three times as high in one direction than in another, especially near the surface. At very deep depths, the stresses have nearly equalized, even in mountainous regions.

Not only is the hoop stress different at different points along the wall, hoop stress changes as we move away from the wall into the formation. Hoop stress is largest at the wall and decreases to zero at a distance of about three radii from the well bore center. (Fig 8-25)

The stress we are interested in is the hoop stress (σ_θ) at the well bore wall along the major and minor directions. This stress can be found from the Kirsch equations in appendix B, but the graphical methods provided by Hoek & Brown[3] are better for our conceptual purposes.

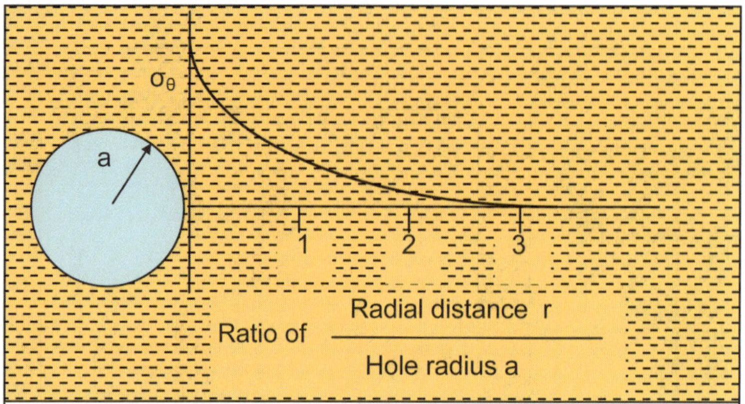

Hoop stress is a maximum at the bore wall and reduces to zero at three hole radius away.

Fig 8-25 Hoop stress away from the wall

Stress Streamlines

Hoek & Brown[3] use an elegant streamline analogy in their book, "Underground Excavations in Rock," for describing the stress field around a borehole. A stress streamline is a principal stress trajectory represented by an imaginary line. The stress streamline is analogous to the streamlines in a smoothly flowing stream.

The circular borehole disturbs stress fields in rock in the same manner that the flow of water is disrupted by a round pier protruding through a stream. (Fig 8-26) Stress must flow around the borehole, just as water must flow around the pier. Immediately up- and down-stream of the pier, the water is slowed and the streamlines spread outward. The flow of water on either side of the pier speeds up because more water now has to flow through a smaller space. The same thing happens with stress around a well bore. The stress streamlines spread apart as they encounter the obstruction, then cram together as they pass around it.

The compressive stress is reduced (and possibly goes into tension) in the region where the stress streamlines spread apart. The compressive stress increases in the region where the streamlines converge.

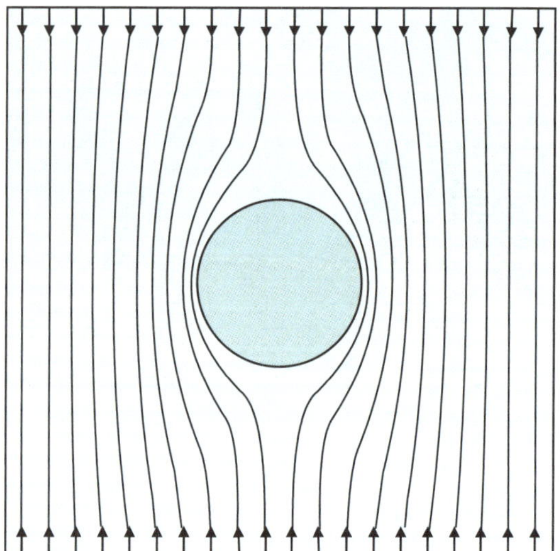

The circular borehole disturbs a stress field in the same manner a pier disturbs the flow of water in a stream. The compressive stress is highest where the lines converge and lowest where they spread apart.

Fig 8-26 Stress stream lines[3]

Stress Contours

Fig 8-27 shows the stress streamlines on one side of the borehole and principal stress contours on the other side. The **stress contours**[3] give the ratios of the principal stresses to the largest applied stress.

Remember, all we are really interested in is the largest hoop stress along the well bore wall. The hoop stress ratios can be read from the point where the stress contours intersect the well bore wall. Note that it is highest at 90° and 270° from the largest stress field and varies according to the difference between the largest and smallest stress fields.

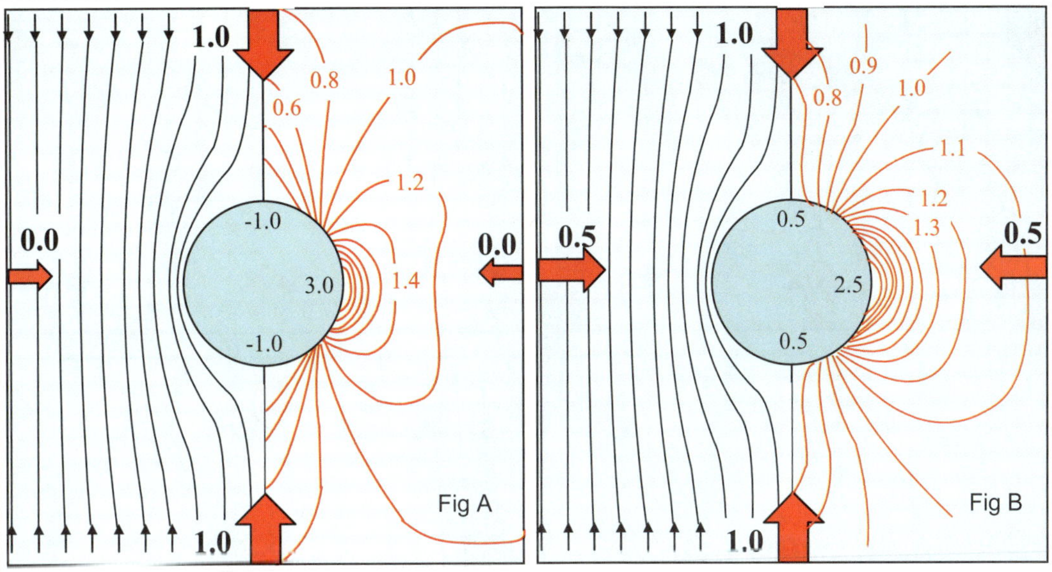

These figures depict the stress streamlines on the left and the stress contours on the right. The stress contours represent the ratio of the principal stress to the highest stress field. In most cases the principal stress will be the hoop stress. Note that as the difference between stress fields increases so does the maximum hoop stress. (Fig A, B, & F)

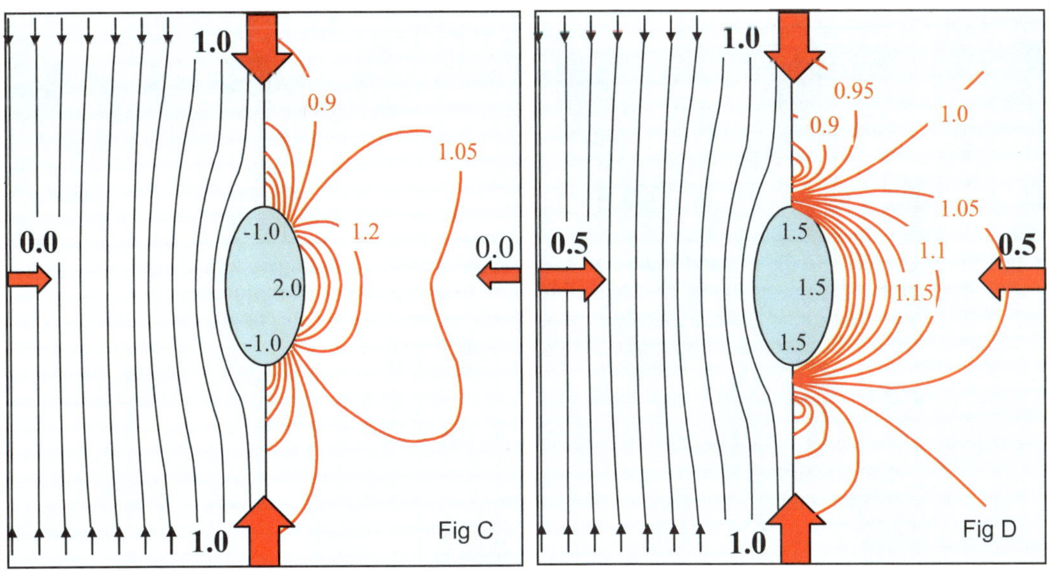

When the well bore is elliptical due to pipe erosion the maximum hoop stress can be less than in a circular borehole _if_ the major axis of the ellipse is aligned with the largest stress field. (Fig C & D) If the minor axis of the ellipse aligned with the largest stress field the maximum hoop stress is much larger than in a circular bore hole. (Fig E, G, & H)

From Hoek and Brown; "Underground Excavations in Rock"

Fig 8-27 Stress contours[3]

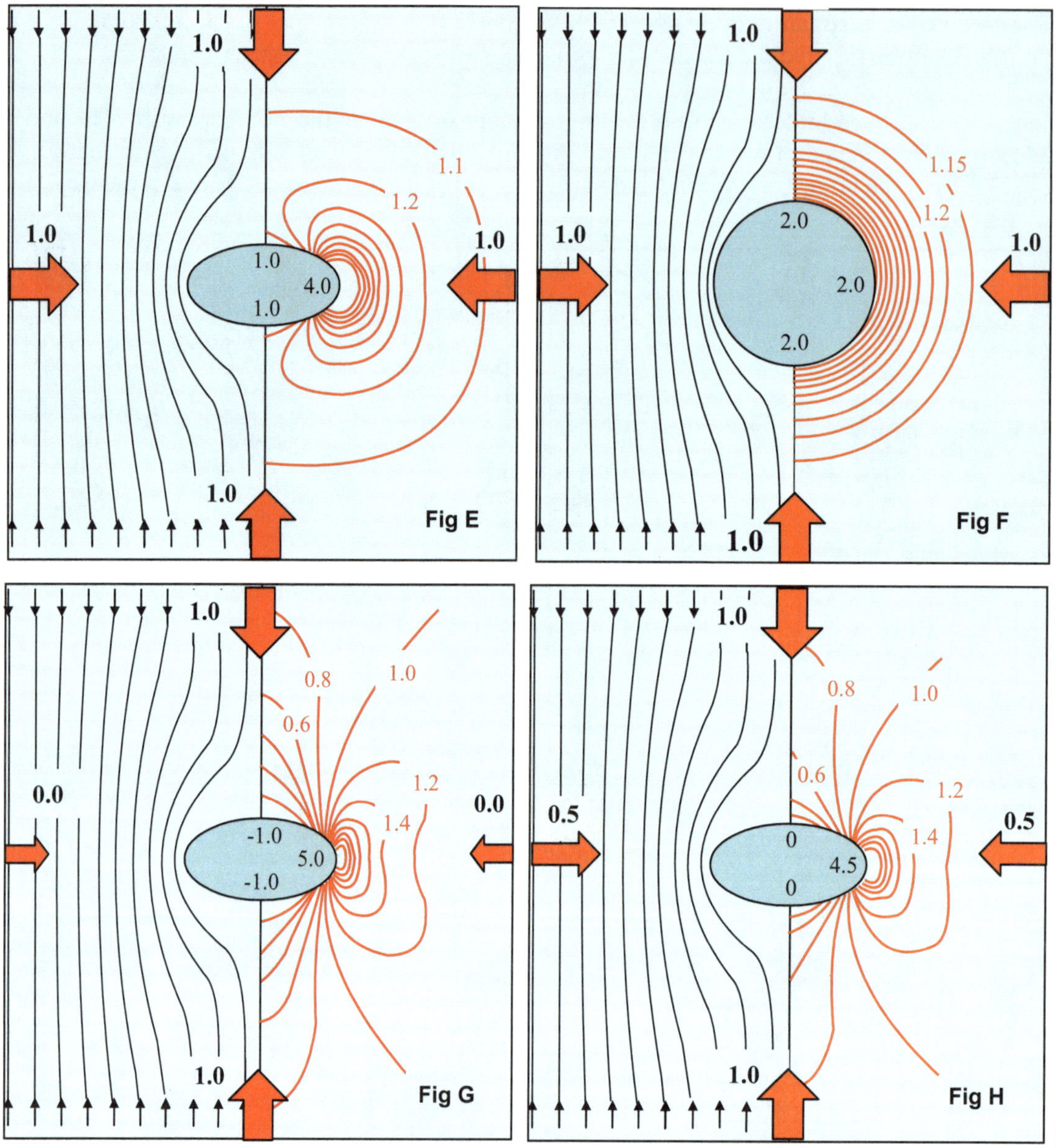

From the stress contours in Figs G and H we can see that as the well bore begins to cave due to instability, the instability worsens. In brittle rocks, this often leads to nearly instantaneous catastrophic failure. In most shale, the freshly exposed rock must weaken from filtrate invasion before it fails. This gives us some time to take corrective action.

Fig 8-27 Stress contours (continued)

These stress contours were prepared by applying stresses to photo elastic material with various stress anisotropy and hole shapes. Note that all of these stress contours were prepared with no radial stress (zero mud weight).

From Figures 8-27A through 8-27H, we see that the shape of the well bore has significant influence on the stress field around it. We can also see how stress anisotropy affects the stress field. Note that we are most interested in the stress at the wall. It is normally a shear stress that causes a well to fail. The shear stress is determined by the difference between the maximum hoop stress at the wall, and the radial stress provided by the mud weight.

The disturbance of the stress field is highest near the well bore wall, then decreases with the radial distance away from the well. Beyond a distance of about three radii from the center of the well, the stress field is undisturbed. (Fig 8-25) Failure from instability will happen where the shear stress is the highest, which happens close to the well bore wall. Deformation in plastic formations such as salt, soft shale, and unconsolidated sands however, occurs up to three radii away from the well bore center.[6]

Note that shale is time dependent. Filtrate invasion causes the pore pressure to increase, so the radial stress is reduced and the hoop stress increases with time.

Radial Stress

The pressure from a column of mud provides radial stress against the well bore wall. This radial pressure reduces the compressive hoop stresses. Hoop stresses develop because the horizontal stress in the material that was removed during drilling has to be replaced somehow. The mud in the well replaces some of this stress and the rest is accounted for with hoop stress. *The more stress that is accounted for by mud weight, the less that needs to be accounted for by hoop stress.* (Fig 8-28)

Total stress redistribution = radial stress + hoop stress   eq. 8.5

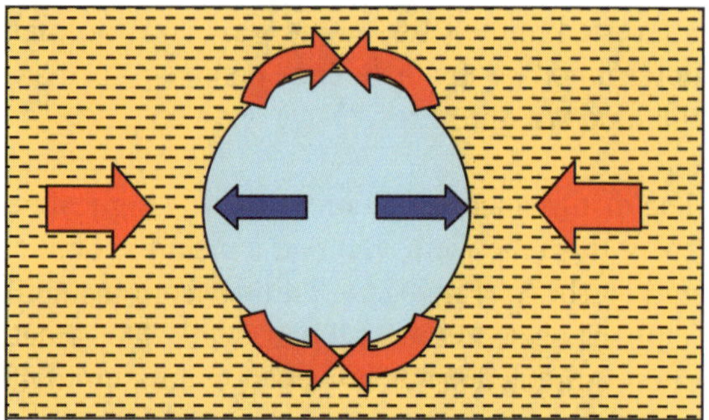

The stress removed by the drilling process is replaced by a combination of radial and hoop stresses. *Stress replacement = radial + hoop*

Fig 8-28 Stress redistribution

Remember the core sample in the laboratory? (Fig 8-5) The little rock element along the well bore wall is compressed by the hoop stress just as the core sample was compressed by the hydraulic press. (Fig 8-29) If the compressive stress exceeds the apparent rock strength the rock will fail. Radial stress from the column of mud reduces the hoop stress and applies a confining pressure. The apparent rock strength increases as mud weight increases and the compressive hoop stress is reduced.

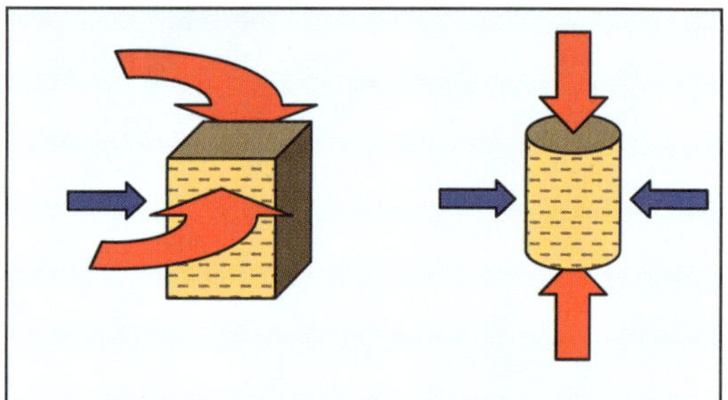

Hoop stress compresses our rock element just as the hydraulic press compressed our core sample. Radial stress becomes a confining pressure that increases rock strength.

Fig 8-29 Radial stress

It is possible to raise the mud weight to a value that reduces the hoop stress to zero. If the mud weight is raised too much the hoop stress will become tensile and the rock may fail in tension. (Fig 8-30) This is what happens when we hydraulically fracture a formation.

Compressive hoop stress is reduced as mud weight increases, possibly to the point of becoming tensile.

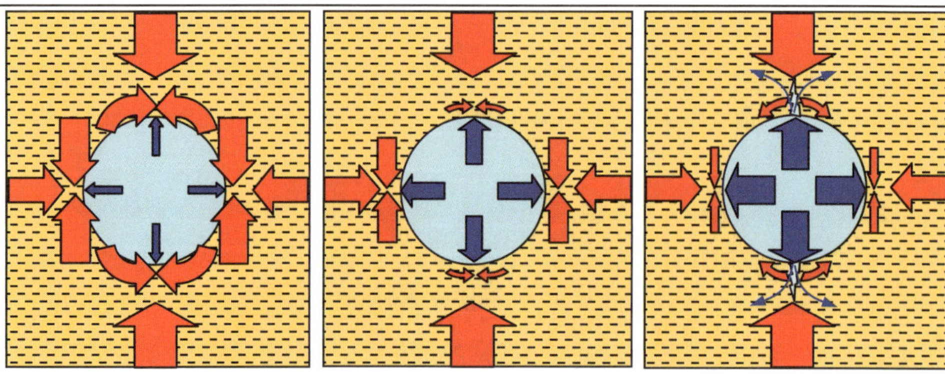

Fig 8-30 Relationship between radial and hoop stress

The rock strength also decreases when mud weight is reduced. When a well blows out of control, large chunks of shale are often blown out of the well shortly before the well bridges over. A common misconception is that the well blew so hard it "eroded" the well and caused it to bridge over. In fact, the radial stress is reduced when the mud is evacuated and the hoop stress increases. This causes the resulting shear stress to exceed the rocks yield point to the extent that the shale fails and collapses into the well.

Axial Stress

Axial stress is the sum of the vertical and horizontal stress components in the axial direction. In a vertical well, the axial stress would be equal to the vertical stress. In a horizontal well, the axial stress would be the sum of the horizontal stress components along the axis of the well.

Note that pipe drag adds to the axial stress.

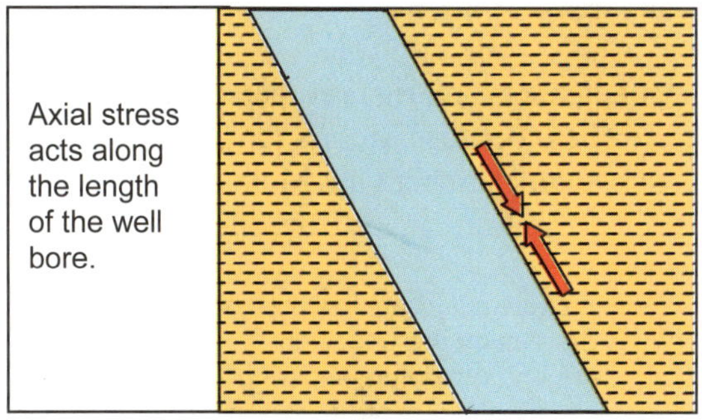

Fig 8-31 Axial stress along well bore

Tri-Axial Stress State Along The Well Bore

The tri-axial state of stress along a well bore wall (Fig 8-32) is represented by these stress components:

- σ_z (axial stress)
- σ_r (radial stress)
- σ_θ (tangential or hoop stress)

We will not bother to calculate stresses in this manual. Our goal is simply to understand what they are and how they affect stability.

The wellbore will fail if the combination of any two of these three stresses exceeds a certain limit.

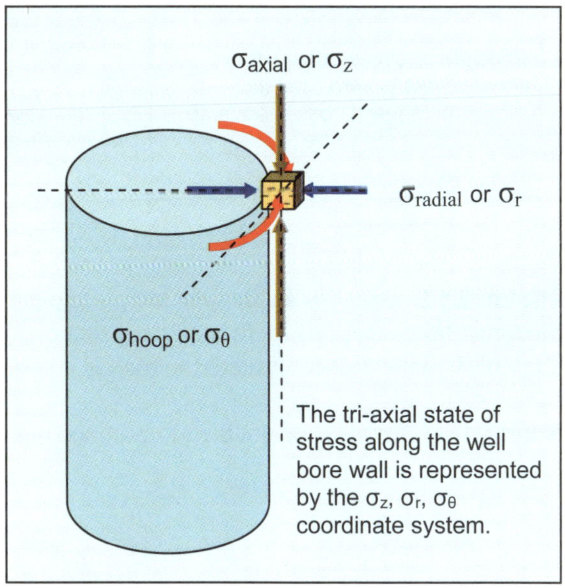

Fig 8-32 Tri-axial stress along well bore

Any two orthogonal stresses produce shear stress on the plane defined by those two stresses. If the shear stress exceeds rock strength the rock will fail and collapse into the well.

Mohr's Circle (Double Angle Theory)

Mohr's circle is often used to represent the stresses in a rock across a plane, at any angle from the direction of the maximum stress. This is useful when we know the stresses in a particular direction and want to know what the stresses are in another direction. It is also useful in finding the principal stresses and maximum shear stress.

Most engineers and rig crew personnel will never have to work with Mohr's circle. A brief explanation is offered here because Mohr's failure envelope is often used to predict the acceptable mud weight window. The German engineer Otto Mohr (1835 –1918) developed this graphical approach, as depicted in Fig 8-33.

Mohr's circle is a graphical method for representing the principal stresses and maximum shearing stress at a point in a stressed rock.

The orthogonal stresses at a point are plotted on a σ, τ coordinate system. Points A and B represent the major and minor stresses across some surface.

Principal stresses occur where shear stress is zero, and the maximum shear stress is equal to the radius of the circle.

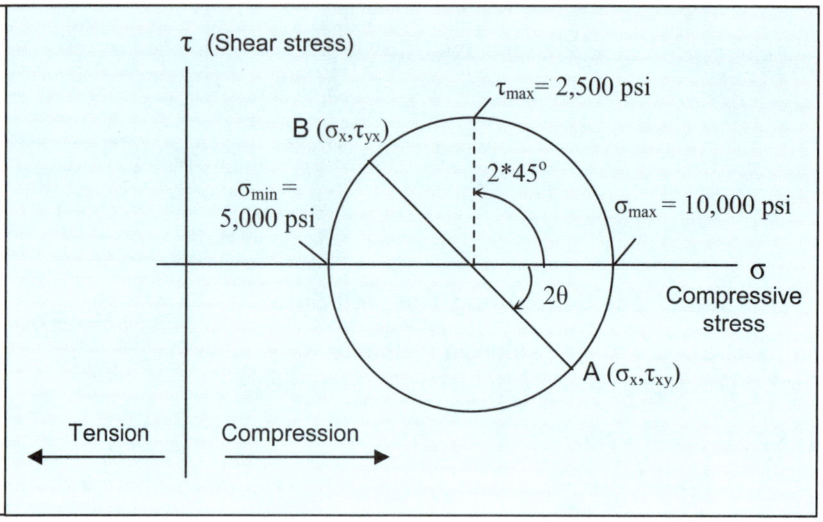

Fig 8-33 Mohr's Circle

The known stresses for an element are plotted graphically on the σ, τ coordinate system, as in Fig 8-33. The fundamental concept of Mohr's circle is that every point on the circle represents the state of stress on an imaginary plane passed through the rock at some angle θ from the plane of principal stress.

Note: All angles must be doubled to produce the circle.

For example, the state of stress in the core sample in Fig 8-34 is plotted in Fig 8-33. Lets assume we wanted to drill a hole through the core at some angle as shown in Fig 8-34. The state of stress on any element across a plane in this direction is shown in the upper right hand corner. The points A and B represent the largest and smallest compressive stresses across this element before the hole is drilled.

The principal stresses in the core sample are 10,000 psi and 5,000 psi. Remember that shear stress is always zero on the planes of principal stress. Thus, the principal stress points are always on the σ axis. The maximum shear is equal to the radius of the circle and exists on a plane that is 45° from the principal plane. In this case, the maximum shear is 2,500 psi.

The principal planes of stress in this core sample are orthogonal to σ_v and σ_h. The state of stress on any element at an angle θ to the principal plane is illustrated on the far right.

The maximum and minimum stresses at this angle, A & B, can be found trigonometrically with Mohrs circle.

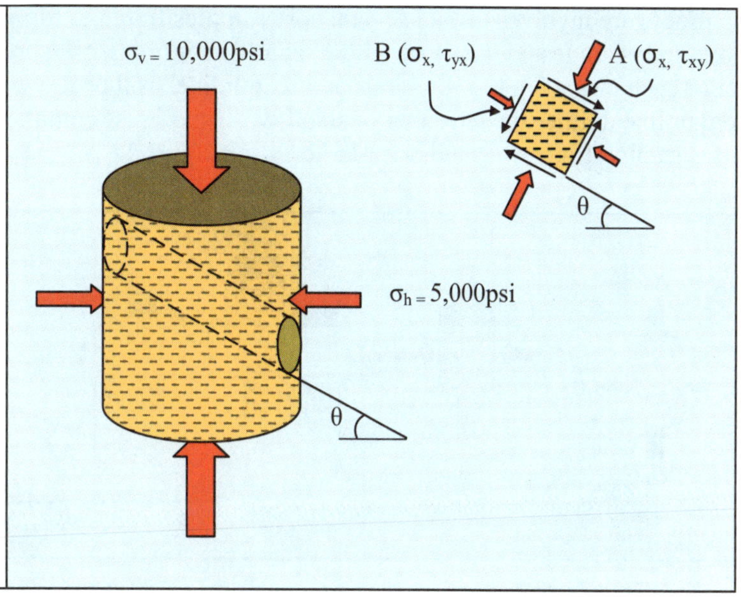

Fig 8-34 State of stress on a plane

From Mohr's circle, we see that *as the difference between the largest and smallest principal stress increases, so does the maximum shear stress*. The maximum shear stress is always equal to one half the difference between the minimum and maximum stresses. (The radius of Mohr's Circle) This is important to note because *well bores generally fail from excessive shear stress*!

The shear stress is always zero on a surface subjected to a principal stress.

Shear stresses cannot exist on an exposed surface, such as the well bore wall. Therefore, once a hole is drilled, one of the principal planes of stress will be parallel to the well path, the others will be orthogonal to it. A new Mohr's Circle will can be drawn to predict the maximum shear in this condition.

Mohr's Failure Envelope

A big part of a successful drilling program is being able to predict the conditions that will provide stability, or lead to instability. If we know the in-situ stress, we need only find a mud weight "window" that will prevent both lost circulation and well bore collapse. A typical **mud weight window** is depicted in Figs 5-1 and 8-40. It represents the range of acceptable mud weights. If the mud weight is too high, the well will suffer lost circulation. If the mud weight is too low, the well bore may collapse. We use a procedure known as Mohr's failure envelope to predict what this mud weight window will be.

The procedure involves testing several core samples from the formation we are interested in, as in Fig 8-35. The samples are tested to failure with a range of confining pressures. A Mohr's circle is constructed from the data of each test. (Fig 8-35) A "shear strength line" is then drawn along the top of the circles. This shear strength line defines the envelope of stability. If the stress condition falls below the line (yellow area), the well is stable. If not, the well is unstable.

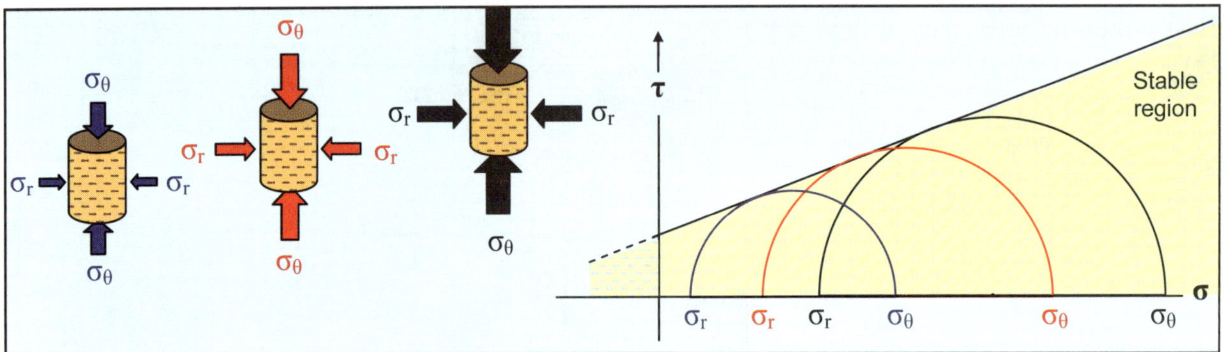

Testing cores at various confining pressures establishes a window of acceptable stress conditions.

Fig 8-35 Mohr's failure envelope

Remember that failure is usually a result of excessive shear stress. Remember also that shear stress increases as the difference between the maximum and minimum principal stresses increases. (The radius of the circle represents maximum shear stress.) The principal stresses represented on the circle are usually *hoop stress,* σ_θ, and *radial stress,* σ_r. As radial stress increases, hoop stress decreases. Thus, an increase in mud weight has an effect on both stresses and will shrink the circle if radial stress is less than hoop stress, or enlarge the circle if radial stress is higher than hoop stress. (Fig 8-39)

We will use Mohr's failure envelope to help understand how a change in factors such as mud weight and temperature affect stability. But first lets look at what factors affect stability.

Factors Affecting Stability

Several factors affect stability, including:

- Mud Weight
- Rock Strength
- Temperature Fluctuations
- Stress and Strength Anisotropy
- Well Path Orientation and Inclination
- Drilling Fluid Filtrate Invasion
- Drill String Vibration
- Well Bore Geometry

Mud Weight (Factors Affecting Stability)

The pressure from a column of mud provides a radial stress against the well bore wall. So let's review what we learned in the section on Rock Mechanics, about how radial stress affects wellbore stability.

The stress that was removed from the well as it was drilled must be replaced with a combination of hoop stress and radial stress. As we see in equation 8.4, the higher the mud weight the lower the hoop stress.

Total stress redistribution = radial stress + hoop stress eq. 8.5

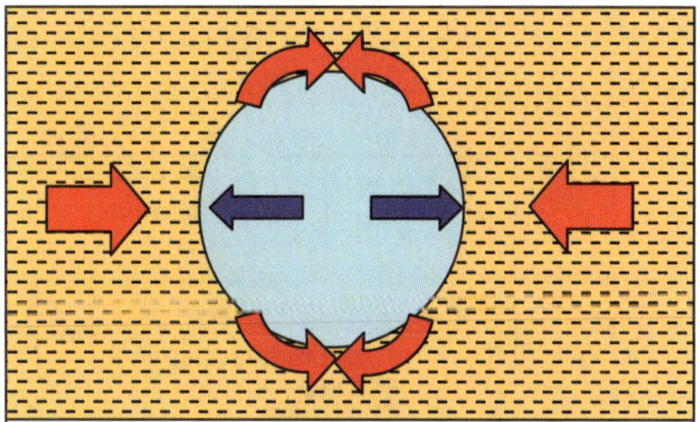

The stress removed by the drilling process is replaced by a combination of radial and hoop stresses. *Stress replacement = radial + hoop*

Fig 8-36 Stress redistribution

Remember the core sample in the laboratory? Mud weight does more than just reduce the compressive hoop stress, it also provides the confining pressure that increases the apparent rock strength.

The combination of reducing the compressive hoop stress and increasing the apparent rock strength makes mud weight a powerful tool for combating wellbore instability.

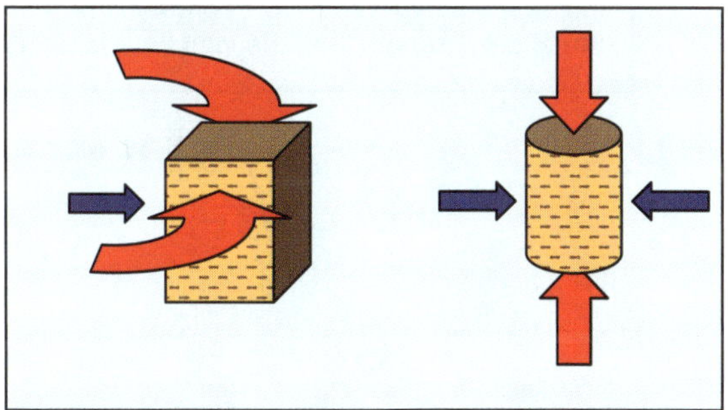

Radial stress becomes a confining pressure that increases the apparent rock strength.

Fig 8-37 Radial stress

Hoop stress decreases as wellbore pressure increases. If the mud weight is raised too much the hoop stress will become tensile and the rock may fail in tension. (Fig 8-38) This is how we hydraulically fracture a formation.

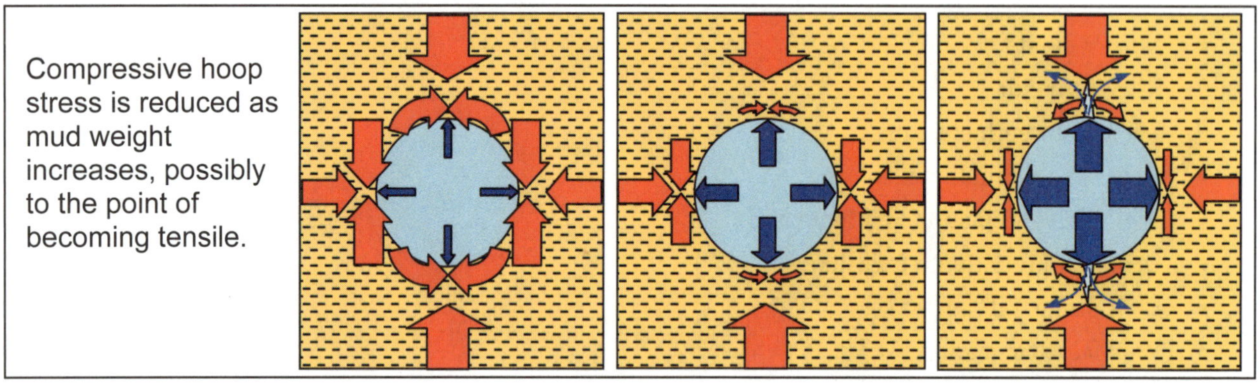

Fig 8-38 Relationship between radial and hoop stress

The effects of mud weight can be graphically illustrated with the Mohr's failure envelope for mud weight in figure 8-39.

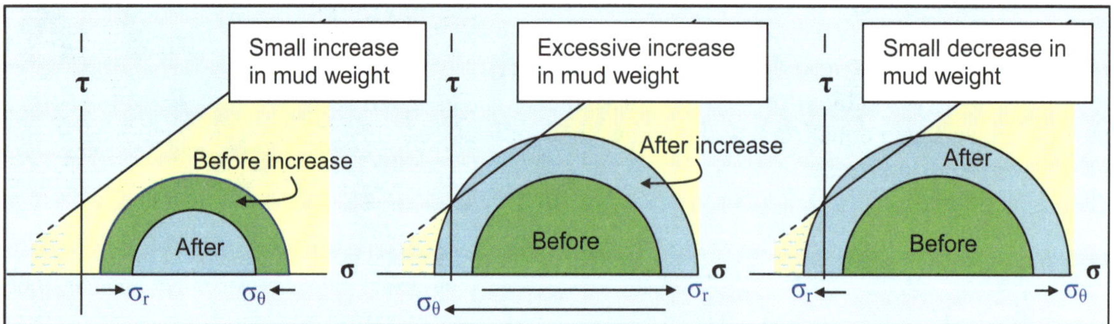

An increase in mud weight increases radial stress and decreases hoop stress. If the increase is excessive, the compressive hoop stress can be changed to a tensile stress. A decrease in mud weight increases shear stress by decreasing radial stress and increasing hoop stress. If the circle shifts above the shear strength line, the well bore is unstable.

Fig 8-39 Mohr's failure envelope for mud weight

There are several approaches to optimizing mud weight while drilling. One approach that attempts to optimize mud weight such that the hoop stress is zero, is the *median line principle* proposed by Aadnoy.[4] The median line principle suggests the mud weight should be half way between the pore pressure and the fracture gradient to bring the hoop stress to zero. Fig 8-40 shows a mud weight vs. fracture gradient with the median line drawn in.

There may be a temptation to keep mud weight as low as possible in order to maximize penetration rate. Unfortunately this often leads to hole enlargement and lost time due to "tight hole" problems. The median line approach sacrifices penetration rate early on in the well but makes up for it by minimizing hole problems. Aadnoy cites record wells that were drilled with the median line approach.[4]

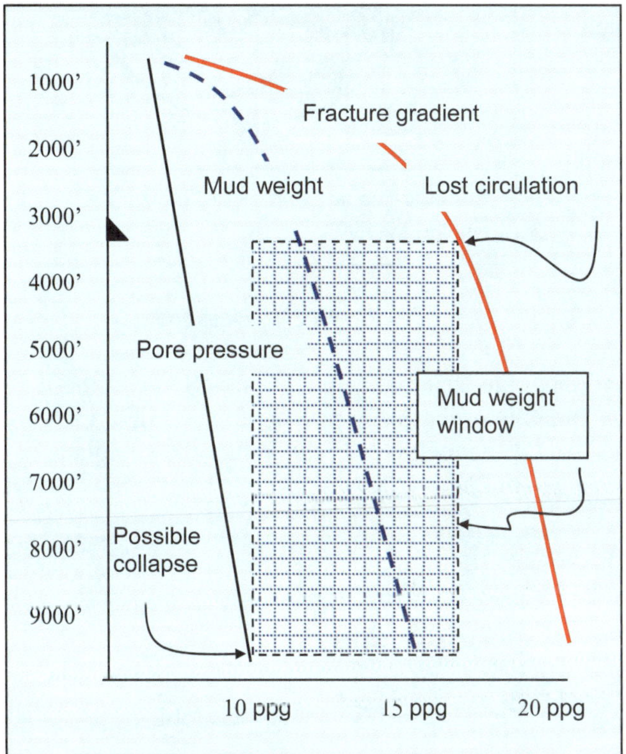

The median line principle attempts to minimize hoop stress with a mud weight that is halfway between pore and facture pressures.

Fig 8-40 Lost circulation and caving

Mud Weight Window

Even with an approach like the median line principle, the mud weight can only be optimized for the well at one depth. An optimum mud weight for the hole at one depth will be too high for the well at shallower depths and too low for deeper depths. This means we can only have an optimum mud weight for a small section of the open hole. The best course of action will be to optimize the mud weight at the drilling depth and continue to raise mud weight as required, but never reduce it. (See Filtrate invasion)

We have what is called an "allowable mud weight window" in the open hole section. (The shaded area in Fig 8-40.) There is a minimum allowable mud weight to hold back formation fluids and prevent well bore collapse in the lower portion of the well, and a maximum allowable mud weight to prevent lost circulation in the upper portion of the well. This allowable range of mud weights is influenced by in-situ stress fields and will be discussed later.

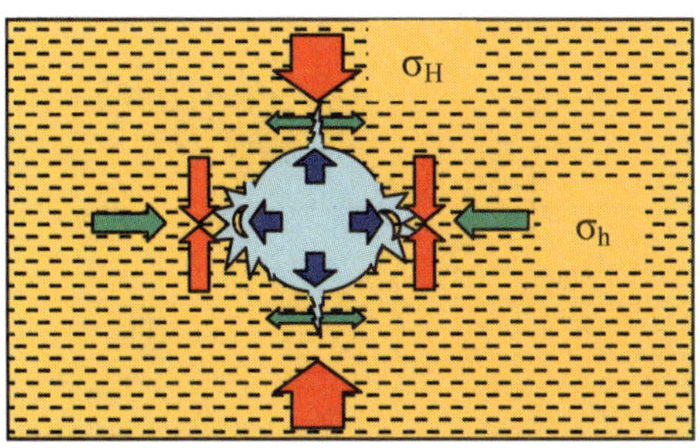

Fig 8-41 Lost circulation and caving

Depth is not the only concern. When there is a considerable difference between the major and minor horizontal stresses, we may find a mud weight that causes lost circulation in a perpendicular direction to the minor stress while the hole continues to cave in the direction perpendicular to the major stress. (Fig 8-41) In this case our mud weight window is too small for the length of open hole. We will have to shorten the open hole sections, run more casing strings, or change the orientation of the well path.

Rock Strength (Factors Affecting Stability)

Obviously, the stronger the rock is, the more stress it can withstand. As we already discovered, rock strength depends mostly on the cementation and friction between the individual grains that make up a rock. The compressive strength and elasticity of the individual grains also contributes to strength.

Look at the rock matrix in Fig 8-42. In order for the rock to break along the slip plane, it must overcome the cementation and friction between the grains that are in contact with each other along the plane.

If the number of contact points is small, the actual stress at these points is very high. Thus, rock strength increases as the amount of rock matrix contact increases.

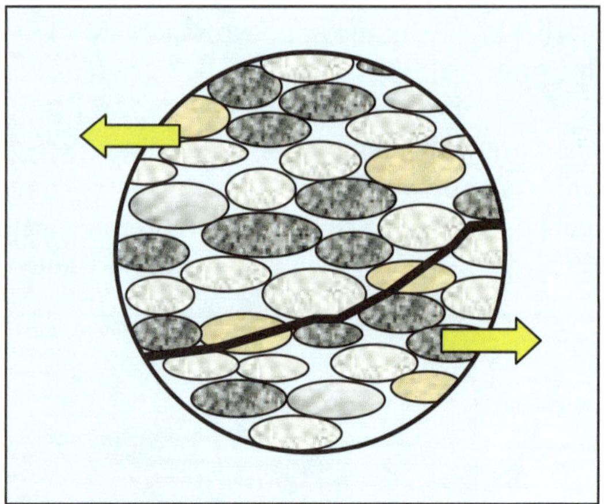

Some grains must break or deform when slippage occurs.

Fig 8-42 Rock strength

Rock strength is also affected by the strength of the individual grains that make up the rock. Some of these grains lie directly across the slip plane and must deform or break in order to slip past another. (Fig 8-42) The stronger these grains are, the harder they are to break. And if the rock is to break, either the slip plane must alter its course around these grains or more stress must be applied.

Weaker formations drill faster and are likely to fail before stronger ones. We can plot the rate of penetration with depth and use it to anticipate hole enlargement. Slower drilling shale is more likely to be in gauge than weaker, faster-drilling shale. (Fig 8-43) This information is even more useful when plotted next to a lithographic plot. Most mud-logging units can prepare a printed chart with this information to present to the drill crew just prior to tripping out of the well. (See Fig 13-1)

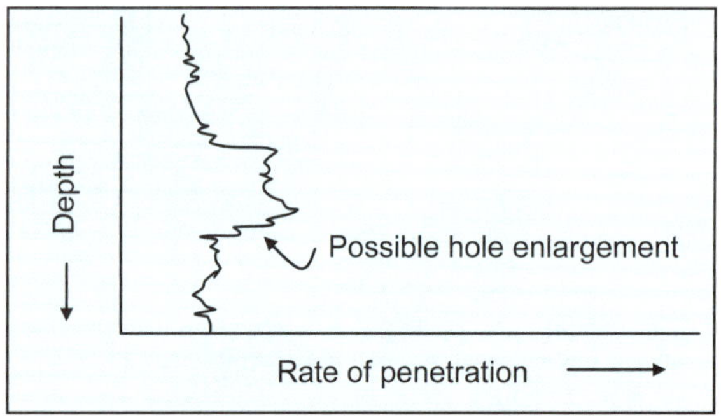

Fig 8-43 Depth vs. penetration rate plot

Temperature (Factors Affecting Stability)

The center of the earth is hot enough that rock under the crust is molten. This heat slowly escapes through the crust, just as heat escapes from the surface of a thick piece of hot steel. (Fig 8-44) As we drill into the crust, we find the temperature generally increases with depth. The average temperature gradient is approximately one degree Fahrenheit for every 100 feet of depth.

As mud circulates through the well, it alters the temperature in exposed formations. Cold mud cools the lower portion of the well; hot mud from depth warms the upper formations. Temperature changes can be detrimental to stability. The change is most pronounced when circulation is first started after a long period of no circulation.

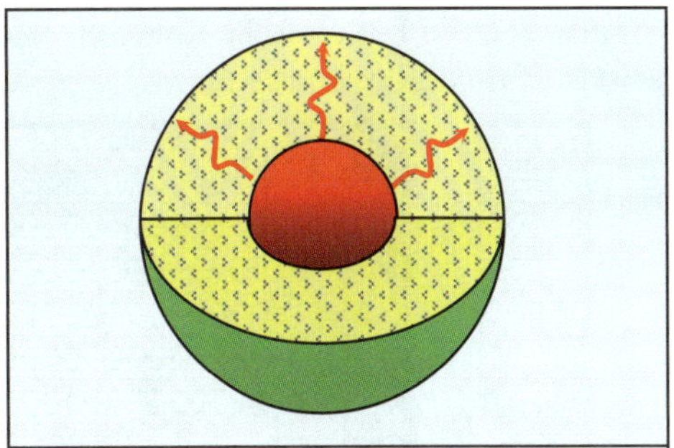

The earth's temperature is coolest at the surface and gets hotter with depth.

Fig 8-44 Earth's temperature gradient

An increase in temperature influences stability in several ways. It can increase the hoop stress in the well bore wall by reducing the radial force provided by the mud pressure. It also weakens the rock by reducing the friction between grains. Water has a larger coefficient of expansion than rock. If there is a temperature increase, the pore fluid will expand more rapidly than the rock matrix. This will increase the pore pressure and reduce the effective matrix stress. An increase in pore pressure also reduces the pressure differential against the well bore wall, thus reducing the radial stress. This raises the compressional hoop stress. (Fig 8-45)

An increase in pore pressure also forces the individual grains in the rock matrix farther apart. This lessens the frictional force between the grains, so rock strength is reduced. The increased volume of pore fluid lubricates and breaks apart cementation as well.

An increase in temperature also causes the rock matrix to expand slightly. Some minerals within the formation expand more than others, which causes shifting and breaking of cementation bonds.

A temperature decrease has the reverse effect. The pore fluid will shrink faster than the rock matrix, thus increasing both the effective stress and the differential pressure felt at the well bore wall. This increases the rock strength, but the higher differential pressure may cause lost circulation in some cases. (Fig 8-45) This is why ballooning is such a common occurance in HPHT wells. (Fig 8-67)

The temperature of the formation changes during trips. While pulling out of the hole, the formation near the well bore heats up or cools down to its original temperature. (Circulation while drilling tends to cool the lower part of the well and may warm the upper part of the well.) Once circulation is resumed, the cool mud traveling down the drill pipe will cool the lower part of the well, while the hot mud circulated from bottoms up will heat the upper portions.

Frequent trips may fatigue the formations and cause failure, even though the stresses were kept within Mohr's failure envelope.

One problem with deep high temperature wells is that while trip gas is being circulated up the well, lost circulation may occur on bottom when cool mud reaches the bit. The expanding gas and reduction in bottom-hole pressure is likely to go unnoticed until the gas is very near surface. By then, a considerable kick may have entered the well bore.

If a well control situation should arise with a high temperature well, the formations near the well bore will tend to warm up while waiting to circulate, and the active volume will have time to cool down. The well bore will also remain hotter at the lower circulation rates. A well that was stable may become unstable after periods of no or slow circulation.

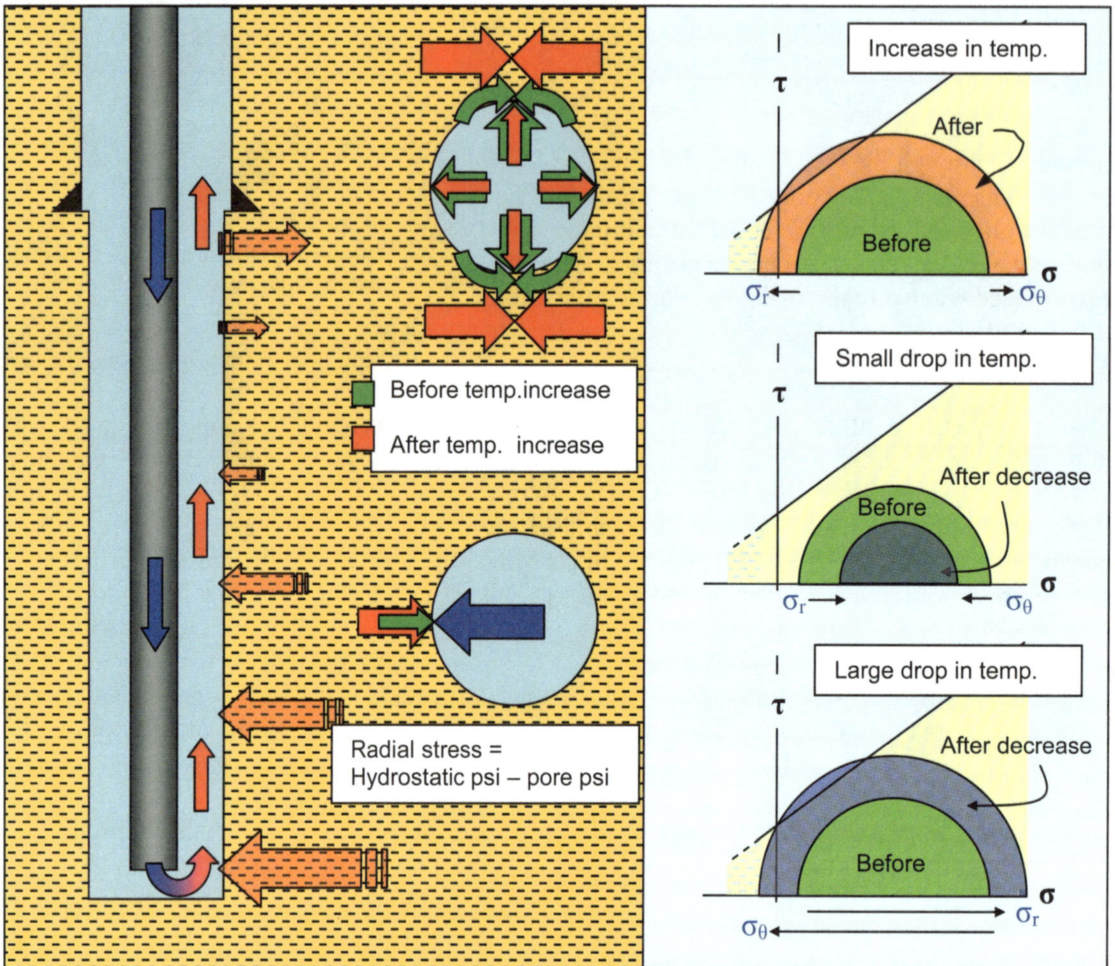

Because formation fluids expand more than the rock matrix, changes in temperature affect the radial stress. Changes in radial stress affect the hoop stress. An increase in temperature will decrease radial stress and increase hoop stress, leading to a less stable well bore. A small decrease in temperature can lead to a more stable well bore. However, an excessive decrease could lead to lost circulation.

Fig 8-45 Mohr's failure envelope for temperature changes

In-Situ Stress Regimes and Stress Anisotropy (Factors Affecting Stability)

The "in-situ stress regime" refers to the regional stress field existing in the drilling location. The stress regime is a result of tectonic forces pushing and tugging on the earth's crust. There are three principal stress regimes, as shown in Fig 8-46.[2] The tectonic stress regimes are defined by the relative strengths of the principal stresses.

A **Normal Faulting** stress regime occurs when $\sigma_v > \sigma_H > \sigma_h$. (Fig 8-46) The largest or principal stress is in the vertical direction and minor and intermediate stresses are in the horizontal direction.

A **Strike Slip Faulting** stress regime occurs when $\sigma_H > \sigma_v > \sigma_h$. (Fig 8-46) In this case, the major horizontal stress σ_H is larger than the vertical stress σ_v and the vertical stress σ_v is larger than the minor horizontal stress σ_h.

A **Reverse Faulting** stress regime occurs when $\sigma_H > \sigma_h > \sigma_v$. (Fig 8-46) In this case, the vertical stress is less than both horizontal stresses.

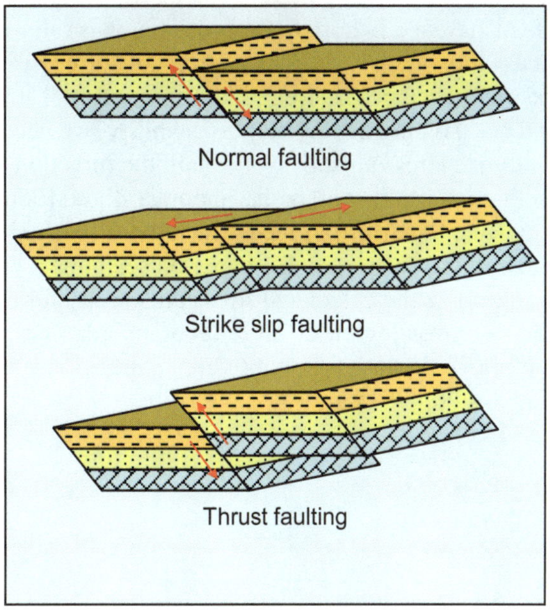

Fig 8-46 Stress regimes

Why are we concerned with the tectonic stress regime when drilling a well? When there is a large difference between horizontal stress fields, the mud weight window is smaller.

Remember, rocks often fail from shear stress and shear stress results from the difference between orthogonal stresses. (Remember from Mohr's circle that the maximum shear stress is ½ the difference between the major and minor principal stresses).

The difference in horizontal stress fields is referred to as **stress anisotropy**. (Fig 47) Stress anisotropy represents the difference in the strength of the horizontal stresses σ_H and σ_h. To maximize well bore stability, we need to minimize stress anisotropy through the direction and inclination of the well path. Shaohua Zhou, Richard Hill, and Mike Sandiford of the University of Adelaide in Australia have presented a paper on selecting a well path to minimize the calculated stress anisotropy for the various stress regimes.[2] Their recommendations are summarized in the following three pages. (Fig 8-49 to 8-51)

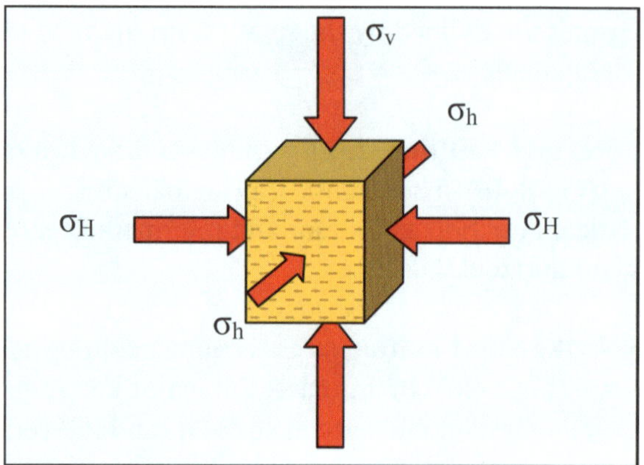

Stress anisotropy occurs when stress fields at right angles to each other are of different magnitudes.

Fig 8-47 Stress anisotropy

As you study the graphs in figures 8-49 through 8-51, keep in mind that the goal is to equalize hoop stress all around the well bore. If this can be accomplished, mud weights can be raised to stabilize the well with less risk of losing circulation. (Fig 48)

The well on the left is outside of its acceptable mud weight window. The mud weight is not high enough to prevent caving, but high enough to cause lost circulation. By equalizing the hoop stresses around the well bore, the well on the right can now be stabilized by increasing the mud weight.

Fig 8-48 Stress anisotropy

For a **Normal faulting stress regime**, $\sigma_v > \sigma_H > \sigma_h$:

- The most stable direction for drilling is along the azimuth of σ_h
- The inclination angle should increase as the difference between σ_H and σ_h increases.
- If the horizontal stresses are equal, $\sigma_H = \sigma_h$, then the inclination angle should be zero. ($\theta = 0^o$)
- If the major horizontal stress is equal to the vertical stress, $\sigma_H = \sigma_v$, then the well should be drilled horizontally $\theta = 90$.
- The inclinations for the various horizontal stress ratios are given in Fig 8-46.

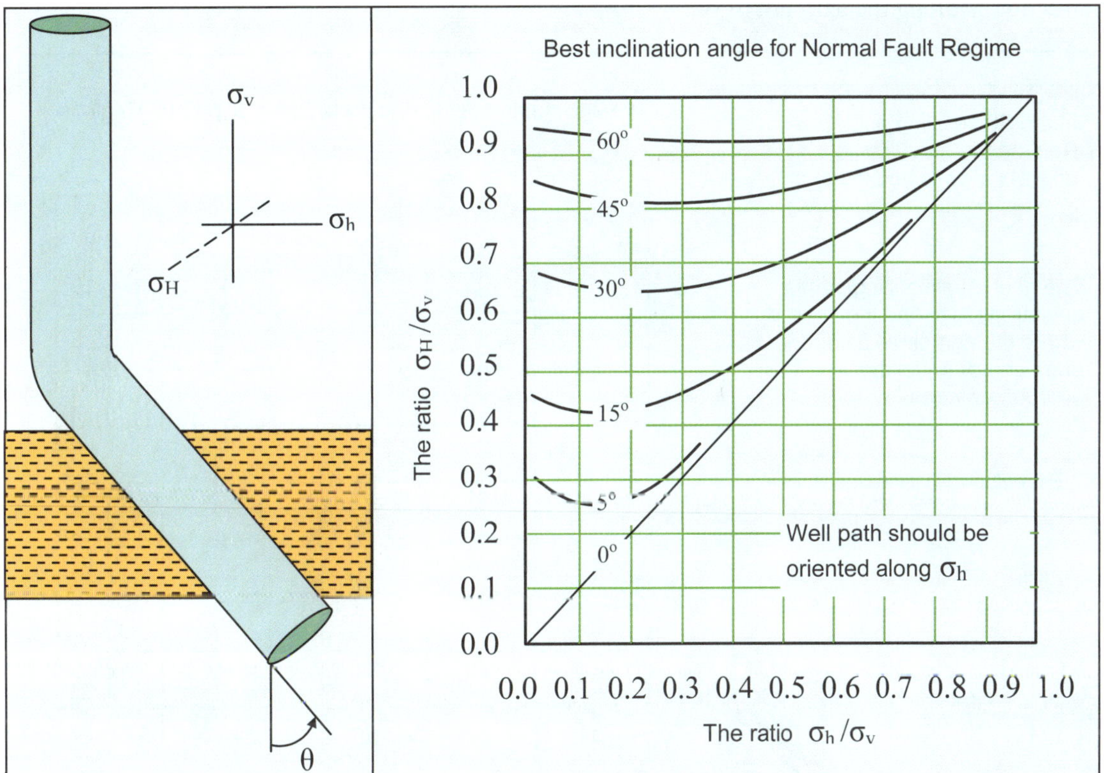

In normally stressed regimes, the most stable direction to drill is along the axis of the minor horizontal stress σ_h.

As the difference between the major and minor horizontal stresses increases, the inclination angle, θ, should increase. In the example above, the ratio of σ_H to σ_v is 0.8 and the ratio of σ_h to σ_v is 0.3. From the chart prepared by Zhou et al., the best inclination angle for the troublesome shale is 45°.

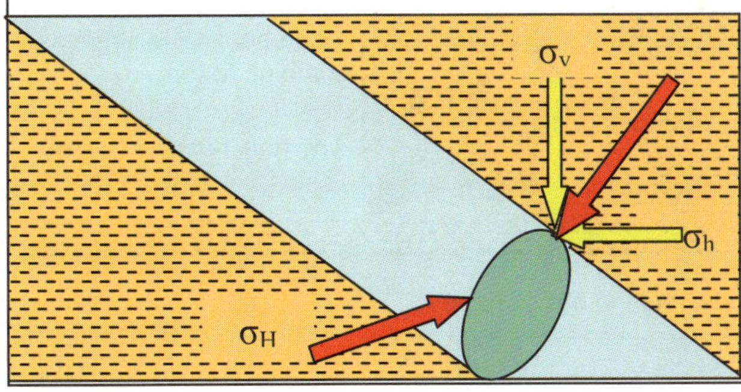

By drilling in the direction of σ_h at some angle of inclination, the component forces of σ_h and σ_v are combined to produce a stress that is closer to that of σ_H. The radial stresses around the well bore are more equal.

Now it is possible to raise the mud weight to offset the maximum hoop stress without losing circulation.

From Zhou, Hill, and Sandiford[2]

Fig 8-49 Normal fault stress regime

For a **Strike Slip faulting stress regime**, $\sigma_H > \sigma_v > \sigma_h$:

- The most stable inclination angle for drilling is horizontally. ($\theta = 90^o$)
- The most stable direction to drill in depends on the ratio of the principal horizontal stresses to the vertical stress. Generally, as the difference between the horizontal stresses increases, the direction needs to get closer to the direction of the major stress σ_H.
- As the ratio between the major stress and the vertical stress increases, the most stable drilling direction gets closer to the direction of σ_H.
- If the minor horizontal stress is equal to the vertical stress, $\sigma_h = \sigma_v$, then the most stable drilling direction is along the azimuth of σ_H.

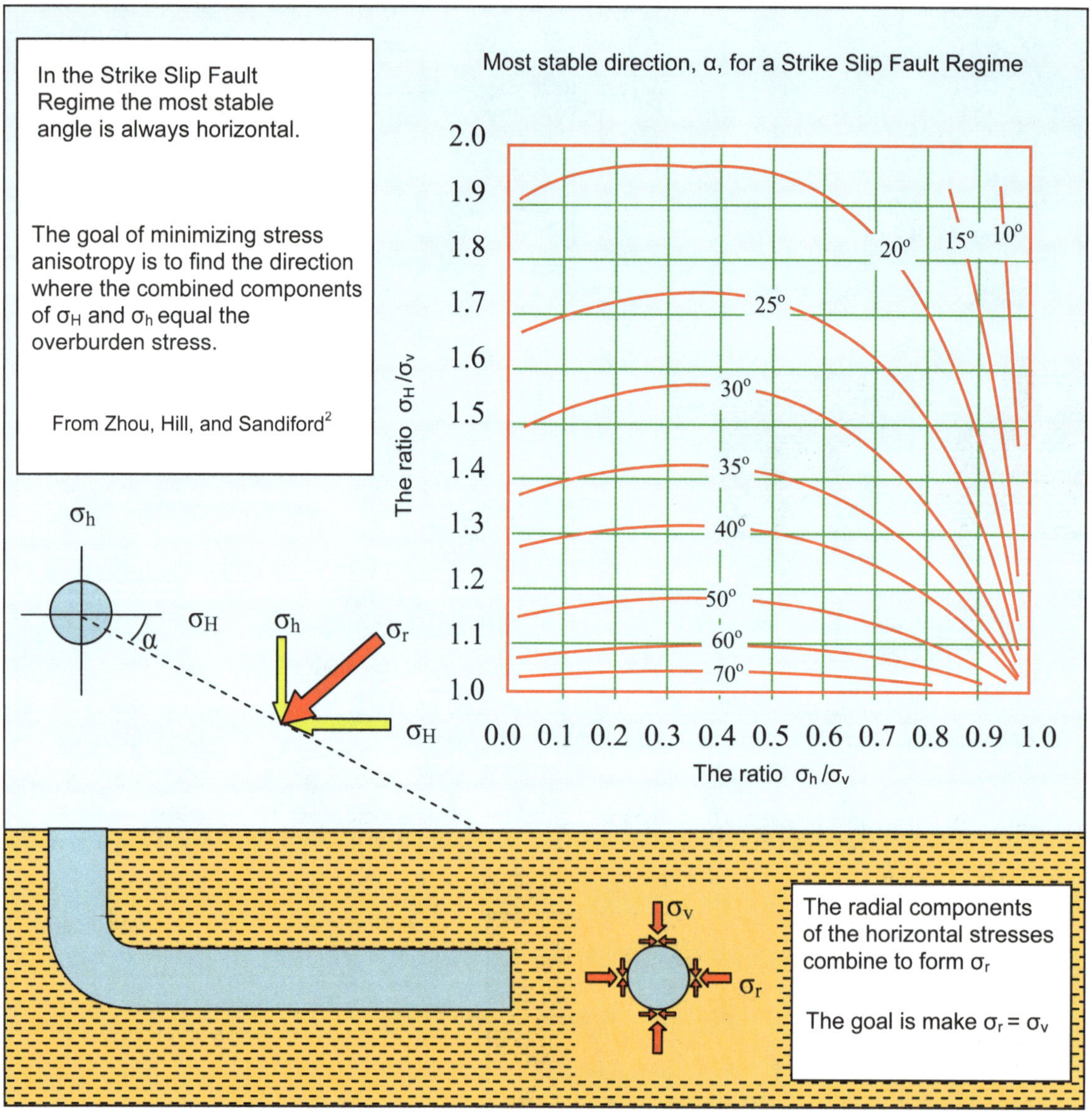

Fig 8-50 Strike slip faulting stress regime

Note from the diagram above that as the direction of the well approaches the direction of σ_H, the contribution to σ_r from the radial component of σ_H is reduced, while the contribution to σ_r from the radial component of σ_h is increased. This reduces the stress anisotropy for the strike slip regime.

For a **Reverse faulting stress regime,** $\sigma_H > \sigma_h > \sigma_v$:

- The most stable direction for drilling is along the azimuth of σ_H.
- The inclination angle should increase as the difference between σ_H and σ_h increases.
- If the horizontal stresses are equal, $\sigma_H = \sigma_h$, then the well should be drilled vertically.
- If the ratio $\sigma h = \sigma v$ then the well should be drilled horizontally. ($\theta = 90^\circ$)

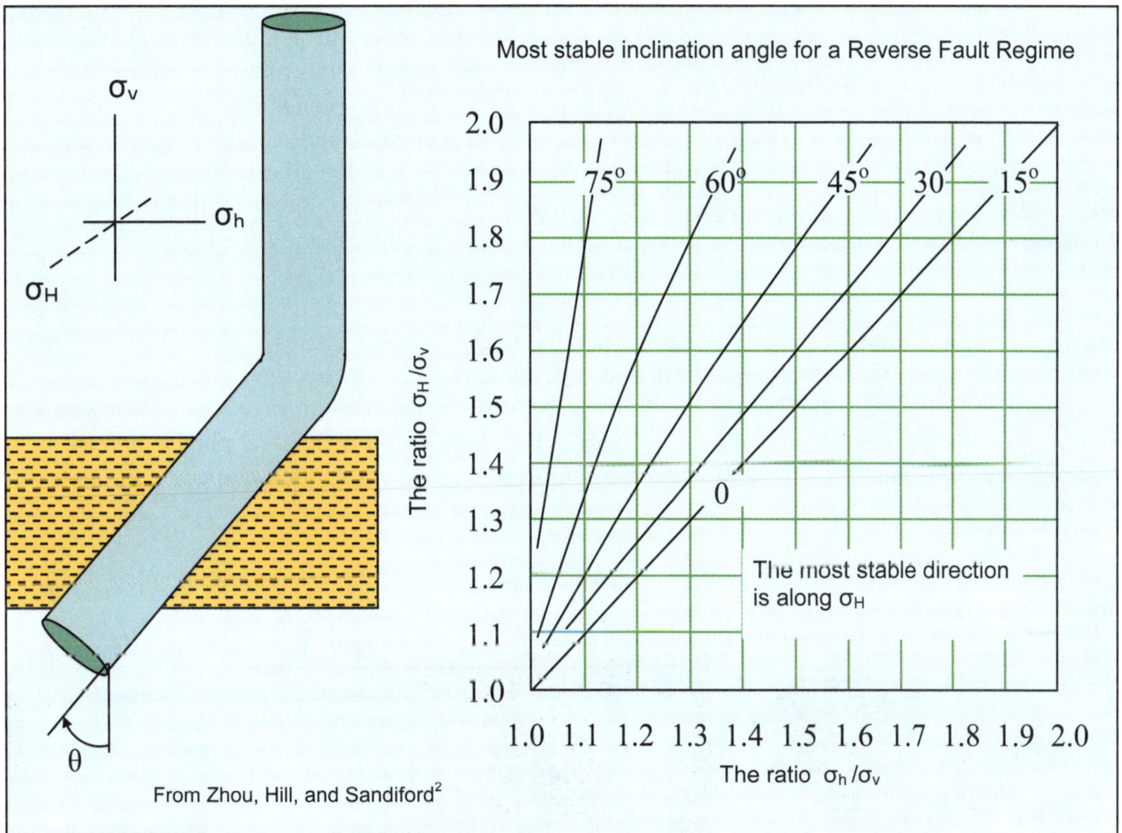

In a reverse fault regime the most stable direction to drill is along the axis of the major horizontal stress σ_H. As the difference between the major and minor horizontal stresses increases the inclination angle, θ, should increase. The goal is to combine the combine the radial components of the overburden and major horizontal stresses in order to equal the minor horizontal stress. As the inclination angle increases the contribution from the major horizontal stress diminishes while the contribution from the overburden increases. The radial stress provided by the minor horizontal stress remains unchanged at any inclination angle as long as the direction is along the major stress axis.

Fig 8-51 Reverse faulting stress regime

From the diagram above, it is clear that as the inclination angle increases the contribution from σ_H is reduced and the contribution from σ_v increases.

Stress anisotropy affects the size of our allowable mud weight window. Remember that the mud weight window is delineated by the minimum mud weight required to prevent well bore collapse and the maximum allowable mud weight that can be tolerated without causing lost circulation. When the anisotropy is high, we will have a smaller window. As we have seen in the examples in figure 8-49 to 8-51, we can reduce the anisotropy by adjusting the inclination and direction of the wellbore. We now have a better chance of stabilizing the well with mud weight.

Rocks cannot support shear stress over time, especially at depth. This is often referred to as **Heim's rule**[3]. The plasticity of rock increases with the increasing confining pressure found at greater depths. Eventually, the rock deforms until the stress anisotropy disappears. This is why stress anisotropy is greatest near the surface and rocks at great depth have nearly equal horizontal stresses.

Bedding Planes (Factors Affecting Stability)

Shale contains bedding planes that give it a plane of weakness. The clay that constitutes shale consists of microscopic bedding planes. Seasonal and geological deposition rates result in layers of shale with varying strength. If we examine a core of shale, we will usually see hundreds or thousands of small bedding planes within a few linear feet of cross section. These bedding planes tend to be planes of weakness that water can penetrate and separate. This gives shale an intrinsic property known as strength anisotropy. **Strength anisotropy**[7] means the shale is stronger in one direction than it is in orthogonal directions. The amount of strength anisotropy varies depending on the type and amount of cementation between the beds and the strength of the shale. The higher the rock strength relative to the bedding planes, the higher the strength anisotropy.

The clay platelets that comprise shale are oriented parallel to the bedding planes. When shale is exposed to water, it swells and produces hydrational stress that is perpendicular to the bedding planes. (Fig 8-52) If the well penetrates the shale at an angle to the bedding planes, this hydrational stress produces stress anisotropy. The swelling shale tends to squeeze and cave in more as the angle with the bedding planes increases. The compressive strength of shale is strongest when the compressive stress is applied perpendicular to the bedding planes.

The effect of these bedding planes on stability is governed by the in-situ stress regimes and the direction and inclination of the well bore.

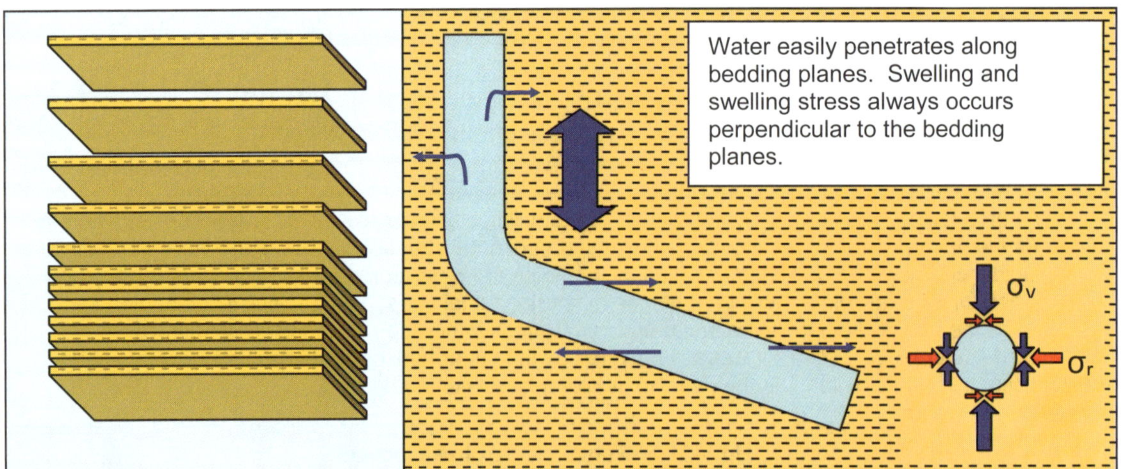

When the well is drilled perpendicular to the bedding planes, the swelling tends to produces more axial stress in the borehole wall. As the inclination to the bedding planes increases, the swelling tends to cause increasingly higher hoop stresses on one side of the well bore wall, but not on its orthogonal walls. By definition, this is stress anisotropy.

Fig 8-52 Hydrational stress in bedding planes

Effect of Well Path and Bedding Planes for the Three In-situ Stress Regimes

Chen, Tan, and Haberfield of Australia have presented guidelines in several SPE papers concerning strength and stress anisotropy analysis and well bore profile design.[8]

The following discussion will seem to be in direct opposition to what Shaohua Zhou, Richard Hill, and Mike Sandiford of the University of Adelaide in Australia have recommended in the previous discussion.[2] This is because Zhou et al were discussing the minimization of calculated stress anisotropy; Chen et al are discussing bedding plane weakness.

We need to distinguish between stress and strength anisotropy. Stress anisotropy refers to a difference in the hoop stress at different locations around the wellbore wall. Strength anisotropy refers to the difference in strength at different locations around the wellbore wall. In either case the degree of anisotropy is influenced by the inclination and direction of the well. Unfortunately for us, the wellpath that minimizes stress anisotropy may not be the path that minimizes strength anisotropy.

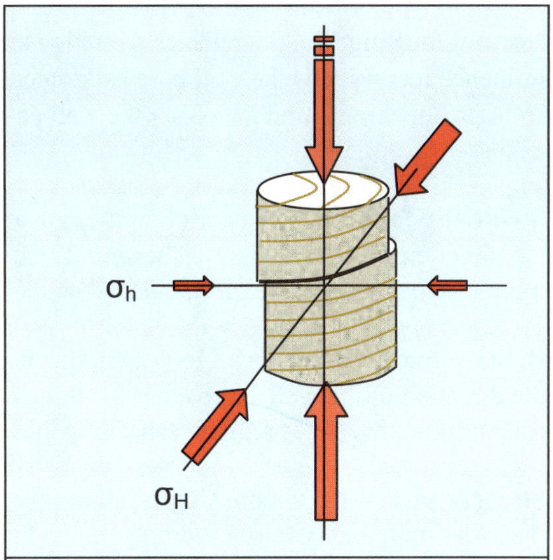

Apparent bedding plane strength is influenced by the confining stresses. The harder the beds are pressed together, the stronger they will be.

Fig 8-53 Bedding plane strength

Remember that the apparent rock strength is influenced by friction provided from confining pressure. The harder the bedding planes are pressed together, the stronger they will be. (Fig 8-53) Thus a shale will be stronger if its bedding planes are perpendicular to the highest stress field. We will also have less bedding plane trouble if we can drill perpendicular to the bedding planes.

The goal or managing the well path is to minimize the effects of both stress anisotropy and bedding plane weakness.

Here are the recommendations of Chen and company regarding strength anisotropy and the well bore profile:

Normal faulting: Bedding planes with high strength anisotropy have significant effect on stability. This influence increases as the well bore is oriented closer to the direction of σ_h. This means that the higher the strength anisotropy, and the closer the well path is oriented in the direction σ_h, the more instability we can expect.

Strike slip faulting: Even with low strength anisotropy, there is significant effect. The influence increases as the direction of the well approaches σ_h. The closer the well path is to the direction of the minor stress, the more instability we can expect, even when the bedding planes are nearly as strong as the shale itself.

Reverse faulting: Even with low strength anisotropy, there is significant effect. The influence increases as the direction of the well approaches the direction of σ_H. Bedding planes have a significant effect on hydraulic fracturing in this stress regime, especially in highly inclined well bores.

Strength anisotropy is more significant in reverse faulting regimes than normal faulting regimes. We are therefore more likely to see bedding related failure in reverse faulting regimes than in normal faulting regimes.

The influence of strength anisotropy increases in this order: **reverse** > **strike slip** > **normal**.

We want to drill perpendicularly into bedding planes if possible and, orient the well path to minimize stress anisotropy. We may not be able to satisfy both conditions. Our well path may have to be compromised in order to satisfy both conditions as much as possible.

As the strength of the shale increases, the strength anisotropy tends to increase and the mode of failure is more likely to occur in bedding planes. When the mode of failure is along the bedding planes, the cavings will have a blocky appearance with parallel faces. If the failure occurred because of excessive shear, the cavings will have a curved geometry. Our field observations may change our guestimates about the optimum well path.

Drilling Fluid Filtrate (Factors Affecting Stability)

Overbalance

We have already talked about the influence of mud weight on well bore stability in the sections on radial stress and Mohr's failure envelope. However, we must differentiate between "overbalance" and the radial stress provided by it. Overbalance refers to the amount of hydrostatic pressure that is in excess of pore pressure. This is **not** the radial force provided by overbalance.

Fluid pressure against the well bore wall provides a radial stress that enhances well bore stability. The radial stress reduces the hoop stress and applies a confining pressure to the rock elements along the well bore wall. This confining pressure increases the shale's apparent rock strength. The radial stress is the result of a differential pressure across the well bore wall. This differential pressure is provided by an overbalance of well bore pressure over formation pressure. However, the differential pressure is not equal to the overbalance! Shale is permeable. Some fluid invades the pores in the shale and raises the pore pressure near the well bore. (Fig 8-54) As filtrate invades the pores in the shale, the differential pressure at the well bore wall is reduced. This process is time dependent.

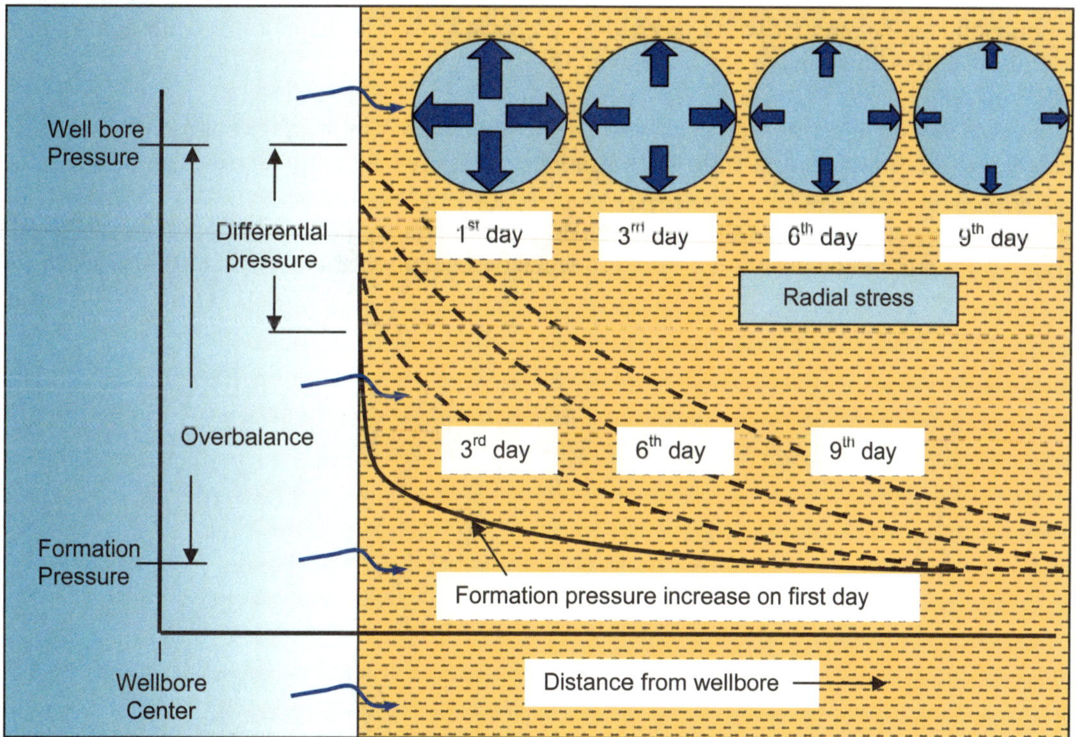

Differential pressure is the pressure across the surface of the wellbore wall. As filtrate penetrates and charges the pores near the wall the differential pressure is reduced.

Fig 8-54 Pore pressure vs. time

To better visualize formation pressures due to filtrate invasion we can compare the filtrate injection curves to drawdown and injection curves in irrigation wells in Fig 8-55 and 8-56.

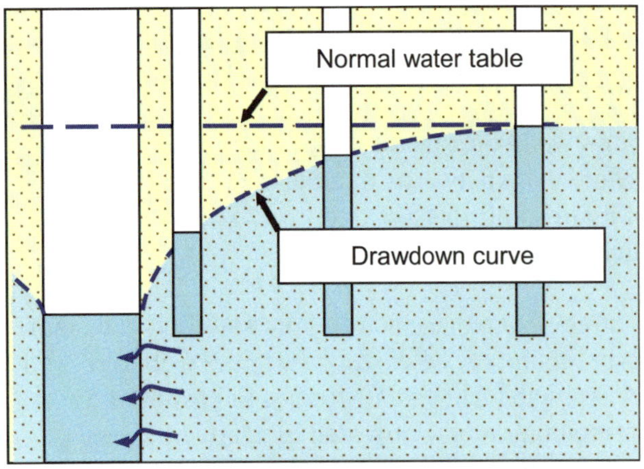

Fig 8-55 Drawdown curve

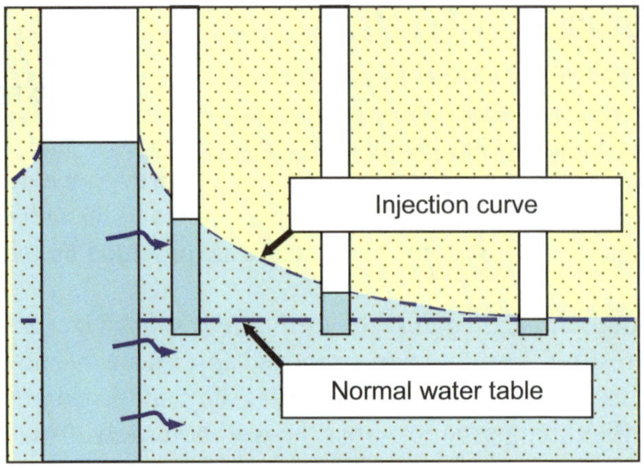

Fig 8-56 Injection curve

A drawdown curve for a large diameter irrigation well often has several monitor wells drilled at various distances away from the production well. When the well is pumping the water level is observed to fall in both the production well and the monitor wells. By plotting the water level in the monitor wells against distance from the well bore a drawdown curve can be drawn. (Fig 8-55)

If water is injected into an injection well, such as in Fig 8-56, the level in the adjacent monitor wells is seen to rise. If the water level is plotted against distance from the well bore an injection curve can be drawn.

The water level in the monitor wells represents formation pressure at that distance from the main well. The curves in Fig 8-54 represent the injection curves produced by injecting filtrate into the formation.

Filtrate invasion

Filtrate invasion is a major cause of well bore instability. Filtrate invasion weakens the rock and alters the stress distribution within the rock. The clay within the shale may chemically react with water, which further weakens the rock. Several mechanisms are involved.

The reduction of radial stress by filtrate invasion reduces the apparent rock strength and increases the hoop stress. As filtrate invades the pores, pore pressure increases. This reduces the effective stress. (Total stress = pore pressure + effective stress.) The fluid breaks the contact between grains, thus lowering the cementation and friction between the grains. The filtrate also acts as a lubricant, which further reduces the internal friction. If all this wasn't bad enough, the filtrate chemically and mechanically reacts with the clay in the shale, causing swelling and dispersion. Adsorption of water onto the clay surfaces causes hydrational stress, which increases the hoop stress and bedding plane weakness.

As filtrate invades the pores, some clays disperse through crystalline and osmotic swelling mechanisms. (See swelling mechanisms) This increases the permeability of the shale and accelerates the rate of filtrate invasion. The amount of swelling and dispersion is dependent on the clay mineralogy, but it is important to note that crystalline swelling will take place with <u>any</u> clay. Non-swelling clays like Illite will not swell or disperse appreciably, but the crystalline swelling will cause hydrational stress and reduce rock strength. With time, all shale will weaken when exposed to water, due to filtrate invasion and the hydrational stress from crystalline swelling.

The increase in pore pressure from filtrate invasion reduces the effective radial stress, which in turn also leads to higher hoop stress. (Fig 8-57)

Filtrate invasion and Mohr's stability envelope

Filtrate invasion increases pore pressure near the well bore wall. This causes the effective radial stress to decrease and the hoop stress to increase. Too much filtrate invasion may cause the resulting shear stress to become excessive and the well to become unstable.

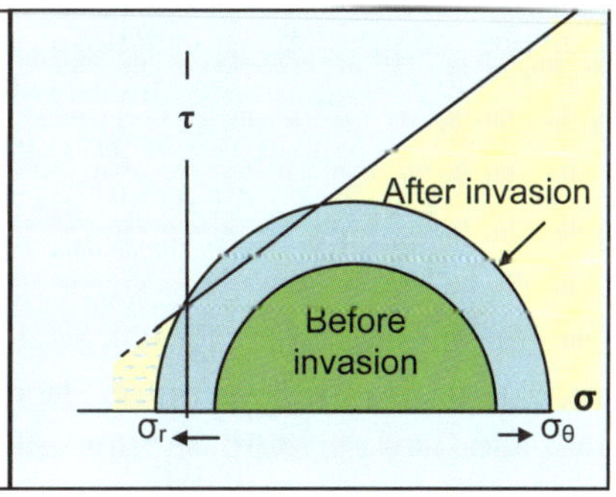

Fig 8-57 Mohr's stability envelope for filtrate invasion

The filtrate invasion makes shale instability time dependent. As filtrate invades the shale, the shear stress increases and apparent rock strength decreases. Some shale will even swell and disperse appreciably when exposed to water. When sufficiently dispersed, the shale can be eroded by turbulent flow and pipe whip. Shale with high smectite contents will suffer the most from dispersion and erosion.

It takes time for filtrate to invade shale. (Fig 8-54, 8-58) The more permeable the shale, and the higher the overbalance, the faster the filtrate will invade the shale. We won't experience a problem with swelling clays as they are drilled. The problem won't develop until the filtrate has invaded the shale and had time to weaken and disperse it. Similarly, non-swelling clays will grow weaker with time.

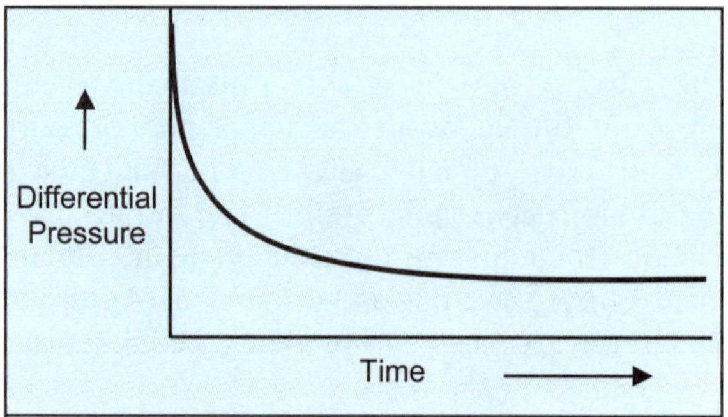

Differential Pressure and thus radial stress decreases proportionally to the square root of time.

Fig 8-58 Filtrate invasion with respect to time

Three mechanisms contribute to filtrate invasion:

- Over balance
- Water activity
- Capillary action

Overbalance is the predominant factor in young, poorly consolidated shale. Overbalance continues to be a major factor in older, more consolidated shale, but water activity becomes increasingly important because the pore throat size decreases. (**Water activity** refers to the attractive and repulsive forces produced by the electrostatic charges on the surface of the water and shale interface.) Capillary action plays a significant effect with fractures, especially when the shale is not completely saturated.[9]

Obviously, filtrate invasion is bad. It leads to instability. One goal of well planners and mud engineers is to reduce the rate of filtrate invasion. Unfortunately, this cannot be done with filter cakes and the filtration controlling additives that work in filter cakes.

In more permeable formations like sandstone, a filter cake against the well bore wall is built to prevent loss of mud into the formation. (See filter cakes) Even with a good filter cake, some filtrate continuously passes through the cake and into the formation. Formations that are permeable enough to accept filter cakes are permeable enough to allow the filtrate to drain away from the well bore. A differential pressure can be maintained across the filter cake to hold it in place.

Shale however, is not permeable enough to deposit dynamic filter cakes. In fact, the filter cakes formed against sandstones are usually several magnitudes more permeable than most shale. Shale pore openings are so small that few, if any, solids can bridge in their openings. Solids are screened out at the well bore wall and only a solids-free filtrate is allowed to enter into the shale. The flow of fluid into the shale is not high enough to hold the solids against the wall. Fluid flow and mechanical erosion from the drill string erode the filter cake off the wall. (Fig 8-59)

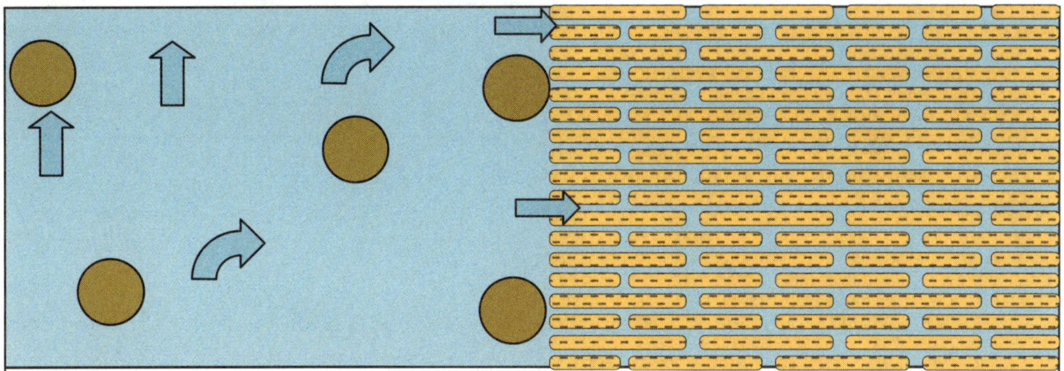

Shale is not permeable enough to build a filter cake. All solids are bridged off at the surface but cannot be "differentially stuck" to the wall hard enough to avoid being swept away by the flowing mud.

Fig 8-59 Filter cakes on shale

To say that no filter cakes are deposited on shale is not entirely true. When pipe and fluid motion is stopped, a static filter cake may develop on the wall, but this cake will be eroded away when the pumps are started again. Some internal filter cake may exist in some pores or fractures that are large enough to allow bridging to take place. However, most of the shale exposed at the wall will not accept a filter cake that will stand up to fluid circulation. Even if it did, the rate of filtration through the shale would be lower than the filter cake.

Several approaches can be taken to minimize or eliminate the rate of filtrate invasion. These approaches include:

- Reduction of overbalance,
- Reduction of the shale's permeability,
- Increasing the viscosity of the filtrate, and
- Creation of a semi-permeable membrane, allowing osmotic pressure to balance the differential pressure.

A **reduction of overbalance** defeats the purpose of having an overbalance in the first place. Some overbalance is necessary to provide radial stress against the well bore wall. However, excessive overbalance may be detrimental, even if it is not severe enough to cause lost circulation. An optimum overbalance should exist to provide adequate radial stress yet minimizes filtrate invasion. One approach for finding this overbalance is the median line principle proposed by Aadnoy.[4]

It would be disastrous to lower an overbalance once it has been established. The filtrate invasion from the original overbalance will lead to a pore pressure that may be higher than the new well bore pressure. This would lead to a low or negative radial stress that would greatly reduce stability. Surging and swabbing while running pipe obviously has this negative effect on well bore stability.

Reduction of the shale's permeability is achieved by chemical reactions between the filtrate and the shale, or by ultra-fine particles that make internal filter cakes. Note that many shale formations are highly fractured, either with naturally occurring fractures or fractures induced while drilling. These fractures are responsible for a large percentage of filtrate invasion. Gilsonite and other "filter cake" forming additives are effective here. (Fig 8-60)

Increasing the viscosity of the filtrate is achieved with additives such as glycol, glycerol, sugars, and silicates. These additives work by interfering with the pseudo-crystalline structure of the water bound to the surface of the clays.

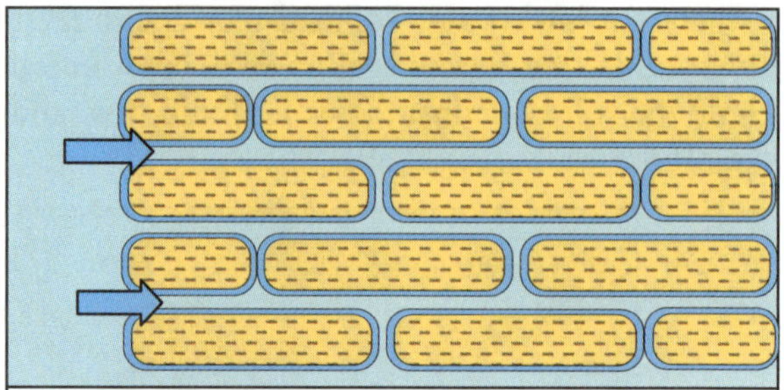

The permeability of shale can be reduced if the filtrate can bond to its surface. The effective porosity is reduced.

Fig 8-60 Shale permeability

A balance between osmotic pressure and overbalance is more readily achieved with oil-base mud than with water-base mud. In oil-base mud, this is accomplished with the use of surfactants and salts. The surfactants are needed to create a semi-permeable membrane across the wall of the well bore. The salts are used to provide the correct salinity of the emulsified aqueous phase of the oil base mud. The creation of a semi-permeable membrane is more difficult with water-base mud. Balancing osmotic pressure with overbalance is still possible, but less likely to happen with water-base mud (WBM).

Osmotic flow of water in or out of the shale is driven by the difference in salt concentrations in the filtrate and pore fluid. Excessive salt concentration in the mud drives water out of shale. (Fig 8-61) Too little salt concentration drives water into the shale. Overbalance also drives water in. It is possible to balance the flow from the overbalance with the osmotic flow from the shale, such that there is no net change in pore pressure.[1]

True osmotic flow requires an ideal, semi-permeable membrane. Shale does not provide an ideal membrane because of the wide variety of pore sizes. Some ions can "leak" through the membrane with the filtrate. When the ion-infested filtrate mixes with the original pore fluid, the potential difference between the drilling fluid and pore fluid is reduced. This reduces the osmotic flow. The success of using osmotic flow to minimize filtrate invasion depends heavily on the quality of the "membrane" formed on the wall of the well bore. The surfactants and quality of the shale determine the quality of the membrane. Many of these surfactants are environmentally unfriendly, as are many oils, so oil-base mud is often not used, even though it makes an excellent drilling fluid. Some success has been made with WBM, but it is difficult to find appropriate surfactants for shale that are also environmentally acceptable.

Fluid moves "osmotically" across a semi-permeable membrane toward the fluid with the higher concentration of salt.

It is possible to balance the osmotic loss of fluid from the shale with the gain of fluid invasion from overbalance.

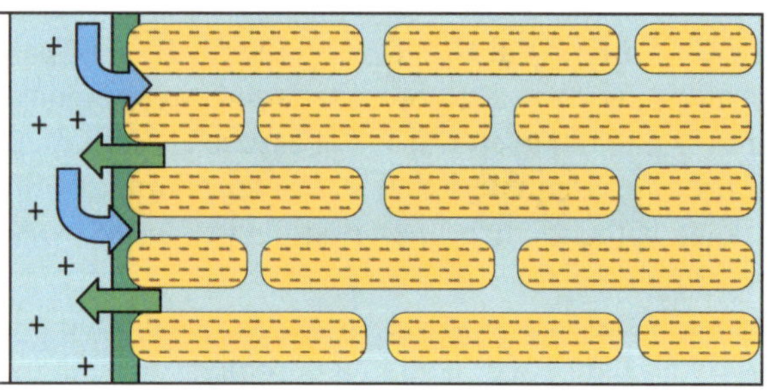

Fig 8-61 Osmotic flow in shale

It is possible for the osmotic flow to be higher than the flow caused by the overbalance. This can lead to dehydration of the shale. This dehydration actually *increases* the strength of the shale for the same reasons that filtrate invasion weakens the shale. However, the risk of lost circulation or failure from tensile stress is now introduced.

If the direction of osmotic flow is into the formation, then the rate of filtrate invasion increases.
It should be noted here that L. Bailey et al. at the Cambridge Research Institute contest osmotic transfer.[11]

Capillary action is another way that filtrate may invade shale. Many shale formations are not saturated with a wetting fluid, such as water, even though the shale is below the water table. When this shale is exposed to a wetting fluid, the fluid will creep into the pores by the capillary process. Air or gasses trapped in the pore spaces will experience an increase in pore pressure that is equal to the capillary pressure.[9] The smaller the pore throat diameter, the higher the capillary pressure. It is capillary pressure that causes apparently hard, dry samples of shale to shatter when exposed to water. Microscopic and visible fractures that appear to be well cemented are actually permeable to the capillary process and become water wet shortly after exposure to water. Capillary pressure spreads the cracks open and breaks the sample apart. Filtrate can enter small cracks and fissures down hole, allowing hydrational stress to develop more rapidly. (Fig 8-62)

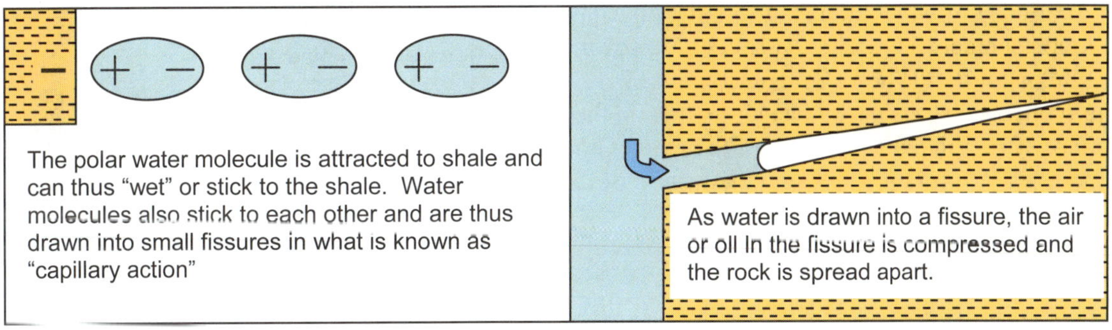

Fig 8-62 Capillary action

Capillary pressure depends on the affinity of the shale for the wetting fluid. The more polar the fluid, the more shale-wetting it tends to be. Water is a highly polar fluid, while oil is non-polar. This explains why the same fractured, but intact shale samples will break apart in water but not in oil.

Shale is not homogeneous. The pore sizes vary, it has many bedding planes, and it is often inter-bedded with sands stringers. Thus, the permeability varies along the length of the well bore and the rate of filtration varies with it. Shale next to permeable sand may receive filtrate invasion from the sand as well as the well bore.

Drill String Vibration (Factors Affecting Stability)

Drill string vibration contributes to well bore instability more than most people realize. Radial stress, axial stress, and hoop stress all fluctuate in the presence of drill string vibration. This fluctuation causes stress fatigue and, in extreme cases, may cause the yield strength of the rock to be exceeded in just one cycle. Lets look at the various types of drill string vibration and how it affects stability.

The drill pipe is almost always in contact with the well bore wall. As it rotates, it may bounce from wall to wall or whirl around the perimeter of the wall. Bouncing is referred to as "pipe whip." The centrifugal force of the rotating pipe forces it to strike the wall. As the pipe strikes the wall, it transfers momentum and applies a radial force. Then, it bounces off the wall and is flung against the wall in a different location. This banging of the pipe against the wall sets up vibrations throughout the length of the pipe. The vibrating pipe contacts the wall at its "nodes" in several places simultaneously. (Fig 8-63) The amount of momentum or radial stress each "bang" imparts on the wall depends mostly on the radial velocity of the pipe. Other factors include tension in the pipe and the weight of the pipe.

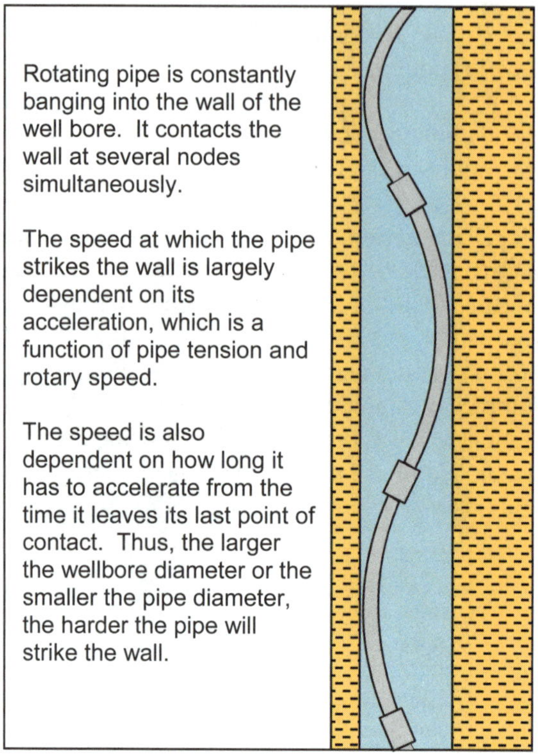

Fig 8-63 Drill string vibration

The radial velocity of the pipe is a function of the distance the pipe has to accelerate after leaving the wall and before striking it again. Thus, the velocity of the pipe increases as the diameter of the well increases and/or the size of drill pipe decreases. The acceleration of the pipe is a function of pipe tension and rotational speed.[11] The acceleration increases as pipe tension and/or rotation increase. *Rotating off bottom at high speeds with small pipe in a large hole is a disastrous combination.*

We are often guilty of rotating at high speeds off bottom to clean the hole. Sometimes, it is necessary to do so. We must be careful not to confuse cavings with cuttings as we assess the benefit of rotating off bottom on a case-by-case basis.

Another type of vibration that occurs is the "standing wave" that occurs when axial pipe motion is suddenly stopped. We sometimes use a standing wave to flip a hose or extension cord around an obstruction so we can continue to drag it. When downward motion is suddenly stopped by the drawworks, a wave propagates down the drill string. Upward motion stopped by a ledge has the same effect, but the wave moves in the other direction. The standing wave travels faster and hits the wall harder when the pipe has more tension in it.

The mud pump creates pressure surges that cause the pipe to vibrate, especially when the valves don't seat right. This vibration can sometimes be seen in the Kelly hose.

The bit also imparts axial and torsional vibrations on the string. Large-tooth, roller-cone bits tend to impart more vibration than small-tooth, PDC bits. These vibrations increase as bit weight and rpm increase. Bit stabilization with packed hole assemblies helps minimize these vibrations. The amount of mass just above the bit serves to dampen these vibrations. The heavier the BHA, the more these vibrations are dampened.

Torsional vibration causes the drill string to form a helical spring that varies in length and diameter with each vibration. This imparts radial and axial stress to the wall. Torsional vibration can also impart tangential or hoop stress to the wall through friction.

Tangential or hoop stress is imparted by the tangential component of the force and by the friction between the pipe and wall. There are two types of friction occurring between the drill pipe and the wall, **static friction** and **dynamic friction**. When the pipe is still, it must overcome static friction to move. When it is moving, the pipe and wall feel dynamic friction. Both types of friction obey the following relationship:

$$F = \mu N \qquad \text{eq. 8.6}$$

Where F is the friction force
μ is the coefficient of friction, and
N is the radial force of the pipe against the wall.

The coefficient of friction is lower for dynamic friction than it is for static friction. When torsional vibration brings the rotational speed of some segment of pipe to zero, the static friction must be overcome to start it moving again. This friction affects the hoop stress more than dynamic friction.

When rotating the pipe very slowly, many parts of the string are temporarily static. The pipe may be turning at the rotary table, but the bit is stopping and starting. The helical spring is coiling and uncoiling and the weight on bit is fluctuating. If we monitor torque against rpm we find that there is a "threshold" rpm that must be reached before the minimum torque can be found. This threshold rpm is necessary to develop enough angular momentum to overcome the friction that brings some portions of the string to a temporary halt. With static friction eliminated, only dynamic friction is experienced, and the torque is lower. As the rpm is increased, the vibration increases, which increases torque.

Generally, the faster we rotate the pipe, the more damaging the vibrations become. In some vertical wells, there may exist a critical rpm that causes resonance in the drill string. This will create horrendous vibrations that are extremely damaging to both the pipe and the well. Contrary to popular belief, *critical rpm's do* ***not*** *generally exist in most wells.*[12] There are too many variations with well bore geometry, vibration, and dampening sources to allow resonance to exist. The occurrence of a critical rpm is most likely to occur in a vertical, in-gauge well. Accurate calculation of the critical rpm requires ideal conditions.

When the pipe is "whirling," the pipe never leaves the wall. Centrifugal force causes the pipe to simply whirl around the wall. While it is whirling, the friction affects the hoop stress and the side load holding the pipe against the wall affects the radial stress. If there is torsional vibration, then the axial stress along the wall is also affected.

Remember that radial and axial stress affect the hoop stress. Any stress imparted by the pipe is cyclical, so the well bore wall is being fatigued. We often talk about how surge and swab fatigues the well, but there are far fewer cycles with surge and swab than there are with pipe rotations.

Well Bore Geometry (Factors Affecting Stability)

The shape of the well bore has a direct impact on well bore stability. High side loads from pipe across a dogleg impart high stress to the well. Keyseats cut into the wall change the stress distribution. As the well enlarges, vibrational impacts from the pipe increase. (Fig. 8-27)

Types of Failures

There are two broad categories of failure for shale — stress-induced failure and plastic creep. Stress-induced failure occurs when the rock strength is exceeded and the rock breaks. Plastic creep refers to slow deformation under stress. When the well bore squeezes in, as with salt and swelling clays, the formation has failed from plastic creep. Plastic creep is stress-induced strain. The stress causing the strain can be of mechanical or chemical origin.

Stress Induced Failure

There are six types of stress-induced failure. Four are shear failure modes and two are tensile failure modes. Furthermore, these stresses can be mechanical or chemical in origin.

Most caving is the result of shear stress. Remember from Mohr's circle that the magnitude of the maximum shear stress is ½ the difference between the largest and smallest principal stresses. The principal stresses along the wall of the well bore are usually the hoop stress σ_θ, the axial stress σ_z, and the radial stress σ_r. (Fig 8-33)

The four **shear failure** modes are: (Fig 8-64)[8]

- Breakout
- Toric shear
- Helical shear
- Elongated shear.

In each case, failure occurs along distinct "shear bands" where slippage initiated and then propagated until a rock chip is detached from the well bore wall.

Breakout shear occurs when the highest stress at failure is the hoop stress, and the smallest stress is the radial stress. $\sigma_\theta > \sigma_z > \sigma_r$ This shear is caused by the difference between the hoop and radial stresses. (This is the most common type of failure)

Toric shear occurs when the highest stress at failure is the axial stress, and the smallest stress is the radial stress. $\sigma_z > \sigma_\theta > \sigma_r$ (This can be caused from drag while tripping)

Both breakout and toric shear occur in the direction of σ_h in the case of vertical wells. Insufficient mud weight or swabbing are the cause of these failures.

Helical shear occurs when the highest stress at failure is the axial stress, and the smallest stress is the hoop stress. $\sigma_z > \sigma_r > \sigma_\theta$

Elongated shear occurs when the highest stress at failure is the radial stress, and the smallest stress is either the hoop stress or the axial stress. $\sigma_r > \sigma_z > \sigma_\theta$ or $\sigma_r > \sigma_\theta > \sigma_z$

Both helical and elongated shear occur in the direction of σ_H and are caused by high mud weight or surging.

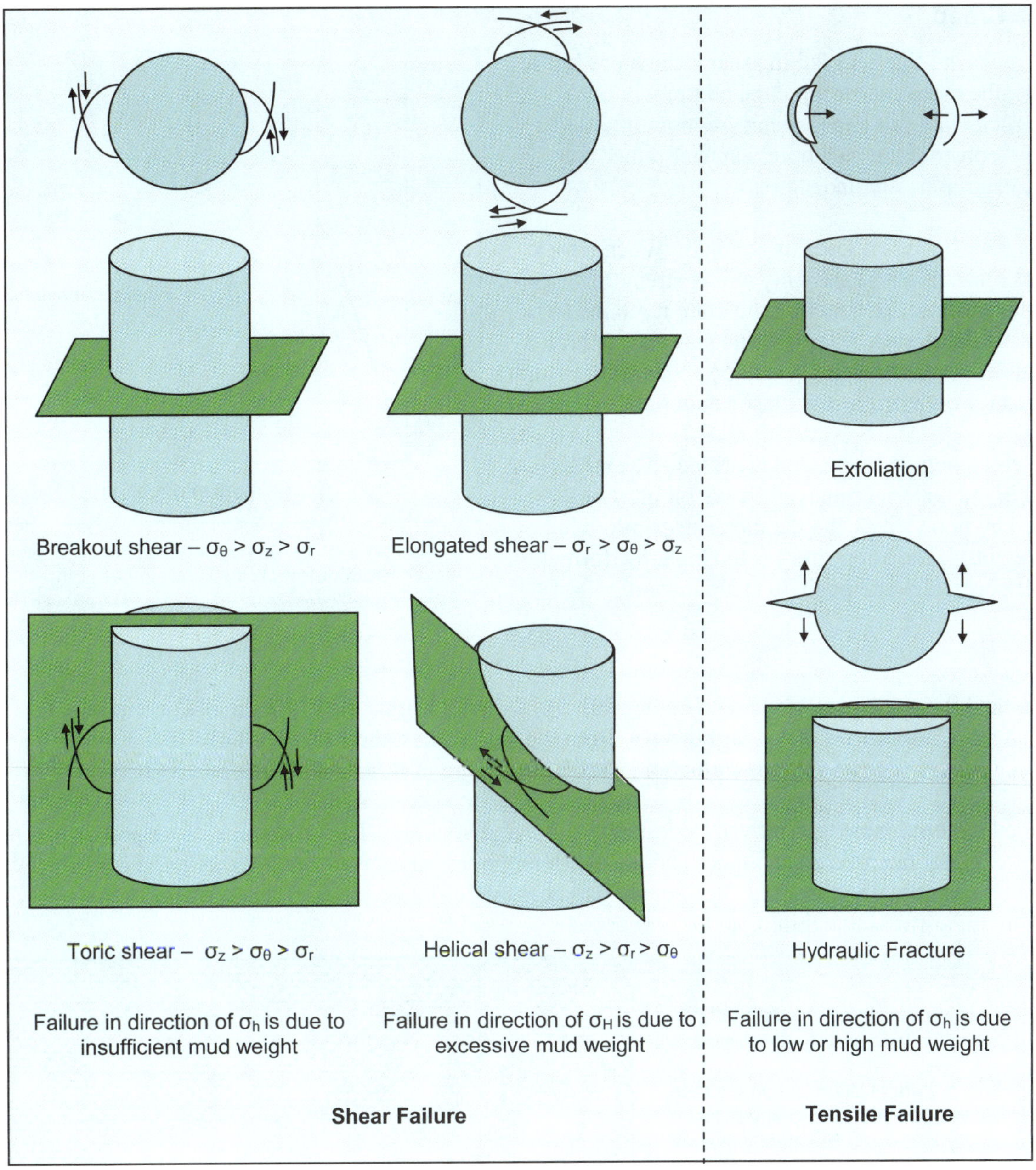

Fig 8-64 Failure modes

The two tensile failure modes are hydraulic fracture and exfoliation. (Fig 8-64)

Hydraulic fracture and ballooning is caused by high mud weight. (See radial stress and Fig 8-30)

Ballooning refers to a type of lost circulation in which mud is slowly lost while circulating, but returns when circulation is stopped. The hydrostatic pressure in this case is very close to the fracture pressure, and the annular friction losses while circulating are high enough to open fissures that have been caused by surging. When circulation is stopped, the fissures squeeze shut, forcing the mud back into the well bore.

Exfoliation tensile failure is caused by low mud weight. This type of failure is common in mine shafts and under-balanced wells. The stress streamlines in figure 8-28A depict tensile stress in the absence of the radial stress provided by fluid pressure.

Plastic Creep

Brittle rocks are likely to fail in shear. The more brittle they are, the more catastrophic the failure will be. Plastic rocks like salt and gypsum are more likely to flow and constrict the well bore diameter rather than break catastrophically under stress.

When a rock breaks, it loses its strength. When a rock deforms plastically, it loses only a little strength. The more elastoplastic the material, the more it retains its strength as it deforms. Elastoplastic materials exhibit a behavior known as "creep." Fig 8-65 shows stress strain diagrams for both brittle and plastic materials.

Plastic creep begins as the well is drilled. The stress that existed in the material removed by the bit must be replaced by hoop stress and the radial stress provided by mud weight. If the hoop stress is too high, the formation will fail plastically and begin to creep inward.

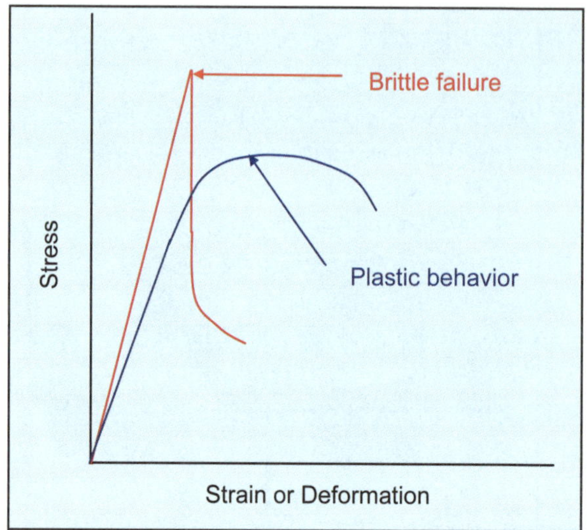

Brittle rocks loose their strength at failure while plastic rocks deform but still retain strength.

Fig 8-65 Plastic creep

At first, the highest hoop stress is right at the wall. As the wall begins to fail plastically, it supports less hoop stress, so more hoop stress is felt farther away from the wall. Thus, the well bore fails first at the wall and then progressively deeper into the formation, up to a maximum of three well bore radii. (Fig 8-66)

The failed material must be removed by reaming to prevent sticking. This material still supports some load, so after reaming, the process continues and the maximum hoop stress is located farther and farther away from well center. Eventually, there may be enough load bearing deformed material inside of the radius of maximum hoop stress that failure no longer occurs.

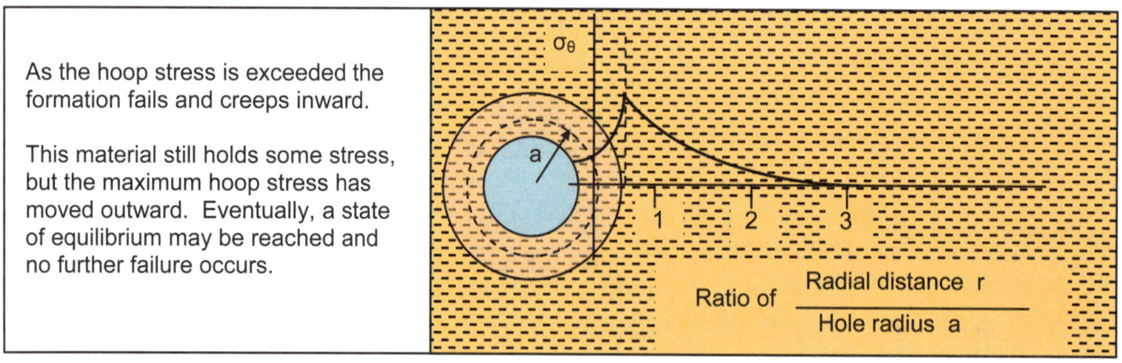

Fig 8-66 Stress distribution in plastic formations

Salts and gypsums exhibit the most creep. Shale and sandstone may also creep significantly. Sandstones tend to creep at great depth or under high tectonic stresses. Young shale with thin bedding planes tends to creep more than older shale. The creeping tendency increases as the well bore intersects the bedding planes at higher angles.

A major concern when tripping is that it is not just the material that is squeezed into the well bore that will be scraped away from the wall as the drill collars pass. All of the failed material may come with it! (Fig. 13.4)

Heaving, Sloughing, and Spalling

The terms, *heaving*, *sloughing*, and *spalling* are often associated with shale instability[13]. Unfortunately, there does not seem to be consensus as to what these terms refer to.

The most widely accepted definition of **Heaving shale** is shale that caves due to low mud weight. Those who use this term are referring to one of the stress-induced shear breakout modes. They might also suggest that this is the type of shale that fails on bottom as soon as it is exposed.

Sloughing shale is generally considered to fail because of filtrate invasion and a chemical reaction with water. This type of failure commonly occurs off bottom after the water has had time to react with the clay. Many people in our industry have the terms sloughing and heaving reversed as to the type of failure. I can find nothing in the industry literature that positively distinguishes one term from the other.

Spalling is not commonly used outside of the mining industry. It generally refers to shale or rock material that explodes off the wall in concentric, concave chips.

These are laymen terms that inadequately describe the type of failure occurring. However, they do offer a broad distinction between chemically stressed shale and mechanically stressed shale. This at least gives rig crews a starting point when addressing the problem.

Lost Circulation and Ballooning

There are two broad types of lost circulation; pressure induced, and naturally occurring. Naturally occurring losses are the result of high permeability and overbalance. If the combination of drilled solids and mud properties are unable to establish an adequate filtercake whole mud is lost to the formation.

Pressure induced losses result from hydraulically fracturing the well with too much pressure. Hoop stress decreases as radial stress increases. If the combination of mud weight and annular friction gets too high the hoop stress will become tensile and the rock may fail in tension. (Fig 8-30, 67) This is referred to as hydraulically fracturing.

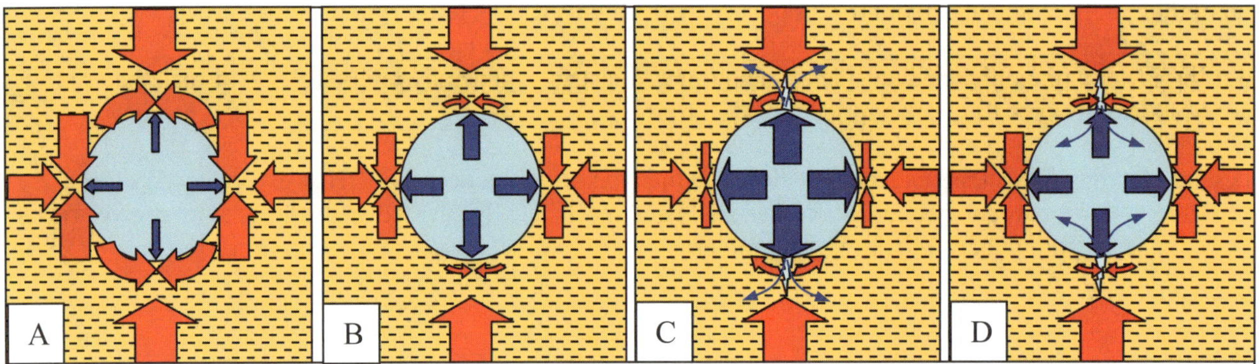

Excessive annular friction puts the wall in tension and initiates hydraulic fracturing. When the pumps are stopped mud is returned to the well as the crack squeezes shut.

Fig 8-67 Ballooning

As depicted earlier in fig 8-30, radial stress and hoop stress combine to transfer the field stresses across the well. A low radial stress results in high hoop stresses. (Fig 8-67a) The hoop stress caused by the minor stress field can be near zero when the wellbore pressure is very high. (Fig 8-67b) The addition of circulation raises wellbore pressure even more. This can put the rock into tension and initiate hydraulic fracturing. (Fig 8-67c) Mud is then slowly injected into the fracture as it extends. When circulation is stopped, hoop stress is returned to compression, and mud is squeezed back into the well as the crack closes. (Fig 8-67d)

Ballooning occurs when the combination of hydrostatic and annular friction pressure cause minor hydraulic fracturing. The back flow of mud from the fracture after circulation has stopped is often confused with an influx of formation fluid. Severe ballooning occurs in many wells, especially HPHT wells, and frustrates well control vigilance.

Determination of Stresses

Overburden stresses are generally one pound per square inch of stress for each foot of depth. (1psi/ft). More accurate formation densities can be obtained from sonic logs and seismic work.

Minor horizontal stress is determined from leak-off data. The **direction** of the major and minor stresses is determined from the breakout orientation. (Fig 8-30)

The **major horizontal stress** can be estimated using Poisson's ratios, the overburden and minor horizontal stresses, and the Kirsch equations.[14] David Woodland of Shell Canada offers an example of how in-situ stresses were calculated in the Canadian overthrust belt.[15]

In practice, stresses are estimated with mathematical models, using as much information as is available. These models are then adjusted until the predicted hole enlargement matches the observed hole enlargement after the well is drilled.

Swelling and Dispersion

Cation Exchange

Before we can understand the swelling mechanisms of clay, we must understand the chemistry of clay. An education in engineering is a prerequisite to thoroughly understand the composition and behavior of clays. Fortunately however, a thorough understanding is not required for most of us in the drilling industry, although some understanding is necessary, and the more we understand the better. The author refers interested readers to the work of Gray et al.[1], and to the industry mud manuals for a deeper understanding of the clay mineralogy and chemistry of mud engineering.

Clay crystals carry a charge on their surface that is compensated for by absorbing an exchangeable cation. A **cation** is an ion with a positive electrical charge. It is attracted to the negative charge on the surface of the clay crystal and sticks to the surface of the crystal like a magnet on a refrigerator. The adsorbed cation alters the physical properties of the clay. The specific properties of the clay depend on the type of ion that is adsorbed.

An **exchangeable cation** is an ion that can be exchanged with other ions in the presence of water. A lot of the magic performed by mud engineers when treating swelling clays is based on exchanging ions in the clay with ions that help hold the clay sheets together. This is possible because one ion can replace another ion if its chemical valence is higher. The order of preferential adsorption generally follows the lyotropic series:

$H^+ > Ba^{++} > Sr^{++} > Ca^{++} > Cs^+ > Rb^+ > K^+ > Na^+ > Li^+$

As can be seen by the above series, Hydrogen, H^+, is strongly adsorbed. This explains the strong influence pH has on the base exchange reaction.

Note that Montmorillonite has a greater selectivity for Potassium (K^+) than Calcium (Ca^{++}) and Sodium (Na^+) ions. This is due to the "non-hydrated" size of the cation. The K^+ ion is just the right size to fit into the hexagonal "hole" in the crystal lattice of the Montmorillonite atomic structure. Other ions appear smaller on the periodic table of elements, but when hydrated, they are actually larger than the Potassium ion.

Swelling Mechanisms

There are two mechanisms of swelling in clays—crystalline swelling and osmotic swelling. Surface hydration, or **crystalline swelling**, refers to the adsorption of layers of water molecules on the surface of the crystal. Its probably called crystalline swelling because the water is so tightly held to the surface of the crystal by hydrogen bonding that the water becomes quasi-crystalline. It assumes the same hexagonal coordination of the hydroxides in the atomic structure of the clay. The water is so tightly bound that it has a higher viscosity and about 3% less volume than free water near the surface of the crystal.

Several layers of molecules will adsorb to the surface of the smectites. The hydrogen in the water is bound to the oxygen in the clay in the first layer. The hydrogen is so tightly bound that it creates a highly polar water molecule. (Fig 8-68) The oxygen on the first layer of water attracts hydrogen from other water molecules, so a second layer of water clings to the first layer. A third and fourth layer of water will also be adsorbed. The fourth layer is bound less tightly than the third, second, or first layers. The first layer is bound so tightly that 80,000 psi of pressure is required to squeeze this water out of the shale. Only 40,000 psi is required to remove the second layer, 20,000 psi for the third, and 10,000 psi for the fourth. These layers of crystalline adsorption will create a hydrational stress equal to these pressures as these layers are absorbed.[1]

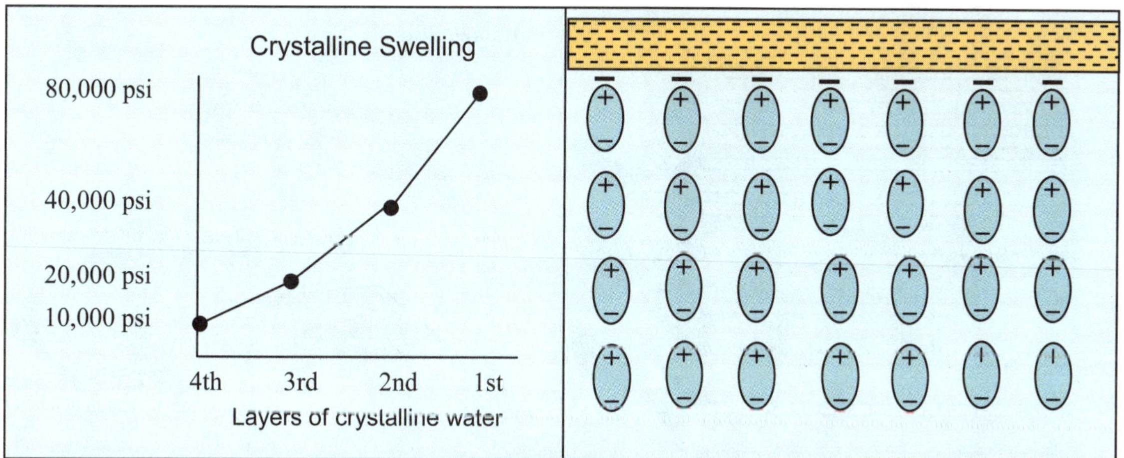

The positive charge in the polar water molecule is attracted to the negative charge on the shale. The water molecule sticks to the shale and becomes highly polar. Additional layers of water molecules become bonded to the layer of water that is bonded to the shale. This water is "stuck" to the shale in a quasi-crystalline state and is very difficult to remove.

Only four layers of water can inject themselves between the layers of clay, so the swelling is not severe. The hydrational stress, however, is very high.

Fig 8-68 Crystalline swelling

Crystalline swelling can cause the smectite clays to swell to twice their size. Crystalline swelling also occurs with Illites and other clays, but to a lesser extent. The water does not penetrate between the layers in Illites and Kaolinites. The water will adsorb on the edges and cause some hydrational stress, however.

Osmotic Swelling

Another type of swelling common with the smectites is **osmotic swelling**. Water is drawn between the clay sheets because the concentration of cations is higher between the clay sheets than in the drilling mud. (Fig 8-69) A much larger volume of water is drawn in between the clay lattice with osmotic swelling than with crystalline swelling. Sodium Montmorillonite can swell 14 to 20 times its size, and completely disperse into colloidal size particles because of osmotic swelling. The hydrational stress is less, however. It is only on the order of about 2,000 psi.[1] This is because the water molecules are not as tightly held to the clay with osmotic swelling, as they are with crystalline swelling. The polar water molecules are attracted to cations, which in turn make them slightly more polar. Other polar water molecules are then attracted to these molecules, and so on.

The water molecules attracted by osmotic swelling are not necessarily bound to the clay as they are with crystalline swelling. They sort of mill around in the vicinity of the cations and frequently change places with other water molecules. With crystalline swelling, the water is bound to the clay and does not exchange places with other molecules. Both crystalline swelling and osmotic swelling occur simultaneously.

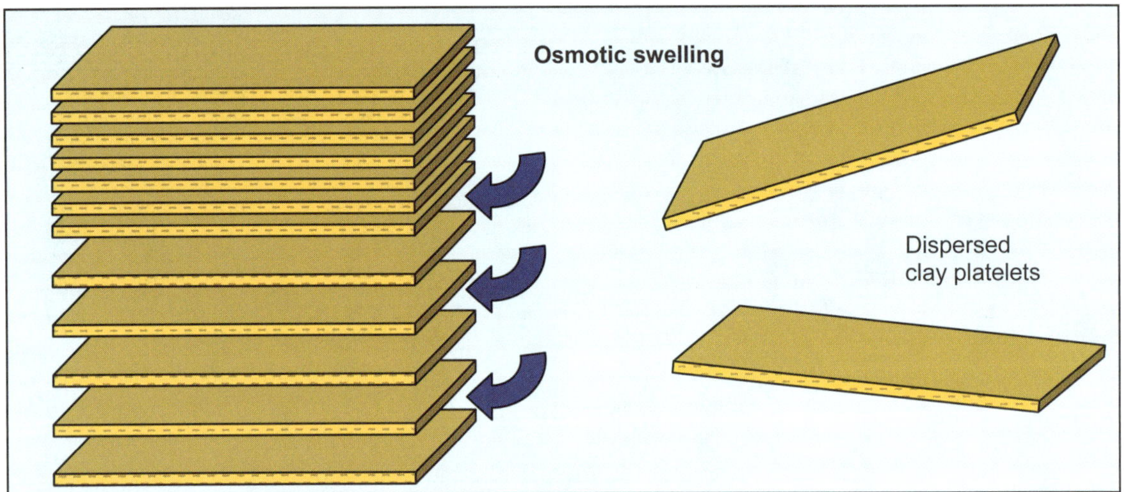

Crystalline swelling can cause bentonite clays to swell to twice their size by absorbing 4 layers of water molecules to the face of each clay platelet. In Osmotic swelling, a large amount of water is drawn in by the waters attraction to the cations between the layers of clay platelets. This causes the platelets to separate substantially and the clay can expand to 14 or 20 times its size and even disperse completely.

Fig 8-69 Osmotic swelling

As shale is buried, the overburden stresses squeeze the water out of the clay structure. First, the osmotically held fluid is expelled, and then, if the overburden is high enough, successive layers of the crystalline held water are expelled.

When the clay is later exposed to water during the drilling process, water is reabsorbed and the clay is subjected to hydrational stress. The smectites will absorb a large quantity of water, allowing them to take on plastic behavior. These swelling clays will squeeze into the well bore, decreasing its diameter at first, then they will disperse, allowing fluid and mechanical erosion to enlarge the hole. When inhibitive mud is used, there will still be some swelling, but the shale may not become quite as plastic. The hydration process lowers the apparent rock strength while increasing the hoop stress. The shale may fail in shear, causing large, hard cavings and hole enlargement. These hard cavings will continue to absorb water and may become plastic and sticky by the time they reach the shale shakers.

It takes time for drilling fluid filtrate to penetrate the shale. (see filtrate). This makes the swelling and caving process time dependent. We normally don't see any problem drilling the shale, but two to twelve hours later we may have severe problems. In some cases, we may not see problems for several days. The higher the smectite content and the more permeable the shale, the sooner we can expect trouble. Young, weakly consolidated shale near the surface tends to have high permeability (for shale) and high Bentonitic content. The lack of confining pressure near the surface also facilitates the re-hydration process.

Fractured shale at depth that does not have a high concentration of smectites may still suffer from hydrational stress if it is fractured. Crystalline swelling will occur on the faces of these fractures, causing an increase in hoop stress and a reduction in rock strength fairly rapidly.

Summary

When to Expect Shale Instability Problems

If shale is exposed, well bore instability should be anticipated. Even if a shale is stable while it is drilled, it will weaken with time because of filtrate invasion. As filtrate invades the shale, the beneficial radial stress is reduced and the harmful hoop stresses are increased. The apparent rock strength is also reduced by the reduction of the confining pressure provided by the differential pressure and the reduction in internal friction and cementation. Rock strength is being reduced and harmful stresses are increasing. Eventually, the shale will fail. It is only a question of time.

Shale is even more troublesome when:

- The well bore is not drilled perpendicular to the bedding planes.
- High stress anisotropy is present such as in a reverse fault stress regime vs. a normal fault regime.
- The shale contains a high Bentonitic content, is young or relatively weak.
- Filtrate invasion is higher due to higher permeability, fractures, sand and shale inter-bedding etc.
- The mud weight is reduced. This causes a sharp reduction in radial stress because the pores will be charged from filtrate invasion from the previous overbalance.
- The temperature increases, such as during a trip.
- The open hole exposure time is long.
- It is subjected to high or long durations of drill string vibration. Drill string vibration increases as the ratio of hole diameter to pipe diameter increases, and as the tension and rotational speed of the string increases.
- The drill string is tripped frequently, especially when surging and swabbing is high, and or when dogleg severity is high.
- The shape of the well bore is not circular. (see Fig 8-27)

Trouble with instability usually occurs while pulling out of the hole or after making connections. In extreme cases, such as well bore collapse, it can occur at any time.

Packoffs are most likely going to occur while pulling the pipe upward. The risk is highest while pumping out of the hole through inter-bedded shale and sandstone sequences. (Fig 8-70) The drill collars tend to pack off where the change in diameter in the drill string first contacts the change in diameter of the well bore. Cavings tend to hold up in the enlarged sections of the well bore and are dragged into the restricted diameters with the drill collars, causing a packoff.

As the well begins to pack off, the pump pressure increases and the drill string is "pumped" into the packoff. *This masks the overpull trend.* If a driller is watching only his weight indicator, the overpull will not increase substantially and may even decrease due to the pistoning effect.

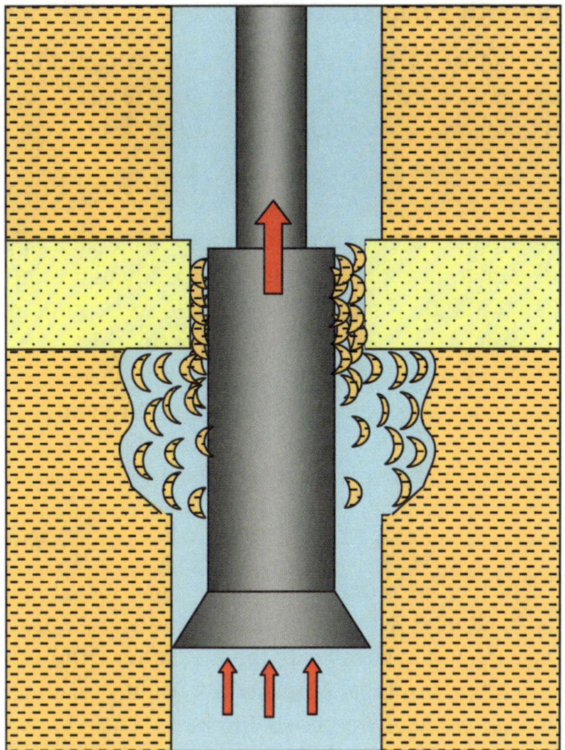

Packoffs tend to occur where the BHA is pulled into full gauge sections of the hole just above hole enlargements. If the string is being pumped out a pistoning effect will mask the overpull trend.

Fig 8-70 Pumping the BHA into packoffs

Preventive Measures

To avoid well bore instability trouble, we must minimize the conditions that lead to instability. Some of these conditions, such as rock strength and the stress regimes, are inherent properties that cannot be changed. The mud properties, well trajectory, drill string design, and drilling parameters are the factors we must focus on.

Trajectory

Choosing the correct well trajectory is our first opportunity to avoid trouble. The well path should be designed to penetrate the shale perpendicular to its bedding planes, whenever possible. This is especially true when the shale is strong compared to its bedding planes. The stress regime must be determined and the well path should be oriented to minimize the calculated stress anisotropy. The well path that minimizes stress anisotropy may not be the path that minimizes strength anisotropy. A decision must be made regarding the relative danger posed by bedding plane weakness vs. that posed by the tectonic forces of the stress regime.

More trouble can always be expected in reverse fault regimes.

Mud Properties

Mud properties probably have the most impact in avoiding stability problems. Of all the mud properties, **mud weight** is the most critical. Mud weight provides the radial stress that minimizes hoop stress. The closer the radial stress is to the hoop stress, the smaller the shear stresses that lead to failure will be. The median line principal proposed by Aadnoy is an approach that aims to do just this.[4] The median line principal suggests the mud weight should be half way between the fracture gradient and the pore pressure gradient. The mud weight should be increased to an adequate level prior to drilling into the shale. *Once it is raised, it should never be lowered.* Raising the mud weight increases the filtrate invasion. If the mud weight is then lowered, the pore pressure will be too high for the mud to apply an adequate radial stress, and failure is likely.

Mud weight should be changed gradually to avoid shocking the well. As the inclination angle increases, the mud weight will probably need to be increased. Excessive mud weight should be avoided.

Another important mud property is **inhibition**. Oil-base mud is the most inhibitive, because oil molecules are not polar and do not have an affinity for the electric charges on the shale. The water phase of oilbase mud must have the correct salinity to limit swelling. Water-base mud will cause swelling in all types of shale. Mud additives such as glycols, potassium salts, etc., will help lower the swelling tendency. Although one type of mud may have been selected in the planning stage, field observations may dictate that a different approach must be taken.

Filtration control is another property that can be beneficial. The filtration control methods used for sandstones, however, are largely ineffective for shale because dynamic filter cakes are not deposited on shale. Filtration control methods for shale include techniques that increase the viscosity of the filtrate, reduce the permeability of the shale, or create a semi-permeable membrane allowing osmotic pressure to balance the differential pressure.

All shale will absorb water and fail in time. Therefore, we should **minimize open hole time.**

Surge and swab pressures vary the radial stress tremendously and must be avoided. Therefore, the **plastic viscosity** of the mud must be minimized. **Good solids control** will help to keep the plastic viscosity low.

Temperature Fluctuations

Fluctuations in temperature cause changes in pore pressure and filtrate invasion. Periods of low or no circulation will cause temperature fluctuations and should be minimized or avoided as much as possible. Increases in temperature can cause caving; decreases can lead to lost circulation.

Drill String Design

The drill string design and the drilling parameters should take instability into account. Drill string **vibrations** should be minimized. High rotary speeds with small diameter pipe, or large diameter wells, causes large vibrational impacts and should be avoided. Large heavy BHAs can resist vibrations, but at high inclinations add stress to the wall and lead to high surge and swabs. The size of the BHA must be carefully selected for the well.

Wiper Trips

Plan regular wiper trips. Wiper trips should be planned, but used only when needed. Swelling clays need to be wiped frequently, but we should avoid tripping through brittle shale as much as possible. The high side loads, surge and swab pressures, and temperature fluctuations are damaging to shale but some times cannot be avoided. Occasional wiper trips to clean the hole, wipe swelling clays, or ream troublesome sections must be taken. Careful thought to what sections may require wiper trips or reaming should be given in the planning stage. The drill crews must also monitor trends and conduct wiper trips as needed. Well designers have a responsibility to ensure the drill crews are aware of the need and dangers of wiper trips in the troublesome shale they are drilling.

Packoffs often occur in doglegs or changes in the well bore geometry. A model of the BHA should be pulled along the lithograph and trend chart depicted in Figure 13.1. The driller can anticipate potential trouble and cautiously pull through theses zones.

Warning Signs

Stuck pipe due to well bore instability tends to be the most severe type of stuck pipe. The well bore sometimes collapses with little or no warning, and we are likely to lose both the well and the drill string when it occurs.

Mechanically Stressed Shale

Failure usually occurs from excessive shear stress at the well bore wall. In the case of brittle rock, once the shear strength is exceeded, the rock may immediately collapse into the well. In more plastic rock, the well may collapse a little at a time. The drill crew must be alert to the warning signs and react quickly to prevent the loss of the well.

The obvious warning signs are:

- **Surface trends**
  - Cavings at the shakers.
    - If the cavings are splinters with curved surfaces, the failure mode is excessive shear stress.
    - If the cavings are angular and blocky, the failure has most likely occurred along the bedding planes.
    - In unconsolidated formations, the cavings may appear like cuttings, or be well rounded and unbroken. The splinters may also break up on the way up the well.
  - Mud tests
    - High filtration rates warn that shale may fail sooner
- **Connection trends**
  - Hole fill after connections or trips.
  - Overpull after connections. (The cavings have settled around the BHA while the pumps were shut down.) If the cavings are bigger and/or more voluminous, overpull may be seen when first picking up for a connection.
  - A pressure surge to start circulation. (The cavings that were hanging up on the walls slip into the center of the well when the pumps are shut down. This partially obstructs flow when the pumps are started up again.) Also, small packoffs around the BHA or drill string may have occurred causing a pressure surge to "break" the packoff.
- **Tripping trends**
  - Swabbing while tripping. Swabbing reduces well bore pressure. This causes an increase in hoop stress and a reduction in radial stress. The result is higher shear stress at the wall and less stability.
  - Pistoning while pumping out. As pump pressure builds up beneath the packoff the drill string is "pumped" out of the well. The overpull trends can become masked because some of the hook load is replaced by the pistoning effect. (Fig 8-70)
  - Excessive and erratic drag.
  - Erratic torque, drag, and pressure trends while reaming.

- **Drilling trends**
  - The torque and drag trends are increasing and erratic.
  - There may be an increase in the rate of penetration followed by a decrease. The ROP increases due to the lower apparent rock strength as formation pressure approaches or surpasses well bore pressure. The higher pressures due to a caving load cause an increase in apparent rock strength, which reduces the penetration rate. (Fig 8-71)
  - Pressure surges indicate minor packoffs due to large quantities of cavings and caving beds.
  - Loss of mud. (Mud is pumped into formation as packoff pressure increases.)

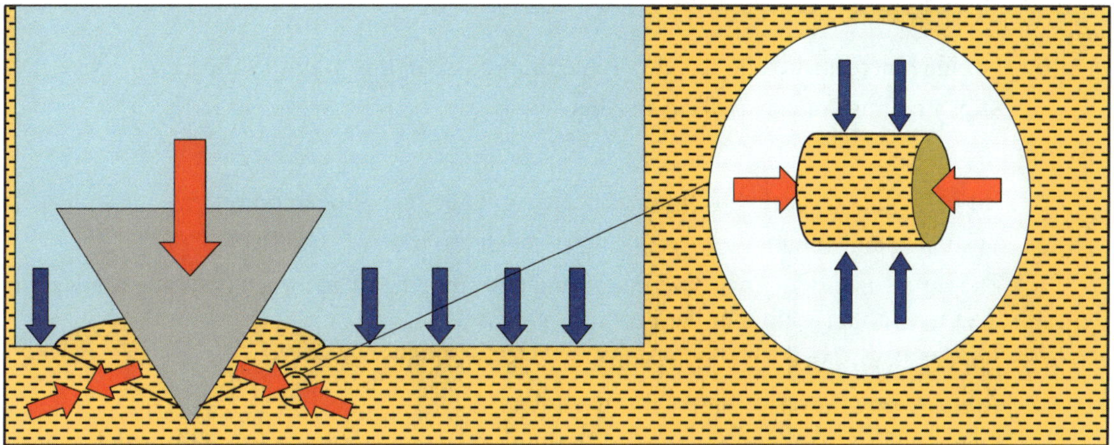

Drillability is influenced by rock strength. As a tooth penetrates the formation the rock must move out of its way. The rock will compress then break just as the core sample did in the laboratory. (See Fig 8-5)

An overbalance applied against a filtercake on the bottom of the well provides a confining pressure that increases the apparent rock strength. The overbalance is reduced as pore pressure increases or well bore pressure decreases. This results in a decrease in apparent rock strength. The penetration rate will increase and wellbore stability will decrease as rock strength decreases.

Fig 8-71 Rock strength and drillability

Chemically Stressed Shale

As with mechanically stressed shale, the failures usually occur because of excessive shear stress near the well bore wall. However, chemical stress takes time to develop, so failure does not occur suddenly. Failure occurs several hours after the formation is exposed, and the failure occurs more gradually.

The most common warning signs include:

- **Surface trends**
    - Increases in funnel viscosity, plastic viscosity, yield point, and cation exchange capacity. (The Bentonitic clays are dispersing, or more eloquently, "The well is making its own gravy.")
    - Possible increase in mud weight. (Due to increase in low gravity solids.)
    - Sticky cuttings or clay balls at shakers.
- **Drilling trends**
    - Bit balling, as indicated by gradual decrease in the rate of penetration
    - Swabbing and surging.
    - The torque and drag trends are increasing and smooth. (The torque and drag are less erratic than with hole cleaning or hard cavings problems.)
    - Loss of fluid. As the well packs off, mud is lost to formation.
    - An increase in pressure and pressure surges
    - Pistoning. (As the bit is pumped off bottom, it becomes harder to find bit weight. In extreme cases, the drill string might literally be pumped up the well.) (Fig 8-70)
    - Problems do not usually occur as the formation is drilled. It normally takes a few hours for symptoms to develop.
- **Connection trends**
    - Increase in torque and drag. Note that drag is very smooth. As the BHA is pulled into a tight spot, the overpull steadily and smoothly increases.
    - Overpull off the slips.
    - There may be a pressure surge to start circulation. (The gel strength and plastic viscosity cause the thixotropic properties of the mud to increase. This means the mud will stiffen and gel up when the pumps are shut down.)
    - Back pressure and backflow through the drill pipe. (Due to trapped pressure in the annulus.)
- **Tripping trends**
    - Smooth but increasing torque and drag.
    - Swabbing. (The most severe swabbing occurs with reactive clays. Well control is a major concern here! Many blowouts begin as stuck pipe problems.)
    - Surging and loss of mud while tripping in.
    - When tripping in, the problems first occur at the depth at which the problem formation is encountered.
    - If pumping out, severe pistoning may occur. Be alert! This tends to mask the overpull trends. (In fact, it may appear that the hook load is actually decreasing as the BHA is pulled and pumped into the packoff!)

Freeing Procedures

First Action

The first action to take for any type of packoff is to **bleed off any trapped pressure and apply 200 psi to 500 psi** to try to re-establish circulation then **torque and slump the** pipe. **Jar down** if jars are in the string.

- Trapped pressure will pump the bit further into the packoff and make matters worse. Also, we will want to move downward, the pistoning effect will reduce the amount of downward force available. Apply low pressure to re-establish circulation once any movement is established.
- Torque is applied to help establish pipe movement and circulation.
- The cavings are moving downwards and are wedged together when the pipe moves upwards. Thus, the best direction for pipe movement is downward to reduce the wedging forces. If downward motion can be established, the packoff may loosen. Circulation can then be re-established and the packoff can be broken up with circulation and pipe movement.
- Jar with the max trip load if jars are in the string. Torque must be used carefully and in accordance with the jar manufacturer's recommendations. Torsional and tensile stresses are additive, so we should *never jar up while applying torque.* It is okay to jar down while applying maximum torque.
- Once circulation is established, the hole must be cleaned before further drilling or tripping can commence.
  - Pump viscous sweeps in vertical wells, and a combination of low and high viscous/high density sweeps in deviated wells.
  - Low viscous sweeps with surfactants and lubricants may be required in both vertical and deviated wells if progress is not made soon after circulation is reestablished.
  - It is important to note that the problem has not gone away once we are freed from the packoff. The cavings causing the packoff must be circulated out of the well or we are likely to become stuck again. The conditions causing the instability must also be addressed.

Secondary Freeing Procedures

If the pipe does not come free with our first action, there are a number of secondary procedures that have proven to be successful.

The **low frequency resonance tool** described in the hole cleaning section may be successful in fluidizing the packoff debris, allowing the drill string to move through it. This tool should be used in conjunction with the first action techniques, as much as possible.

Backing off and **washing over with wash pipe** may be successful if the conditions causing the instability have been addressed. Washing over is more successful with settled cuttings and when the cavings are small, such as with unconsolidated formations. Large hard shale splinters may become bridged around the larger wash over pipe.

Other Types of Well Bore Instability

Shale causes most of our well bore instability problems, but well bore instability is not limited only to shale formations. Well bore instability problems are also common in unconsolidated sands and conglomerates, fractured and faulted formations, graded salt formations, and around ledges formed in hard/soft inter-bedding.

Unconsolidated Formations and Conglomerates

Unconsolidated formations present a unique type of well bore instability. These formations tend to flow into the well bore, just as sand caves into the holes we attempt to dig in the beach. They can cause so much hole fill that it is difficult to make connections. Large, cavernous washouts occur that make it difficult to bring larger cuttings to the surface.

The well can experience cave-ins and packoffs very quickly once circulation is stopped. Worst of all, a large enough cavern may be washed out beneath the rig, causing the rig to fall into a crater. Many rigs have been lost this way!

Unconsolidated formations tend to be recent sand deposits at, or very near, the surface. They have not had time to become cemented, and usually have never been exposed to high overburden stress.

Occasionally, older formations at depth can be loosely consolidated or unconsolidated if they have high pore pressures and are filled with hydrocarbons instead of water.

The strength of an unconsolidated formation depends entirely on the friction between the grains that make up the sand. If the formation is dry or damp, the friction can be high enough for the well bore to remain intact.

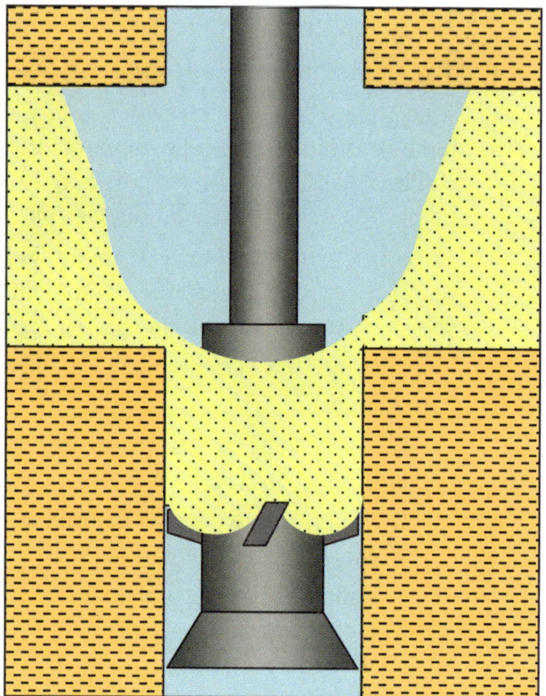

Without an effective filter cake for an overbalance to work against, unconsolidated material will slough into the well when circulation is stopped.

Fig 8-72 Unconsolidated sand

All of the overburden will be carried across the grain-to-grain contacts, resulting in high friction. As fluid enters the well, it lubricates the grain-to-grain contacts and helps support the overburden stress. Thus, the effective stress is reduced and the overall friction between grains is reduced.

The sand on a beach is a classic unconsolidated formation. When the sand is dry, we can dig a hole, but the sides will slope. The friction between grains is not strong enough for the walls to stand vertically. If we pressed a flat plate with a hole in it on top of the sand, we would increase the grain to grain contacts beneath the plate and could dig more successfully beneath the plate. If we tried to dig this hole with water, or if we reached the water table, the grains would become too lubricated and the sand would flow into the well. (See Fig 8-7 concerning rock strength.)

To successfully drill through unconsolidated formations, we must apply a filter cake across the formation and maintain an overbalance against it. The same principles of rock mechanics apply to unconsolidated formations as to shale. One difference, however, is that the maximum hoop stress is felt further inside the formation in unconsolidated formations than in consolidated shale.[7] The radial stress applied against the filter cake reduces the hoop and maximum shear stresses, and thus increases stability. (Remember that the maximum shear stress is equal to ½ the difference between hoop and radial stress.)

As long as an adequate differential pressure can be applied against the filter cake, the well bore can remain stable. However, filter cakes are not impermeable, so fluid is flowing through the cake and into the formation. If this filtrate is able to charge the pore pressure, the differential pressure will reduce with time, and the well bore will become less stable. The increase in pore pressure also reduces the grain to grain contact forces providing the internal friction. This further reduces rock strength.

Obviously, an impermeable filter cake is desired. However, filter cakes against unconsolidated sands tend to be more permeable than those deposited on consolidated sands.

Annular friction losses cause the well bore pressure to be slightly higher while drilling, so the pore pressure just inside the formation is highest while circulating. When circulation is stopped, the differential pressure against the filter cake may no longer be high enough to support the well bore wall.

Vibrations from drilling can lower the friction between the grains and lead to instability. In the absence of any movement, the friction between the grains is a result of the static coefficient of friction. Once movement between contact points is established, the coefficient of friction shifts from static to dynamic. The dynamic coefficient of friction is much lower. Vibration from drilling can cause just enough movement between grains to allow this shift to the dynamic coefficient of friction. The lower friction results in less strength, so the formation becomes less stable.

When to Expect Trouble with Unconsolidated Formations

Trouble from unconsolidated formations can be expected when the BHA is against or below the unconsolidated formation. The problem will be even more pronounced if a dogleg is across the formation or the well is out of round. Often, the problems don't appear until circulation is stopped for a connection.

Anything that charges the pore pressure of the formation increases its instability. If loss of circulation occurs in this zone, trouble can soon be expected. The higher the pore pressure, the harder it will be to apply a differential pressure across the filter cake. Packoffs above the sand, excessive cuttings load, and circulating out a gas kick are some methods of charging the pore pressure of unconsolidated sand.

The younger and shallower the unconsolidated formation, the more likely we are to have trouble with it. The longer the sand is left exposed, the more likely it is to cave.

When drilling in permafrost, unconsolidated formations are frozen year round and can never become consolidated. They behave as consolidated formations when cemented with ice, but as they thaw out with the heat from the drilling mud, they become unconsolidated again. Mud coolers are sometimes employed to limit the amount of thawing in these cases.

Warning Signs of Unconsolidated Sands

- **Large quantities of sand coming across the shale shakers.** If more material is being removed than was drilled, the material may be running in from an unconsolidated formation.
- **Hole fill on connections.** The sand may only run when the pumps are shut off, as in a connection. In more severe conditions, the sand may be running while circulating, but it is being carried out of the hole unnoticed.
- **Backflow on connections.** The sand flowing into the well causes an effective density in the annulus that is higher than in the drill pipe.
- **Pressure surges.** If the sand is flowing hard, it may be momentarily packing off, causing torque, drag, and pressure surges.
- **An increase in standpipe pressure** may occur if the annulus is loaded with sand.
- **Lost circulation.** Unconsolidated sands are normally very permeable. They also provide the most permeable filter cakes. Lost circulation can warn us that we have entered unconsolidated sand. If the unconsolidated sand has been exposed, a loss of circulation can indicate that the filter cake is sloughing off the wall. The sands may have begun flowing, or soon will.
- **Torque and drag after a connection that disappears significantly once circulation is established.** Caving sands may settle around the BHA and bit when the pumps are shut down, but become loose and fluid once circulation is re-established.
- **Settling or shifting of the rig.** As a cavern begins to collapse, early signs of cratering may appear.
- **Swabbing while tripping.** Flowing sands are easily swabbed in.
- If a **gas kick** is taken at a deeper depth, it will charge the pore pressure of the unconsolidated sand as it is circulated out of the well. Instability is almost guaranteed.

Prevention

- The best protection against unconsolidated sands is steel wall cake. The sooner casing can be set, the better. It is extremely risky to get the bit and BHA below the potentially flowing sand.
- If the program calls for open hole below the unconsolidated formation, a drilling jar should be placed near the top of the BHA and all stabilizers removed.
- Viscous pills can be placed across and below the sand prior to stopping the pumps for a connection or trips. A weighted pill could be placed below the sands prior to running casing.
- A thin, impermeable filter cake is desired so graded carbonates, LCM, and fluid loss agents can be added to the mud.
- The bit should never be left in one place if circulating off bottom in unconsolidated sand. The jets can easily erode these loose formations.
- Never leave the bit on bottom with pumps off. Keep the pipe moving slowly.

Freeing Procedures

Torque and downward movement are required to un-bridge and break free from a packoff. Any trapped pressure should be removed and 200 to 500 psi pressure applied in an attempt to re-establish circulation. Once circulation is established, viscous sweeps should be pumped to clean the sand from the well.

If downward movement is not possible or successful, it may be possible to jar up through the sand. Low frequency vibration tools should be successful in fluidizing the sand, but may cause further caving. The use of vibration tools should be carefully considered on a case-by-case basis.

Backing off and washing over may be successful if conditions leading to the instability have been addressed.

Fractured and Faulted Formations

Large chunks of formation can break loose from naturally fractured and faulted formations and fall into the well. These can cause bridging and potentially a packoff if a large quantity of debris breaks off at once.

Brittle formations, such as limestone, tend to fracture more than pliable formations, such as clay. The formation is more likely to contain fractures if it has been subjected to regional or local tectonic forces, such as faults, salt domes, or mountain building. Ledges that are formed from graded salts and hard soft inter-bedding can also break loose and fall into the well.

The problems will be exacerbated by doglegs and drill string vibration.

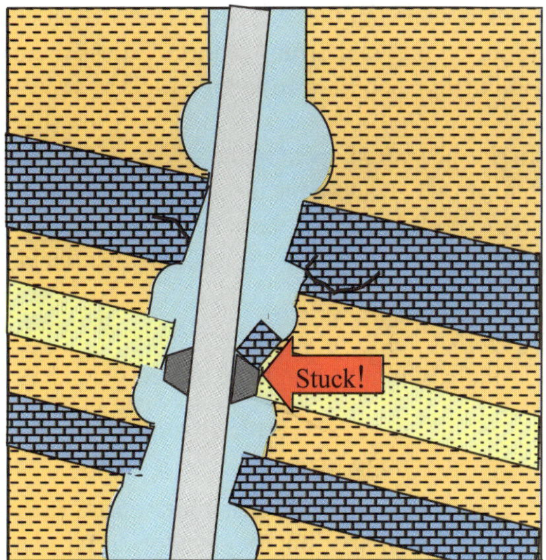

Broken ledges or chunks of rock fall into the well and wedge against the drill string.

Fig 8-73 Fractured and faulted formations

Warning Signs

The warning signs of fractured formations and broken ledges include:

- Sudden and erratic torque and drag.
- Hole fill on connections.
- It occurs as the formation is being drilled or while moving the drill string up or down.
- The problems may appear and disappear and may be hard to pin down to a specific depth. This is because the bridging material is moving up or down the well.
- Cavings or cement chunks on the shakers.

Freeing Procedures

If the pipe has become stuck, the first action should be to apply torque and jar down. If there are no pressure restrictions, circulation can be maintained at the full rate.

Secondary Methods

Acid pills may be useful in carbonate formations. Spotting pills may be of some use to reduce friction. A reduction of friction may be helpful at the stuck point, or in reducing torque and drag along the entire string (or at doglegs), so that more force is available at the stuck point.

Low resonance frequency tools should be successful in breaking up debris and lowering friction.

Once the pipe is free, the large chunks of rock must be broken up and swept from the well. It is best to get the material beneath the bit where it can be drilled up. Viscous sweeps will help clean the well and viscous pills will help keep the debris from settling around the drill string while the pumps are off.

Junk in the Hole

Junk dropped in the well often leads to bridging and stuck pipe. Far more junk goes into a well than ever gets reported on the morning report. Unless there is a high degree of earned trust, both on the rig and between rig and office, people tend to hide their mistakes.

In the 17 years I spent at the rotary table, I have seen a lot of junk go into the well. Two things always happen when someone drops something in the well. First, the person who dropped the junk immediately goes over and looks down the well. (As if there were something to see.) The next thing he does is look around to see if anyone saw him. If not, he must decide whether to tell anyone or not. (See section on earned trust.)

When to Expect Trouble with Junk

Trouble with junk can be expected when:

- There is very little clearance between the well bore wall and the pipe in the hole, such as when running liner.
- When there are hard formations present. Junk can be wedged into the wall in softer formations, such as gumbo, but may wedge and bridge in harder shale, sandstone, and carbonates.
- The well is uncovered and/or poor housekeeping is practiced.
- Tools or equipment on the rig floor is unaccounted for. Missing tools at the rotary table often lead to the symptoms of junk in the hole, even though no one believes the tools have fallen in the well.
- The rig is suffering from a rash of bad luck. “When it rains, it pours.” If morale is suffering or things are going bad, we must be more alert to this type of problem.

Warning Signs

The symptoms of junk in the hole are very similar to fractured formations, sudden and erratic torque, and drag. Usually, there are no pressure restrictions. Unlike cavings from fractured or broken ledges however, junk is much harder to break up. It may have to be milled with wash over tools.

Freeing Procedures

The freeing procedures for junk are similar to faulted formations. The first action should be to apply torque and jar down.

Acid pills may be useful if the pipe is stuck against a carbonate formation. Low frequency vibration tools may also be successful. If all else fails, a backoff and installation of fishing jars may be successful.

Once the pipe is free, the junk must be removed from the well. The density of steel rules out viscous sweeps. A junk·basket will most likely have to be used, perhaps in conjunction with milling tools.

Nomenclature

F = Friction force
N = Normal force ε = Strain
α = Direction or azimuth
θ = Inclination angle
σ = Normal stress
σ_θ = Hoop stress
σ_h = Minimum horizontal stress
σ_H = Maximum horizontal stress
σ_{max} = Maximum stress
σ_{min} = Minimum stress
σ_r = Radial stress
σ_v = Overburden stress
σ_z = Axial stress
τ = Shear stress
μ = Coefficient of friction

Test your understanding of Wellbore Instability

1. What is the difference between rock strength and apparent rock strength? (Fig 8- 4, 8-5, pg. 121)
2. Where does rock strength come from? (Fig 8-6 & 8-7 pg. 123)
3. What is effective stress? (Fig 8-9 pg. 126)
4. What is the relationship between effective stress and pore pressure? (Eq 8.2, pg 126)
5. How is stress distributed around a well bore, and away from the well bore? (Figs 8-22 to 8-27, pg 134 to 138)
6. How far into the formation will we have to go to see the hoop stress reduced to zero? (Fig 8-25, pg. 136)
7. How does and increase in mud weight affect rock strength and hoop stress around the well bore? (Fig 8-28, 8-29, 8-30, pg. 140)
8. What does Mohr's Circle represent? (Fig 8-33 pg. 142)
9. What is the median line principle? What is it used for? (Fig 8-40, pg 147)
10. What is meant by stress regime? (Fig 8-46 pg. 151)
11. What is stress anisotropy? (Fig 8-47, 8-48, pg. 152)
12. How much bigger is the hoop stress than the horizontal stress in a vertical wellbore in a region where the maximum and minimum horizontal stresses are equal? (Fig 8-27, pg 138)
13. What happens to stress in the wellbore wall as the hole enlarges in one direction? (Fig 8-27, pg 137,)
14. What would happen to a brittle rock like granite once it started to elongate from caving? Why doesn't this happen with shale? (Fig 8-27 & fig 8-54)
15. How is a Mohr's failure envelope prepared? (Fig 8-35, pg. 144) What is it used for?
16. How does a change in mud weight affect wellbore stability? (Fig 8-29, 8-45)
17. Name 8 factors that affect wellbore stability (Pg 145)
18. How and why does a change in temperature affect wellbore stability? (Fig 8-45, pg. 150)
19. How does direction and inclination affect wellbore instability? (Fig 8-49 to 8-51) (pg. 153-155) What are we trying to accomplish when we choose an inclination and direction?
20. Why are bedding planes sometimes considered to be a weakness? (Fig 8-52, pg. 156)
21. How does the well path affect bedding plane instability? (Fig 8-53, pg. 157)
22. How does overbalance affect stability? (Fig 8-54, pg. 159)
23. How does filtrate invasion affect stability? (Fig 8-54 to fig 8-58)
24. What affect does pipe diameter, hole diameter, rpm, and drill string tension have on drill string vibration impact forces. (Fig 8-63, pg. 166)
25. What are the six types of stress induced failure? (Fig 8-64, pg. 169)
26. Why does plastic creep slow down with time? (Fig 8-66, pg. 170)
27. What are "ballooning shales"? (Fig 8-67, pg 172)
28. Why is potassium an effective inhibiter of some shale? (Pg. 174)
29. What are the two swelling mechanisms in clays? (Fig 8-68, 8-69, pg. 175)
30. Which mechanism causes the most swelling? (Fig 8-69)
31. List ten conditions that should alert us to expect shale instability. (Pg 178)
32. When is a packoff most likely to occur? (Fig 8-70, pg. 179)
33. How can a packoff trend be masked when pumping out? (Fig 8-70)
34. How does a dogleg affect packoff? (Fig 7-62, pg. 108)
35. How can we prevent packoff due to wellbore instability? (Pg. 180)
36. What are the obvious warning signs of an impending packoff due to shale instability? (Pg. 182)
37. Why does overbalance affect penetration rate? (Fig 8-71, pg. 183)
38. What are the obvious warning signs of swelling clays? (Pg. 184)
39. What are the first actions for freeing pipe that is becoming stuck or has become stuck from packoff due to wellbore instability? (Pg. 185)
40. What secondary freeing procedures are available? (Pg. 177)

Bibliography

1) Gray, George R. & Darley, H. C. H.: "Composition and Properties of Oil Well Drilling Fluid's" fourth edition, Gulf Publishing Company (1980)
2) Shaohua Zhou, Richard Hillis, and Mike Sandiford, Dept. of Geology and Geophysics, university of Adelaide Australia: "On the Mechanical Stability of Inclined Wellbores" SPE 28176 (1994)
3) E. Hoek & E.T. Brown: "Underground Excavations in Rock" The Institution of Mining and Metallurgy, London UK (1990)
4) Bernt S. Aadnoy: "Modern Well Design" Gulf Publishing Company (1977)
5) Jaeger, J.C. and Cook, N.G.W.: "Fundamentals of Rock Mechanics" Chapman and Hall, London (1976)
6) Yarlong Wang, Maurice Dusseault, University of Waterloo: "Borehole Instability and Fluid Loss in Poorly Consolidated Media" Petroleum Society of CIM/Society of Petroleum Engineers, paper no. CIM/SPE 90-25 (June, 1990)
7) D. Økland, J.M Cook: "Bedding-Related Borehole Instability in High-Angle Wells" SPE/ISRM paper 47285, 1998 SPE/ISRM Eurock, Trondheim Norway (July, 1998)
8) X. Chen, C.P. Tan, and C.M. Haberfield; Australian Petroleum Cooperative Research Centre, CSIRO Petroleum & Monash University: "A Comprehensive Practical Approach for Wellbore Instability Management" SPE paper 48898, presented at the 1998 SPE International Conference and Exhibition in Beijing China (Nov 1998)
9) Thierry M. Forsans & Laurent Schmitt, Elf Aquitaine Production, Pau, France: "Capillary Forces: The Neglected Factor in Shale Instability Studies?" ISBN 90 5410 502 X, Eurock '94 Balkema, Rotterdam. 1994.
10) L.Bailey, P.I. Reid, and J.D. Sherwood, Sclumberger Cambridge Research, "mechanisms and solutions for Chemical Inhibition of Shale Swelling and Failure" SPE# 627043
11) Helio Santos, Petrobras, Adel Diek and Jean-Claude Roegiers, Rock Mechanics Institute, U. Oklahoma: "Wellbore Stability: A New Conceptual Approach Based on Energy" SPE paper 49264, presented at the 1998 SPE Annual Technical Conference and Exhibition in New Orleans. (Sept 1998)
12) Authors personal experiments with real time drill string vibration analysis in three degrees of freedom while drilling off the coast of Gabon in 1991.
13) Adam T. Bourgoyne Jr., Martin E. Chenevert, Keith K. Millheim, F.S. Young Jr.: "Applied Drilling Engineering" SPE, (1986)
14) E.R. Leeman, National Mechanical Engineering Research Institute, CSIR, Pretoria, South Africa,: "The Determination of the Complete State of Stress in Rock in a Single Borehole – Laboratory and Underground Measurements" Int. J. Rock Mech. Min. Sci. (Feb 1967)
15) Woodland, David C., Shell Canada: "Borehole Instability in the Western Canadian Over thrust Belt," SPE paper 17508, SPE Drilling Engineering (Mar 1990)

Chapter 9
Differential Sticking

History

The problem of differential sticking was first recognized by Hayward in 1937[1]. However, numerous technical papers have recently been written suggesting that differential sticking was not recognized until 1955. These papers generally give credit to Helmick and Longley. This is probably because Helmick and Longley referred to the phenomena as differential-pressure sticking, where previously it was called "wall sticking" or "frozen drill pipe". Anyone researching differential sticking must include the terms "wall sticking" and "frozen drill pipe" in their search strategy to capture the work done prior to 1957.

Helmick and Longley were the first to demonstrate the phenomena of differential sticking in the laboratory. They wrote the first technical paper dedicated to the mechanics of differential sticking in 1957[2]. Outmans performed a thorough analysis of the mechanisms of differential sticking the following year[3]. The progress made by these gentlemen was so complete that very little information has been added since then.

Sadly, however, the industries understanding of differential sticking is becoming distorted. Contrary to popular belief the wall cake does not become thicker to increase the area of contact, it shrinks in the area of contact. The differential pressure responsible for sticking the pipe is not equal to overbalance. Also, most differential sticking occurs high in the hole in the drill pipe and not at the bottom hole assembly.[6]

The first part of this chapter focuses on the mechanics of differential sticking, with emphasis on the factors affecting it. The last part summarizes these findings by explaining the preventive measures, warning signs, and freeing procedures adopted by the industry.

The Mechanisms of Differential Sticking

I often use the example of hanging wet coveralls across the grill of an engine radiator to explain the mechanics of differential sticking. However, H. D. Outmans offered a more accurate and descriptive explanation in 1974[4]. Outmans used the example of a rubber sink stopper slid over a drain hole in a basin of water. As the stopper blocks the flow of water through the drain, hydrostatic pressure forces the stopper against the drain opening, causing it to stick. (Fig 9-1) The friction between the stopper and the drain is high enough to hold it in place. There is very little friction between the stopper and the bottom of the basin when the stopper is not over the drain opening. This is because there is a thin film of water between the stopper and the basin that eliminates any differential pressure that might hold the stopper against the basin. Hydrostatic pressure acting up on the stopper balances the pressure acting down. This thin film of water also acts as a lubricant. When the stopper is against the drain opening, the lubricating layer drains away, allowing a differential pressure to develop and force the stopper against the drain. The presence of this differential force and a lack of lubrication results in a relatively high friction force.

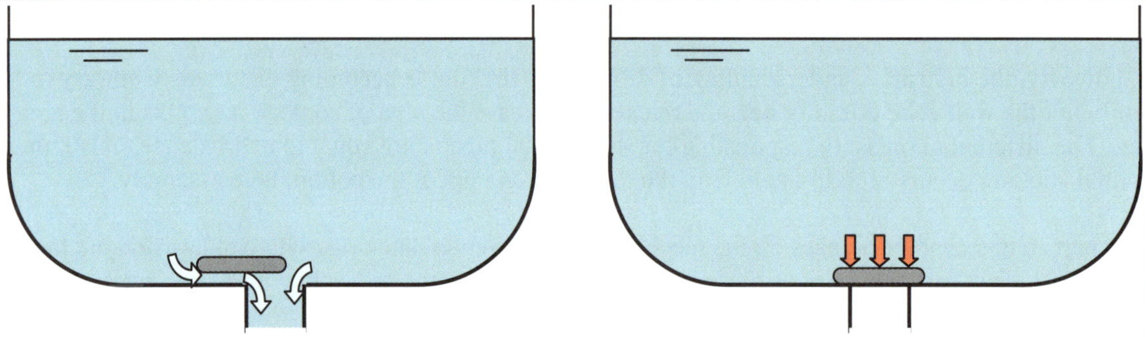

The layer of fluid under the stopper balances the pressure, such that no differential pressure exists. When the stopper blocks the drain, the fluid under the stopper drains away, allowing the differential force to develop.

Fig 9-1 Differential pressure

Now, let's examine the mechanics of differential sticking in the well bore.

Consider a well bore with permeable sand exposed. If the well bore pressure is greater than formation pressure, drilling fluid will enter the formation and deposit a filter cake. Filter cakes are somewhat permeable, so fluid will continue to pass into the formation. However, most of the solids are filtered out of the fluid at the filter cake. Only a clear filtrate passes through the filter cake. The filter cake will grow in thickness as new solids are deposited, until the rate of deposition equals the rate of erosion. (See filter cake formation. Fig 9-7)

The drill string is almost always in contact with the wall of the well bore. While rotating, it will drag a thin layer of fluid between it and the filter cake. (Fig 9-2A) This thin film serves three purposes:

- It lubricates the drill string.
- It provides a means of transmitting pressure between the tubular and the filter cake.
- It provides the filtrate that continually passes through the filter cake in the area of contact between the pipe and cake. Outside the area of contact, the filtrate is supplied directly from the mud. Note that filtrate will continue to pass through the filter cake as long as any overbalance exists.

When the drill string becomes motionless, new mud is no longer dragged into the lubricating layer, so the flow of filtrate through the cake in the contact area is interrupted. (Fig 9-2B) Fluid from the thin lubricating layer will continue to supply new filtrate until this source is exhausted. The filtrate in the filter cake will continue to drain into the formation until no filtrate remains in this part of the filter cake.

As the filtrate is drained out of the filter cake in the area of contact, this part of the filter cake collapses and becomes thinner than the rest of the filter cake. (Fig 9-2C) The shrinking of the filter cake allows the pipe to penetrate deeper into the cake, thus increasing the area of contact between the drill string and filter cake. The compressed filter cake also has a higher coefficient of friction than it had previously, when it was full of filtrate.

If circulation is stopped a static filter cake will be deposited on top of the dynamic filter cake. This also serves to increase the area of contact.

As the lubricating layer disappears, the hydrostatic pressure between the steel and the cake disappears with it. This allows a differential pressure to develop, which presses the string against the filter cake. This causes a frictional force that resists movement of the pipe. The friction force resisting pipe motion may eventually become high enough to prevent movement of the pipe. At this point, the pipe has become differentially stuck!

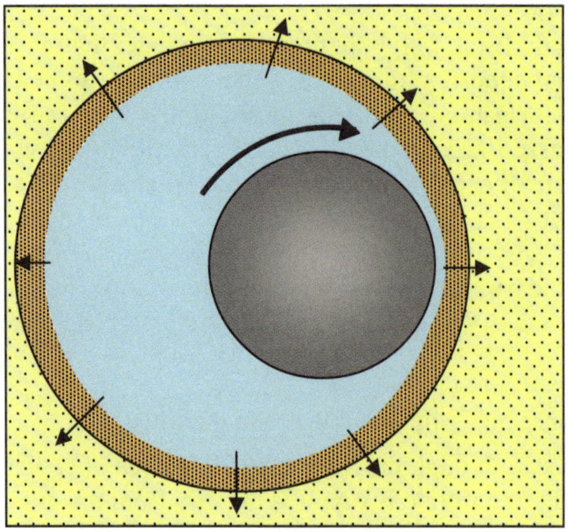

Fig A

A thin film of fluid is drawn between the pipe and filter cake while the pipe is moving.

This thin layer of fluid equalizes the pressure around the pipe.

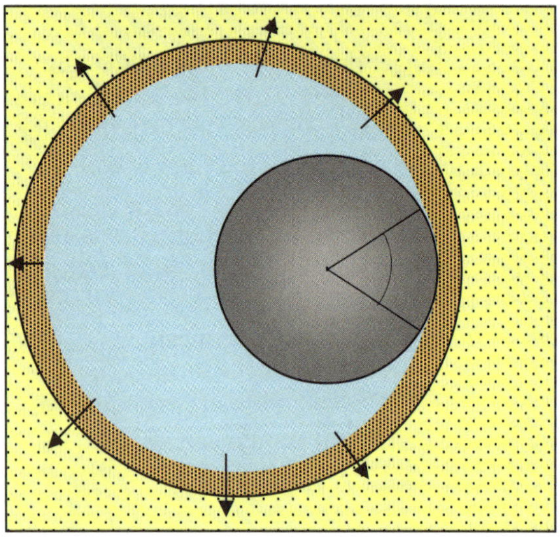

Fig B

The “lubricating” layer drains away when the pipe is stationary.

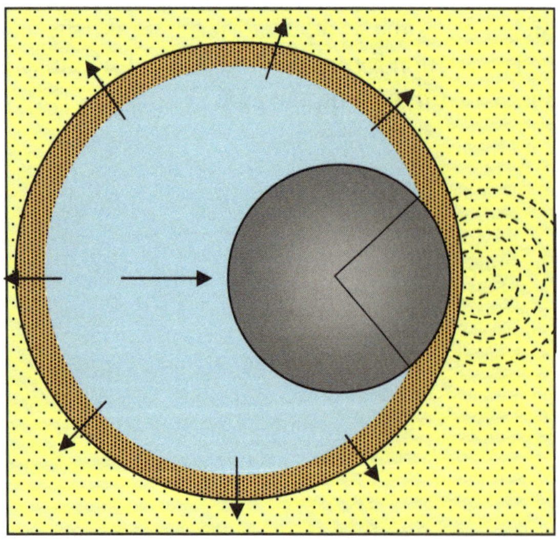

Fig C

Filtrate continues to drain out of the filter cake causing the cake to collapse and increase the area of contact.

As the filtrate drains out of the filter cake, a differential pressure begins to develop across the area of contact.

Fig 9-2 Collapse of the filter cake

Factors Influencing Differential Sticking

There are several factors that influence differential sticking. These factors include:

- Permeable formations
- Overbalance
- Filter cake
- Wall contact
- Lack of pipe movement
- Time
- Side loads

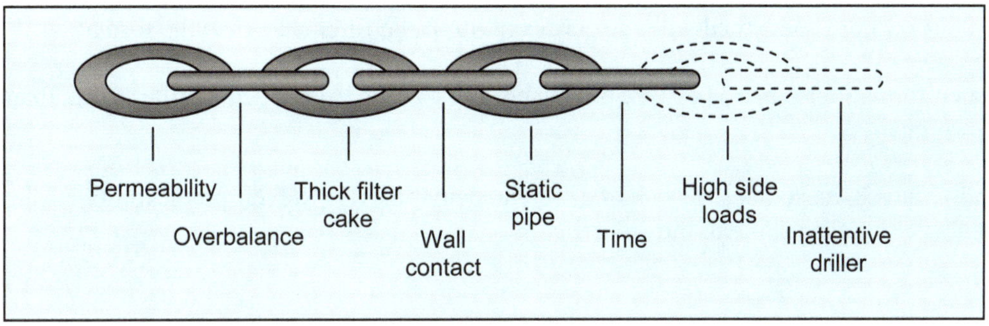

Fig 9-3 The chain of events leading to differential sticking

We normally need to have all of the first six factors present to get differentially stuck. If we have just five of them, we aren't likely to get stuck. The seventh factor, side loads, isn't necessary to get stuck, but it greatly contributes to getting stuck. All of these factors contribute to the differential force that holds the tubular against the filter cake. Let's examine each of these factors separately.

Permeable Formations (Factors Affecting Differential Sticking)

A permeable formation is required if the drill string or casing is to become differentially stuck. We don't get differentially stuck inside casing, unless it has become permeable from perforations or wear. Permeable formations include sandstones and fractured formations. We can become differentially stuck against shale if it is fractured and permeable. We occasionally get stuck against the perforations in casing. It is also possible to become differentially stuck in casing when it has lost its integrity from internal wear.

If we don't have a permeable formation, there will be no filter cake and no development of differential pressure.

The formation does not have to be very permeable to cause differential sticking. The formation need only be permeable enough to allow a filter cake to be deposited on it. The filter cake is the "slow drain" the filtrate is flowing through, and the formation need only be permeable enough to allow the filtrate to drain away from it. Thus, we should be more concerned with the permeability of the filter cake than the formation.

Unconsolidated formations tend to be more permeable and have more permeable filter cakes than consolidated formations.

As permeability increases, so does the risk of differential sticking. However, permeability has less influence on differential sticking than some of the other factors.

Overbalance/Differential Pressure (Factors Affecting Differential Sticking)

Overbalance is the most significant factor influencing differential sticking. This is because it has the largest influence over the differential force holding the tubular against the filter cake. The differential force is the product of the differential pressure times the area of contact.

$$\text{Differential Force} \approx \text{Differential Pressure x Area of contact} \qquad \text{eq. 9.1}$$

Obviously, a higher differential pressure produces a higher differential force. It is important to note that the differential pressure that holds the drill string against the filter cake is **not** equal to the overbalance. Having an **overbalance** means that our well bore pressure is greater than the formation pressure. **Differential pressure** refers to a difference in pressure across some surface.

When the drill string is moving, we may have a substantial overbalance, but there is **no** differential pressure holding the pipe against the filter cake. (Fig 9-2A) This is because of the thin film of fluid the pipe drags between itself and the filter cake as it moves. This thin layer of fluid is capable of transferring pressure according to Pascal's principle, so the force against the tubular is equal in all directions. The thickness of this film is on the order of a couple of microns.[5]

It isn't until the pipe motion has stopped, and the fluid in the thin lubricating layer has filtered into the filter cake, that we see a development of differential pressure. (Fig 9-2B) Once the thin lubricating layer is gone, a seal forms between the steel and cake. At this point, the differential pressure is the difference between the pressure in the well bore and the pressure of the filtrate in the pore spaces of the filter cake. Initially however, the pressure at the surface of the filter cake is nearly equal to the well bore pressure.

As the filtrate drains out of the filter cake in the area of contact, a pressure differential between the cake and steel can develop. (Fig 9-2C) Eventually, enough filtrate will drain out of the filter cake to reduce the pore pressure in the filter cake to the formation pressure, immediately adjacent to the filter cake. (Fig 9-3 and 9-4)

It is doubtful the entire projected area of the contact area will reduce to formation pressure. Some filtrate from the filter cake immediately surrounding the contact area will likely drain into the contact area filter cake as its pore pressure reduces. Thus, the pressure would be lowest in the center of the contact area and highest at its perimeter.

Several statistical studies show that most of the stuck pipe occurring in the GOM occurs high in the hole around the drill pipe[6]. This is because, as the well is deepened and the mud weight is increased, the overbalance in the upper part of the hole increases. This is not the only explanation for pipe sticking in the upper part of the hole, but it does support the claim that overbalance is the single most important factor affecting differential sticking.

Fig A

The lubricating layer balances pressures around the pipe.

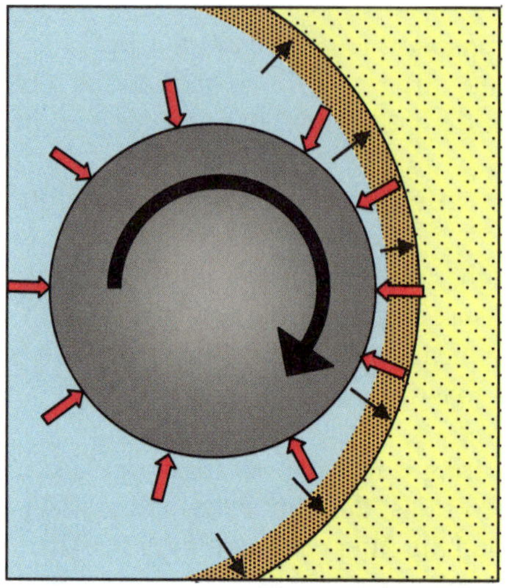

Fig B

Once the lubricating layer has drained away, a differential pressure begins to develop.

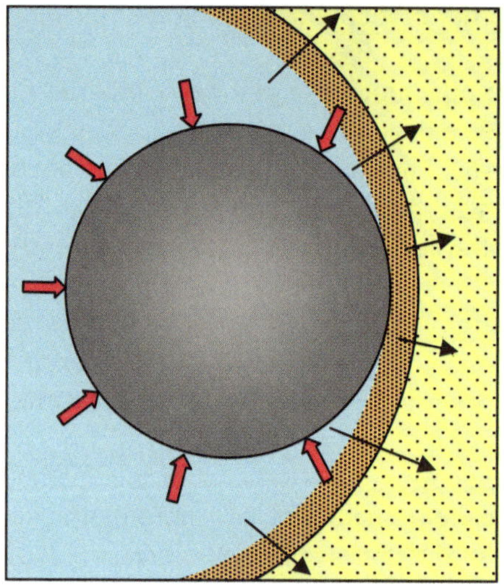

Fig C

As the filtrate drains out of the filter cake in the area of contact the differential pressure grows.

The differential pressure will eventually reach a maximum value that is proportional to the overbalance.

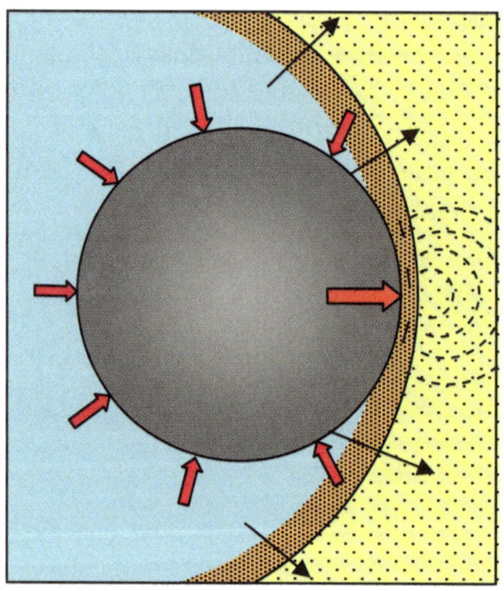

Fig 9-4 Differential pressure behind the contact area

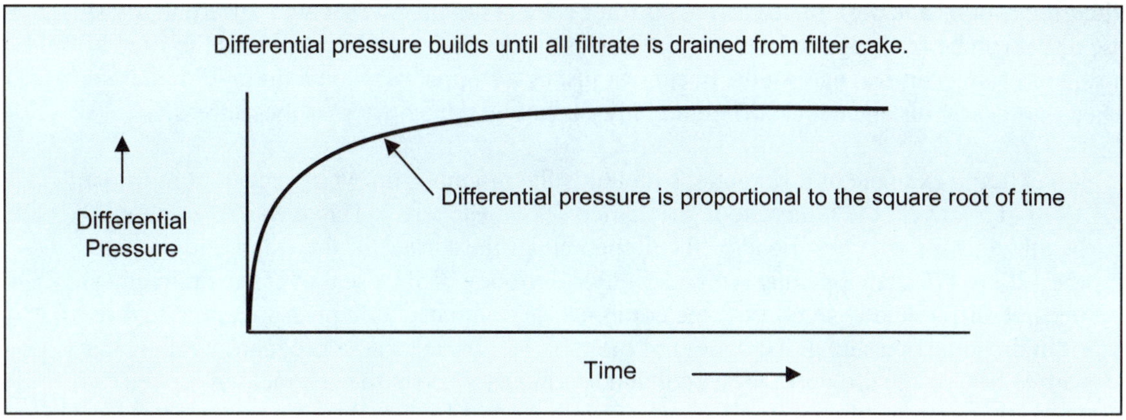

Fig 9-5 Differential pressure and time

Filter Cake (Factors Affecting Differential Sticking)

So far, we have only talked about differential sticking on a filter cake against a permeable sand. With sand, a filter cake is necessary to become differentially stuck. If not for the "slow drain" of the filter cake, there would be no differential pressure across the steel and formation face. The pressure at the well bore wall would be very close to the well bore pressure. (Fig 9-6) The injection curve for fluid entering the formation would be similar to a draw down curve, only in reverse. (Fig 8-55) The slow drain effect of the filter cake allows the formation pressure against the filter cake to be nearly the same pressure as the rest of the formation.

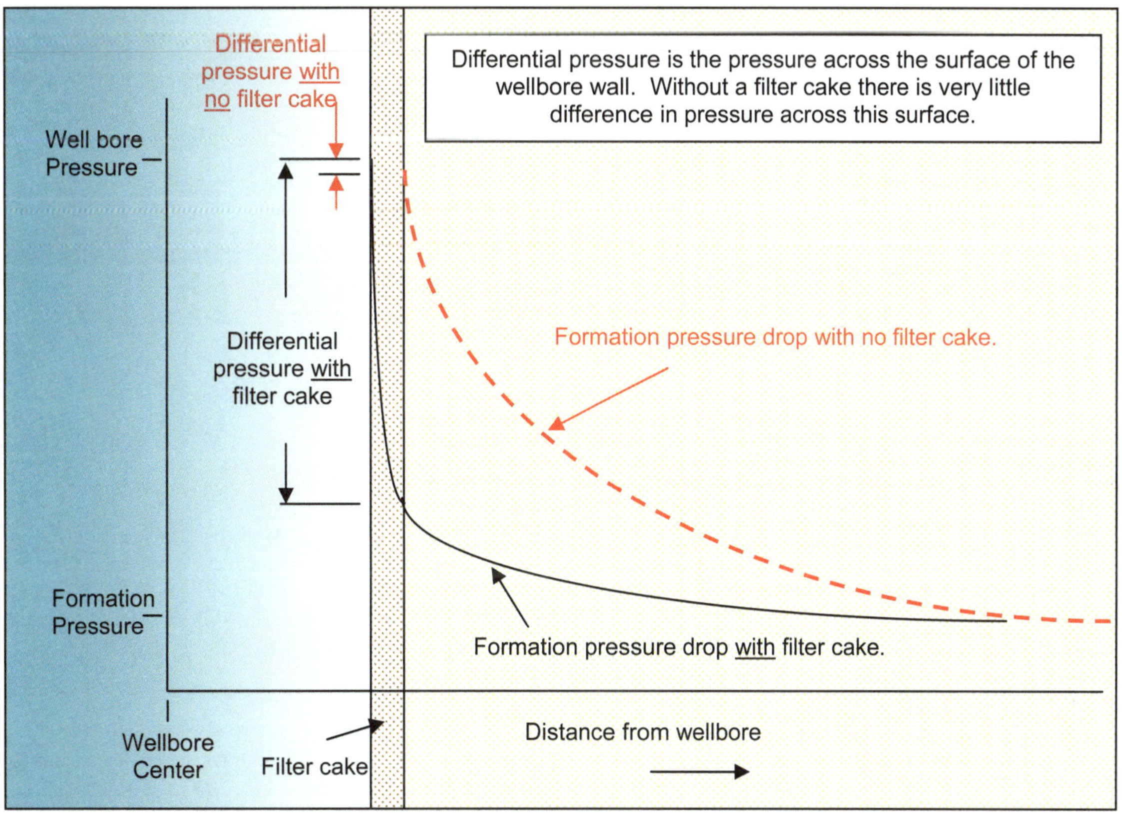

Fig 9-6 Effect of filter cake on formation pressure drop

In faulted formations and perforated casing, no filter cake is required to become differentially stuck. This is because a seal can be made against the perforations or cracks that completely blocks the flow of fluid. The drill string or casing can seal against the openings, just as a stopper seals on a drain. Smaller contact areas may exist, but nearly instantaneous differential pressures will develop across these areas.

> An extreme example of differential sticking is the infamous drowning accident in May of 1996 of a 16 year old high school girl named Tanya Nickens[7]. Tanya was relaxing in a health club spa with her friends. She dipped under the surface of the water and covered the 12" by 12" drain opening with her body. Her body made a seal over the drain and she became differentially stuck from the combination of atmospheric pressure and 3 to 4 feet of hydrostatic pressure at the bottom of the tub. Her friends and a lifeguard were unable to free her, so she drowned. The pool and spa industry refers to such incidents as "suction entrapment", but it is, in fact, an example of differential sticking. No filter cake was necessary to create the seal that allowed the differential pressure to develop.

A thick, permeable filter cake leads to differential sticking. *To prevent differential sticking, we want a filter cake that is thin, hard, and impermeable.* What we mean by a "hard" filter cake is one that is relatively incompressible.

A filter cake is a layer of solids that is filtered out of the mud as it flows into a permeable formation. To better understand the basics of filter cakes, let's consider some filter cakes as they are deposited.

Filter Cake Formation

Consider a permeable formation as it is drilled with a slight overbalance. Drilling fluid will pass into this formation as it becomes exposed. Drilled solids smaller than the openings in the pore spaces of the formation will pass into the formation with the drilling fluid. Particles larger than 1/3 the diameter of these openings will become wedged together and form a bridge that prevents similar sized particles from passing. Smaller particles will then bridge in the spaces between the larger particles. Eventually, even the colloidal size particles are unable to pass through the filter cake, and only a clear filtrate can enter the formation. (Fig 9-7)

The larger particles are necessary to bridge in the pore spaces and fractures in order to provide a means of trapping smaller particles. If not for the larger bridging material, whole mud would be lost into the formation, as is the case with lost circulation.

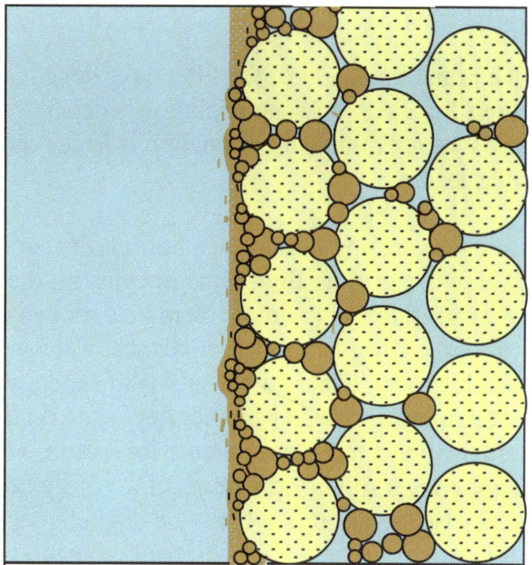

A filter cake is formed as solids bridge across the pore openings in the formation.

Fig 9-7 Dynamic filter cake

The filter cake is generally recognized as having three zones, or layers[8]. There is the **invaded zone**, which consists of whole mud, and extends a couple of inches into the formation. Then there is an **internal filter cake**, consisting of bridging material. It extends only a few grain diameters into the formation. Finally, there is the **external filter cake**, consisting of mostly colloidal size particles. (Fig 9-7) The thickness of this layer varies with time and annular velocity.

Note that the filter cake is permeable. As long as there is an overbalance, fluid will continue to filter through the cake of solids. This means that solids will continue to be deposited on the surface of the filter cake, and it will become thicker with time.

While the drilling fluid is circulating, it is eroding the filter cake. Equilibrium is reached where the rate of erosion equals the rate of deposition, and the cake does not get any thicker. This type of filter cake is referred to as the **dynamic filter cake**. If circulation is stopped, no erosion takes place and a **static filter cake** will continue to thicken. (Fig 9-8) The static filter cake is thicker, and has softer surface layers that make it difficult to determine where the mud ends and filter cake begins. It is also less permeable than the dynamic filter cake. When circulation is started again, some of the static filter cake that was deposited on top of the dynamic filter cake is eroded away.

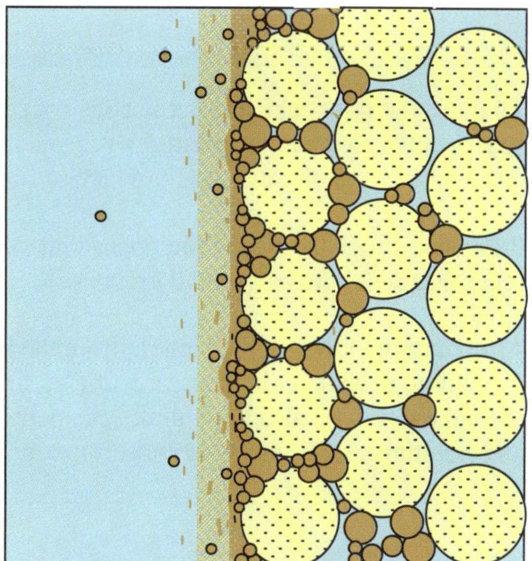

The static filter cake is thicker and less permeable than the dynamic filter cake.

Fig 9-8 Static filter cake

If circulation is stopped and started several times, there will be several layers to the filter cake, layers that are similar to the rings in a tree. Mechanical erosion from rotating drill pipe limits the size of this composite filter cake.

The final thickness of the filter cake depends on the amount and type of solids in the mud. The rate at which the filter cake grows in thickness is dependent on the permeability of the filter cake. The filter cake grows most rapidly when the formation is first exposed, and then decreases with time as the filter cake becomes less permeable.

Imagine drilling through sand of uniform grain size, using clear water as a drilling fluid. Imagine that all the grains are exactly the same size and none of them gets broken up in the drilling process. If we have an overbalance, water will flow into the sand above the bit. As the grains of sand are washed up the well, a layer of sand becomes deposited on the wall as some of the drilling fluid flows into the formation. (Fig 9-9) But a single layer of uniform grains of sand is extremely permeable. Because the grains are the same diameter as the grains in the formation, the well bore will behave as though there is no filter cake whatsoever. It would be as though the well bore were smaller in diameter by the thickness of two grains of sand.

Fluid would continue to flow into the formation and additional layers of sand grains would be deposited. If all the sand grains are the same size, the filter cake is essentially as permeable as the single layer, no matter how many layers deep it gets. Additional layers will be deposited until the rate of deposition equals the rate of erosion.

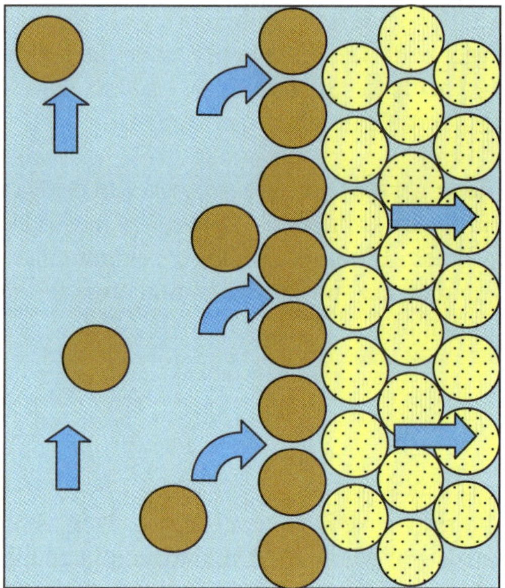

If all the bridging material is of a uniform diameter the filter cake will be highly permeable and thick.

Fig 9-9 Thick, permeable filter cake

To make the filter cake less permeable, we could add a variety of grain sizes. The smaller grains nest in the spaces between the larger grains. Even smaller grains can nest into the pores between the small grains, and so on. The mixture of grain sizes produces a filter cake that is much less permeable.

It is not so much the size of the grains that matter, but more the variety of the sizes that makes the cake impermeable. Even very small grains will make a highly permeable filter cake if they are all the same size.

A mixture of particle sizes ranging from the largest bridging size required to block the pore opening in the formation, down to colloidal size particles, are needed to make an effective cake. An abundance of the colloidal size is needed to minimize the permeability of the cake. Deformable colloidal solids like asphalts and Bentonitic clays make the most impermeable cakes.

When a formation is freshly drilled, the first filter cake laid down will be laden with large drilled solids. It will be thick and somewhat permeable. When a freshly drilled sand is wiped during a connection, we often see some extra drag as this fresh filter cake is wiped off. A new filter cake gets deposited, but by now an internal filter cake has already been established in the formation. There is little or no mud spurt lost to formation as we apply the new filter cake. When this section of hole is wiped later on, such as after the next connection, we notice the extra drag is no longer there. Only freshly drilled and un-wiped sands produce this extra drag, because the new filter cake is not as thick. It probably is not as permeable, either. Our chances of becoming differentially stuck are greater against the original un-wiped filter cake than one that is more established and conditioned.

Filter cakes are generally not deposited on shale. Shale pore openings are so small that few, if any, solids can bridge in their openings. Solids are screened out at the well bore wall and only a solids-free filtrate is allowed to enter into the shale. (Fig 9-10) The flow of filtrate through the shale is much slower than the flow would be through a filter cake, if it existed. This is because shale is a couple of magnitudes less permeable than the usual filter cake deposited on sandstones. The flow of filtrate into the shale raises the pore pressure near the well bore wall, such that there is no differential pressure to hold a filter cake against the formation. Fluid flow and mechanical erosion from the drill string erode the filter cake off the wall faster than it can be deposited.

The only way for a filter cake to be deposited on shale is for the shale to be highly fractured, such that it is permeable. Differential sticking can occur against shale if they are sufficiently fractured and permeable.

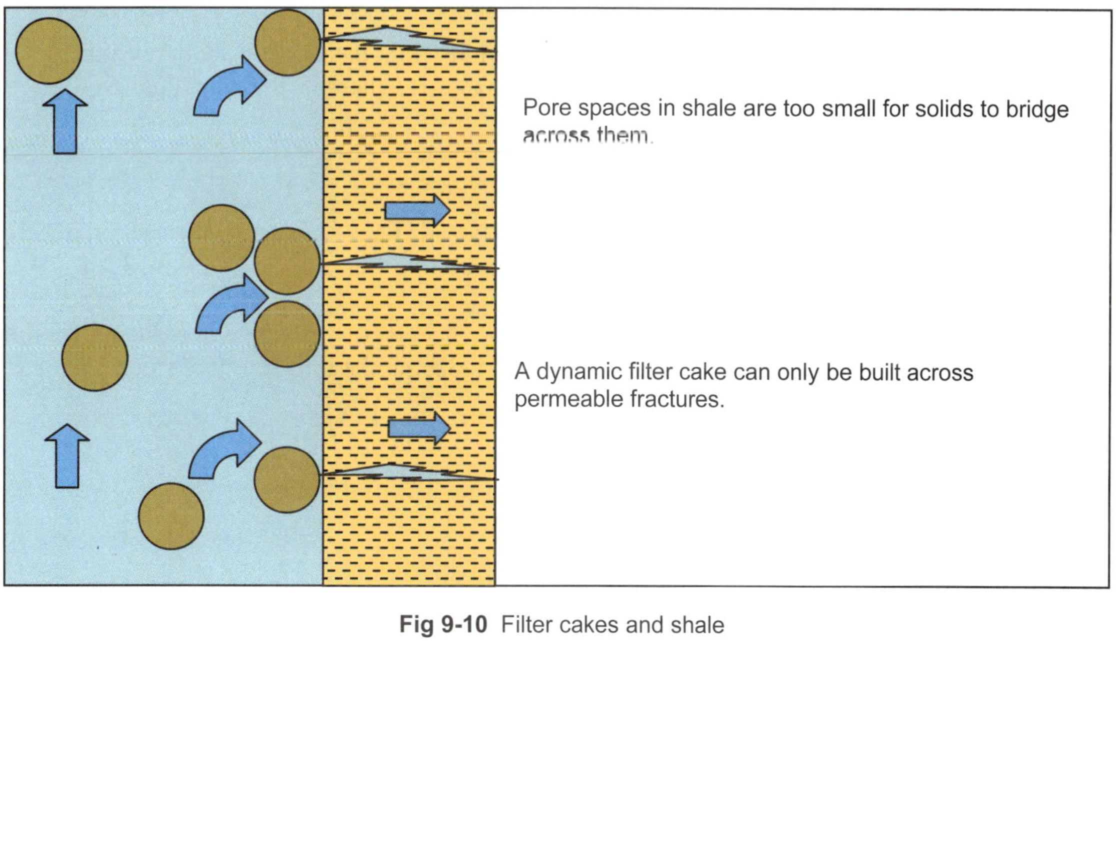

Fig 9-10 Filter cakes and shale

Filter Cake Quality

A high quality filter cake is one that is *thin, hard, and impermeable.* Several factors influence the quality of the filter cake:

- Solids
- Lubricants
- Overbalance
- Temperature

Solids (Filter Cake Factors)

The level of solids in the mud and filter cake has a detrimental effect on differential sticking. Several studies have shown that, as the level of solids in the mud increases, the harder it becomes to free differentially stuck pipe.[9] These solids include weighting agents, as well as drilled solids, though drilled solids are more detrimental than commercial weighting agents.[9]

There are several reasons for the detrimental impact of solids in the mud:

- Non-deformable solids, such as drilled solids and weighting agents, increase the permeability and thickness of the filter cake.
- Solids increase the coefficient of friction between the steel and the filter cake. A higher concentration of solids in the mud will result in a faster deposition of both static and dynamic filter cakes.
- Solids interfere with mud additives meant to condition the filter cake. Mud additives attach themselves to the surface of the solids. The more surface area there is, the more additives are required.

Lubricants (Filter Cake Factors)

Lubricants are added to the mud to reduce torque and drag. They also have a beneficial effect on differential sticking. Lubricants in the filter cake can typically reduce the force to free stuck pipe from 33% to 70%.[9, 10, 11]

Lubricants that are effective in reducing torque and drag are not necessarily effective in reducing the torque to free stuck pipe. David Krol has suggested the mechanism for reducing the torque to free stuck pipe is a combination of:

- Reducing the fluid loss of the mud,
- Coating the solid particles in the mud, and
- Wetting of the metal surfaces.[10]

Effective lubricants adhere to the surface of the steel and the solids in the filter cake. When lubricants wet the steel, they reduce the coefficient of friction between the steel and the cake and thus lower the sticking force. Coating the steel surface reduces the adhesion of the filter cake to the pipe. This film of lubricant may also make it easier for fluid to penetrate between the steel and the cake as the pipe is worked. *To be effective however, these lubricants must be in the filter cake prior to becoming differentially stuck.* Once the pipe is stuck against a filter cake, no fluid can get between the steel and the cake.

When lubricants are in the filter cake prior to getting stuck, the pipe generally breaks free between the steel and the cake[10]. When no lubricants are present, the point of failure may be between the filter cake and the formation, or possibly in the cake itself. The filter cake will remain stuck to the pipe and may even be present when the pipe is removed from the well.

Lubricants also reduce the friction between the particles in the filter cake. This lowers the yield strength of the filter cake, making it easier to tear free from the pipe. (See rock strength.) To be effective, the lubricant has to be able to coat all of the solids in the mud. The total surface area must be accounted for. As the amount of solids increases, or as the solids break into smaller pieces, the total surface area increases and additional lubricants must be added.

The lubricant can decrease fluid loss in the filter cake by several mechanisms:

- One method is by *reducing the flow area in the pore spaces within the cake.* As a lubricant coats a particle, it increases its effective diameter. The shell of lubricant around the solid is deformable, so it allows close nesting of the coated solids in the filter cake. The lubricant film surrounding the particles will partially block the pore openings, thus reducing the flow of filtrate through the filter cake.
- Another mechanism of lowering fluid loss is by *increasing the viscosity of the filtrate.*
- A third method of reducing fluid loss is by *deflocculating the colloidal clays in the cake with thinners.*

When fluid loss is reduced, the amount of time the pipe can be stationary before it becomes stuck is increased. (Fig 9-11)

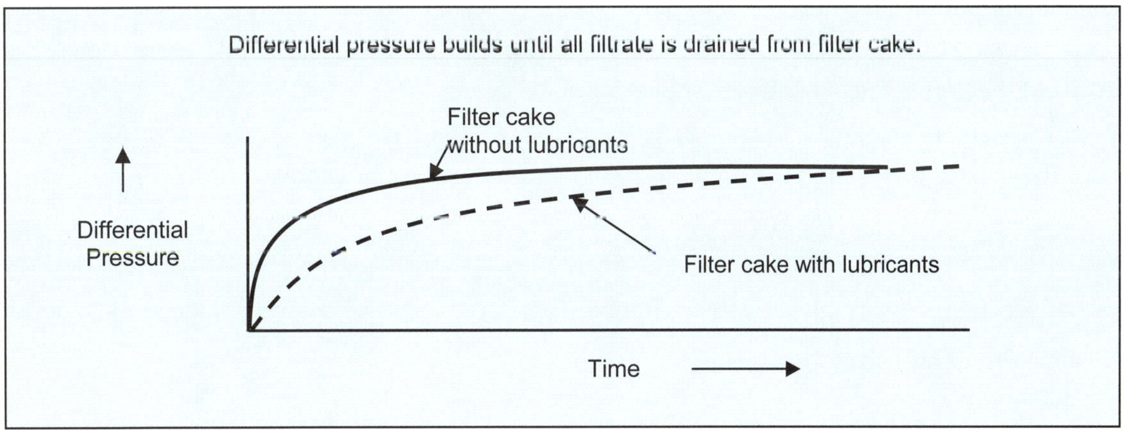

Fig 9-11 Differential pressure with lubricants

Some lubricants, like mineral and diesel oil, also produce a thinner filter cake. Some fluid loss control additives for water-base mud actually increase the thickness of the cake.[8] Generally however, if fluid loss is reduced, the rate of filter cake deposition is reduced and the dynamic filter cake is thinner.

Overbalance (Filter Cake Factors)

Pressure has two affects on the filter cake:

- It helps drive the filtrate through the cake.
- It compresses the filter cake, thus making it thinner and less permeable.

These two effects offset one another. If the filter cake has an abundance of deformable colloidal size particles, such as bentonite clay particles, the rate of filtration through the filter cake might actually decrease with an increase in pressure. Flocculated filter cakes may also be compressed with increases in pressure. If the filter cake is made up of nearly spherical sand grains, the filtration rate will increase with an increase in pressure.

The filter cake, compressed by pressure, will have a higher coefficient of friction, but will give less contact area. A high overbalance will lead to faster drainage of the filtrate into the formation once the pipe becomes stuck. Higher pressures will lead to higher sticking forces, but the quality of the filter cake can influence how much this increase will be.

Temperature (Filter Cake Factors)

An increase in temperature lowers the viscosity of the filtrate. As filtrate moves more readily through the filter cake, solid deposition rates increase. The erosion of the cake by a less viscous fluid will also increase. The degree of flocculation and aggregation of the clay in the cake is also affected by temperature. Investigations into the effect of temperature on filtration rate have demonstrated that the filter loss at high temperatures cannot be predicted from lower temperatures. This is why mud is occasionally tested at the temperature of interest in a high temperature cell.

Wall Contact (Factors Affecting Differential Sticking)

The differential force holding the tubular into the filter cake is obviously affected by wall contact. The differential force is the product of the differential pressure times the area of contact.

$$\text{Differential Force} = \text{Differential Pressure x Area of contact} \qquad \text{eq. 9.1}$$

If the area of contact increases, the differential force will increase. Several factors influence wall contact:

- Tubular fit to hole size
- Tubular size
- Doglegs, ledges, keyseats, and other well bore geometry
- Inclination
- Cutting beds
- Thickness and compressibility of filter cake

The more closely the tubular fits the hole size, the greater the angle of contact, and therefore the more wall contact. The larger the tubular, the greater the contact. Note however, that a small tubular can have more wall contact than a large tubular, if their respective hole sizes dictate so. (Fig 9-12) *Statistical studies show that most stuck pipe occurs in the smaller well bores.*[6]

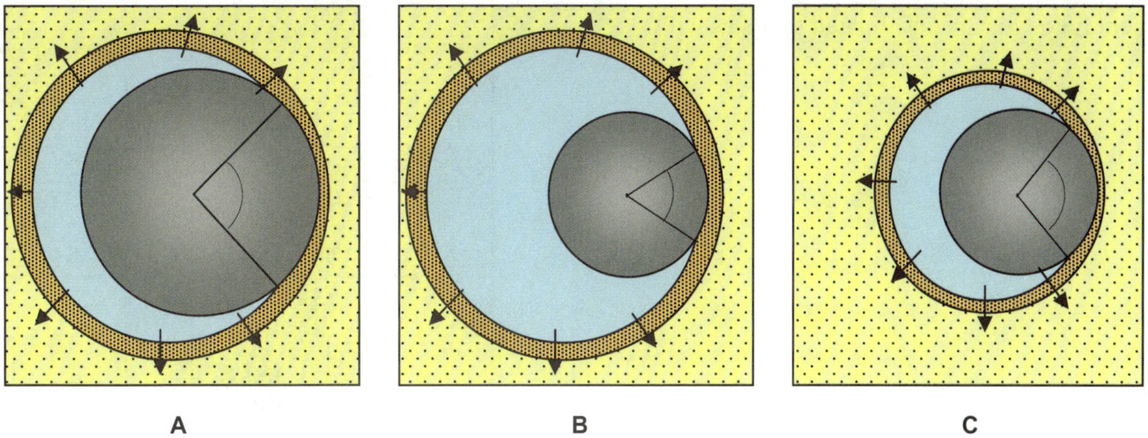

The wall contact area increases as hole and pipe size converge.

Fig 9-12 Contact area vs. pipe size

The drill pipe may cut small keyseats in doglegs or ledges. If so, the drill pipe very closely approximates the hole size and we get very high contact angles. The pipe in figure 9-13 has cut a keyseat only 1/4th the diameter of the pipe but the wall contact is nearly the full diameter of the pipe!

Keyseats are more readily cut into a dogleg or ledge when there is a high side load. The side load is a function of tension in the string, at the depth of interest. The side load, and thus the tendency to cut a keyseat, is greater as the length of open hole below the dogleg increases.

Wire line is exceptionally susceptible to this type of sticking.

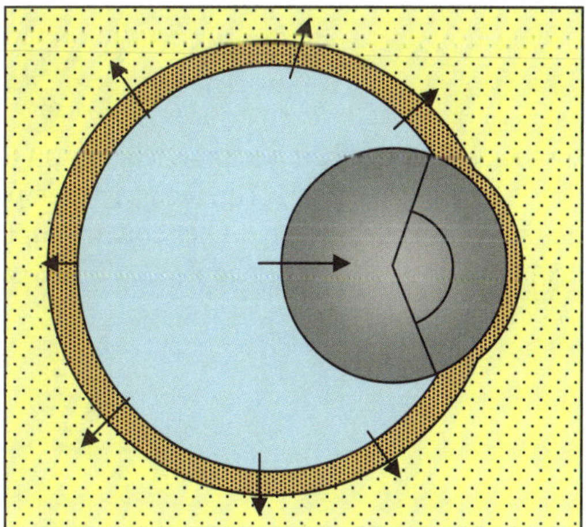

Small key seats provide high contact areas.

Fig 9-13 Keyseat contact areas

If alternating formations of sand and shale are exposed, the harder sandstone formations will be full gauge, while the softer shale will have enlarged. Thus, the permeable sands stick out proud. Stabilizers may not effectively keep the collars off the sand in this case. (Fig 9-14) The same argument holds for unconsolidated sands as well.

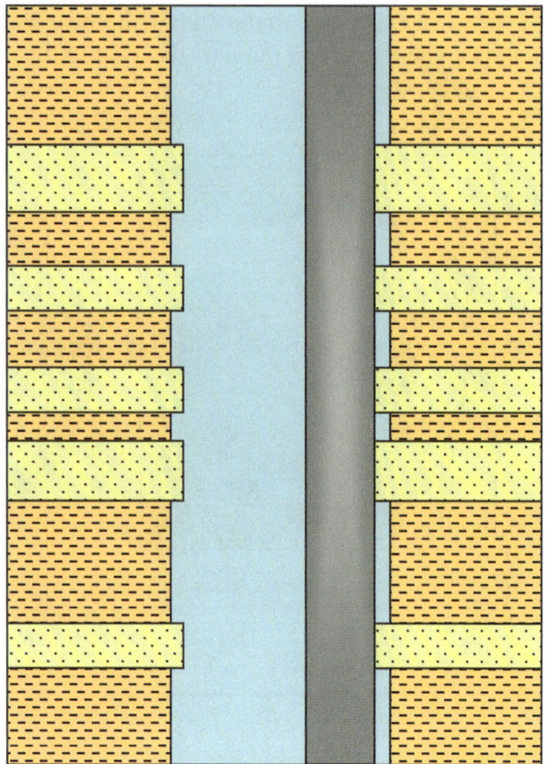

Proud ledges allow full contact against sand.

Fig 9-14 Proud ledges

When the well is inclined, the pipe will lay on the low side of the hole. (Fig 9-15) The weight of the pipe forces it into the filter cake, giving higher contact angle than in vertical holes. Often, the pipe cuts small keyseats into the low side.

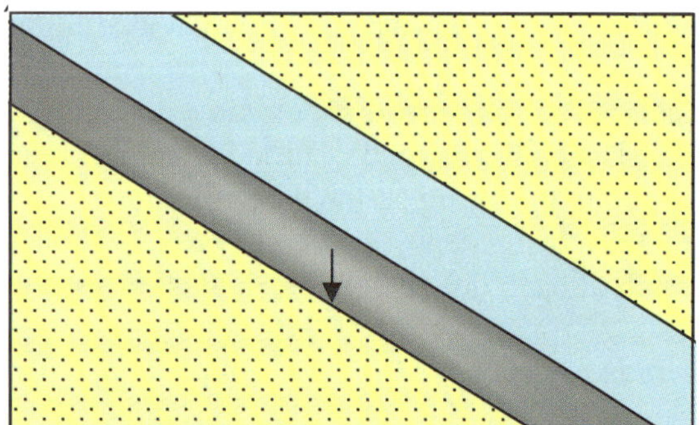

Gravity forces the pipe into the cake on the low side of the hole.

Fig 9-15 The drill string lays on the low side

There are also cutting beds that the pipe settles into and that settle around the pipe. The cutting beds behave as a very thick filter cake. When the pipe is buried in cutting beds, the angle of contact can reach or exceed 180°. (Fig 9-16)

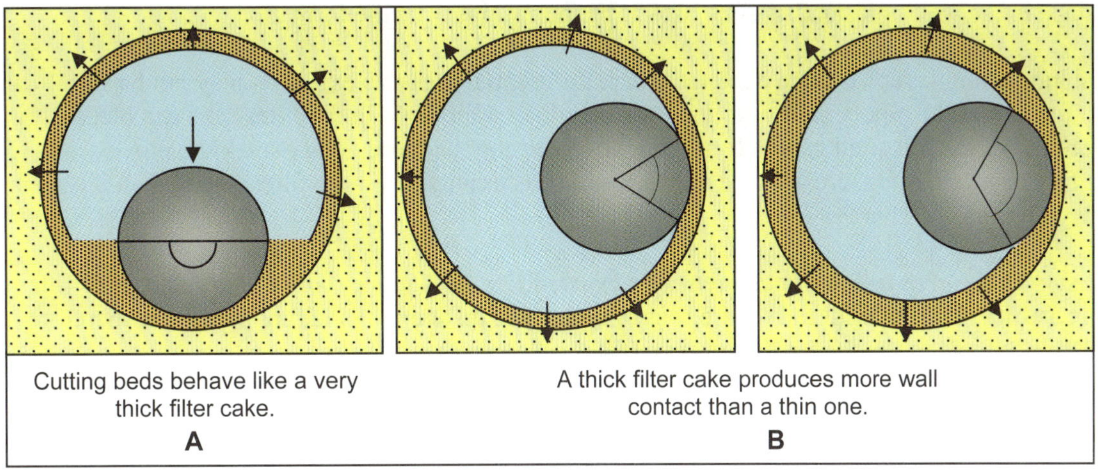

Fig 9-16 Thick filter cakes and cutting beds

The thickness of the filter cake affects angle of contact and thus the wall contact. The thicker the filter cake the higher the wall contact. (Fig 9-16B) Bentonite filter cakes tend to be spongy and compressible. The bentonite particles contain a high percentage of trapped water that makes them deformable. These filter cakes are thick and spongy, even though they are relatively impermeable.

Lack of Pipe Movement (Factors Affecting Differential Sticking)

As mentioned previously, pipe movement is necessary in order to drag a thin film of fluid between the tubular and the filter cake. Without this thin film of fluid, the pressure will not be balanced all the way around the tubular. The thin lubricating layer of fluid also provides filtrate to the filter cake. Without this layer of fluid, filtrate cannot be replaced in the filter cake as it drains into the formation. This allows a lower pressure to exist in the filter cake in the area of contact than exists in the well bore. A differential pressure will develop across the filter cake.

Pipe rotation is the best form of pipe movement. However, even the slight movement caused by pumping helps to wet surface of the pipe against the filtercake. The pressure surges from a triplex pump cause the pipe diameter to fluctuate. The pipe also moves axially due to small changes in length. This "snaking" action may or may not be enough to prevent differential sticking, but it will influence the amount of wall contact and the time required to get stuck.

Note that one method of freeing differentially stuck pipe is the crippling of the pump to cause severe pressure surges in the pipe.

Time (Factors Affecting Differential Sticking)

The development of the low-pressure area takes time. The amount of time for the pressure in the filter cake to reach formation pressure depends on the permeability of the filter cake, the viscosity of the filtrate, the overbalance in the well bore, and to some extent, the permeability of the formation.

After the lubricating layer has been depleted, some differential pressure exists. It may not be significant at first, but the longer the pipe remains motionless, the longer the filtrate in the filter cake can bleed off into the formation, the closer this fluid gets to formation pressure, and the more firmly stuck the pipe becomes. The sticking force continues to increase until all the filtrate has drained from the filter cake—then it remains relatively constant.[2, 4, 9] (Fig 9-17)

The pressure in the area of contact decreases proportional to the square root of time. The static filter cake also increases in thickness with the square root of time. The differential sticking force will, therefore, also increase proportional to the square root of time (filter cake cycle).

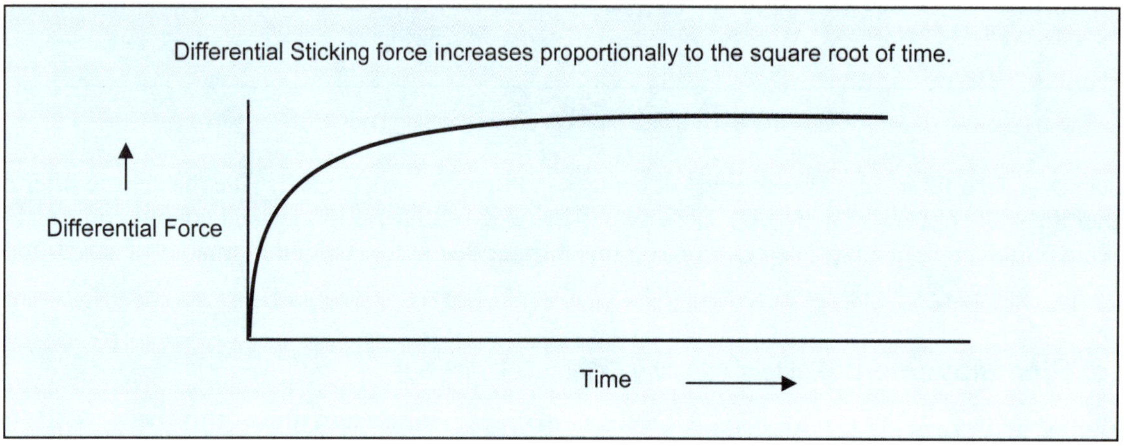

Fig 9-17 Differential sticking vs. time

Side Loads (Factors Affecting Differential Sticking)

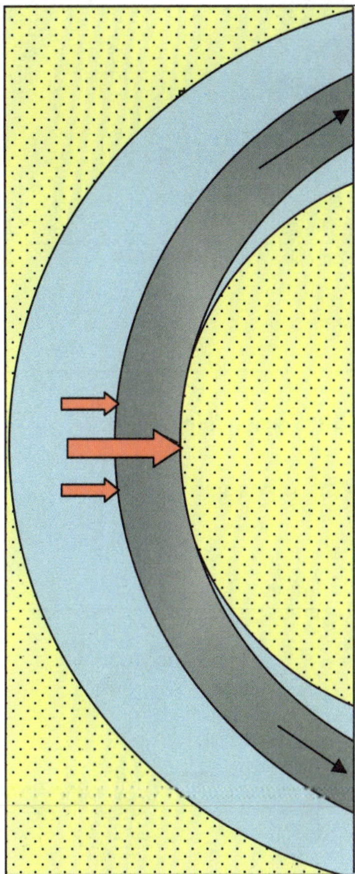

The side load is added to the differential force.

Fig 9-18 Side load

Side loads contribute to differential sticking, but do not necessarily cause it. A side load can be caused by high tension in the pipe when it is against a dog leg, or by the weight of the tubular against the low side. (Fig 9-18) Remember, differential sticking begins when the lubricating layer drains away, such that a differential pressure begins to develop across the filter cake and pipe contact area. The peak differential force isn't attained until the filtrate has drained out of the filter cake and it has been compressed. *A side load speeds up this process.*

The side load is also added to the force caused by differential pressure. *The total friction force is a combination of both forces.*

Recent statistics indicate that more than half of all differentially stuck pipe in the Gulf of Mexico occurs off bottom, up in the area around the drill pipe[6]. Increased overbalance in the upper sections as the well is deepened offers one explanation for this. The added weight of drill pipe to reach the deeper depths increases the tension and thus the side load in the upper section. The combination of higher overbalance and extra side load explains why differential sticking does not occur around the drill collars in this section while it is drilled, but will occur with the drill pipe when the hole is deepened.

All of these factors influence the differential force that holds the tubular against the formation. This is not the sticking force, however. This is only the force holding the tubular against the formation. The sticking force is the friction felt between the tubular and the filter cake.

Friction Force

The equation for friction is usually described by:

$$F = \mu \bullet N \qquad \text{eq. 9.2}$$

Where:

F = friction force resisting motion.
μ = coefficient of friction between the two surfaces
N = normal force

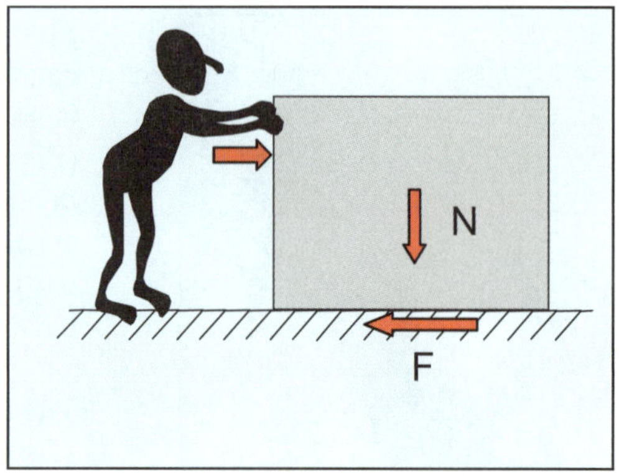

Fig 9-19 Friction force

The "Normal force" is the total force, perpendicular to the formation, that holds the pipe against the formation. This force is sometimes called the *side load.*

In the case of differential sticking in a straight vertical well, it would follow that the normal force would be the differential pressure times the area of contact. One might then conclude that the frictional force resisting pipe movement would be defined by the following equation:

$$F = \mu \bullet [A \bullet (p_m - p_f)] \qquad \text{eq. 9.3}$$

Where

A = the area of contact
p_m = the pressure of the mud in the well bore
p_f = the pressure of the filtrate in the filter cake
μ = The coefficient of friction between the steel and filter cake

If this were the case, then the sticking force in Fig 9-21 would be:

F = 0.3 x [(2" x 30' x 12 in/ft) x (1,400 psi)]
F = 302,400 lbs!

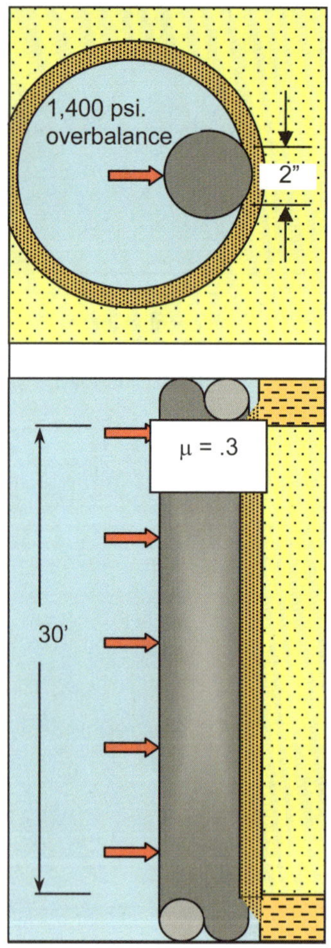

Fig 9-20 Differential sticking force

However, this is only the theoretical maximum value of the frictional force, and it is not usually reached in the field. As discussed in the section on differential pressure, the differential pressure is not constant across the entire area. The pressure difference is a maximum in the center of the pipe and a minimum at the edge of the contact area.

A more practical assumption would be to assume the average differential pressure is about ½ of maximum. The practical equation for the force required to initiate motion when differentially stuck against a sand formation would be:

$$F = \tfrac{1}{2} \bullet \mu \bullet [A \bullet (p_m - p_f)] \qquad \text{eq. 9.4}$$

This is the friction due to differential pressure <u>only</u>. There are additional friction forces when the drill string is laying on the low side of a deviated well, or held against a dog leg by tension. This additional side load must be added to the normal force to determine the actual frictional force to overcome.

$$F = \tfrac{1}{2} \bullet \mu \bullet [A \bullet (p_m - p_f)] + \text{side load} \qquad \text{eq. 9.5}$$

Sticking Force Due to Filter Cake Adhesion

Another factor to consider is the *adhesion of the filter cake to the drill string.* As the drill string presses against the filter cake, the fluid is squeezed out from between the steel and the cake, and the cake becomes stuck to the steel. This is partly due to the adhesive properties of the colloidal size particles in the cake, and partly due to the filter cake becoming differentially stuck to the steel. This is similar to a suction cup stuck to a window. A seal exists between the steel and the cake and, because all the fluid has been expelled, any attempt to increase the volume of space between the cake and steel will result in a drastic reduction of pressure to near zero. This can be demonstrated with a ball of gumbo clay stuck to a wall—a suction sound is sometimes heard as the ball of clay is pulled away from the wall.

Several studies have demonstrated the extent of the adhesive impact on differential sticking.[9,11] Adhesion becomes a significant component of the friction force at very low differential pressures, but becomes less significant at higher pressures. (Fig 9-21) This is because the adhesive force remains relatively constant, while the friction due to differential pressure increases substantially with higher pressures.

Evidence of filter cake adhesion to the pipe is often found when differentially stuck pipe is recovered from a well. In fact, filter cake stuck to the drill string is a warning sign that differential sticking has occurred.

The effect of adhesion is greatly reduced when lubricants are added to the mud.

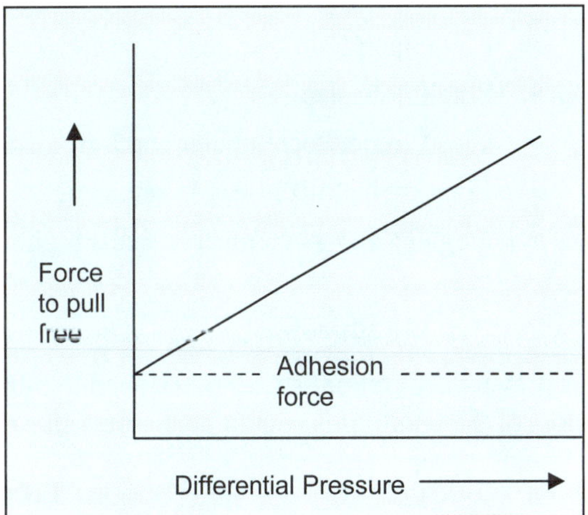

The sticking force due to filter cake adhesion dominates at very low differential pressures, but looses significance at higher differential pressures.

Fig 9-21 Filter cake adhesion

Summary

When to Expect Differential Sticking

Differential sticking occurs when a tubular is motionless across a permeable formation long enough for a differential pressure to develop across the steel and filter cake interface. If the well is inclined, or there are doglegs or other well bore geometry features that create a side load that forces the tubular into the filter cake, then differential sticking may happen more quickly and become more severe.

Conditions that should cause us to be alert for differential sticking are:

- Permeable formations
- High overbalance
- Flocculated filter cakes
- Solid laden, water-base mud
- Rapid drilling
- Long open hole sections
- Small hole size
- Close fit between tubular and hole size
- Long un-stabilized BHAs or casing
- Doglegs across permeable sands high in the hole

Preventive Measures

To prevent differential sticking, we need to minimize the seven conditions that lead to differential sticking. Most of the conditions cannot be controlled, so we must focus on the ones that can.

Permeable Formations

We don't have much control over this condition. We can however, case off permeable zones and limit open hole length to minimize overbalance in these zones.[6]

Overbalance

We may be forced to live with a high overbalance because of the well plan. We can limit the overbalance somewhat, by controlling the mud weight and circulating cuttings out of the vertical part of the well prior to surveys or long connections. We can also pay close attention to solids control to minimize excessive overbalance with light weight mud. We might also consider adjusting the casing setting depths to minimize overbalance[6].

Filter Cake

We have more control over the filter cake than any of the other conditions that contribute to differential sticking. Remember that filter cake should be thin, hard, and impermeable.

Drilled solids should be kept to a minimum in order to limit the cake thickness and coefficient of friction of the cake. If flocculation is observed, it should be chemically treated. Chemical additives such as thinners, lubricants, and deformable colloids will help condition the cake to lower the coefficient of friction and the thickness of the cake. Lubricants must be in the filter cake prior to sticking to have any benefit. To be effective at preventing differential sticking, the lubricants must also be able to coat all the solids in the mud and cake, wet the steel surfaces, and reduce fluid loss through the filter cake.

Freshly drilled formations may have a thicker, more solids laden cake than one that has been wiped once. It can be beneficial to condition the cake with a short wiper trip prior to taking a lengthy survey. Static cakes will build on top of dynamic cakes. If several static periods have occurred without pipe rotation, the cake will be thicker. Occasional rotation to mechanically erode the filter cake can help reduce the sticking tendency.

Wall Contact

Smaller tubulars and larger hole size will give less wall contact. The contact angle, and thus contact area, increases as the tubular approaches the hole size. Stabilizers, spiral drill collars, and heavy weight help to minimize wall contact around the BHA. Centralizers help minimize wall contact with casing.

Keyseats, ledges, and cutting beds lead to increased wall contact. It can be beneficial to wipe out keyseats and ledges in permeable formations. The well plan should take these wall contact issues into consideration during the design phase.

Balled up collars and tool joints may closely fit the well bore in the full gauge sections of the well. In alternating sands and clays, it will be the sands that are full gauge. The pipe should be worked until drag disappears before making a connection.

Static Pipe

Differential sticking does not occur until the pipe remains motionless long enough for the lubricating layer to drain into the filter cake. Motionless pipe is unavoidable, as connections and surveys must be made. We must try to avoid any unnecessary static pipe and plan surveys carefully.

If the risk of differential sticking is high, static time should be avoided until the risk is reduced. If a long static period is expected and we are unsure of the sticking potential, we can check the differential sticking signature with a couple of short static periods prior to sitting motionless for the longer period (Fig 9-22).

If the string must remain motionless due to an unscheduled repair, the mud should be conditioned and the string positioned to minimize wall contact, if at all possible.

Since downward motion is desired to free stuck pipe, we must ensure downward movement will be possible before allowing the string to become static.

Note: The author does not recommend moving the pipe during a well control operation to prevent stuck pipe. Many blowouts occur from BOP failure as a direct result of this practice.

Time

It takes time to develop the differential pressure necessary to cause a sticking force. The lubricating layer must first drain into the filter cake to create a seal. Then, the filtrate in the cake must drain into the formation in order to develop a differential pressure. As the filtrate drains into the formation, its pressure is reduced, so there is less pressure to drive the filtrate into the formation. The filtrate drains slower and slower as time goes by. The rate at which filtrate drains into the formation, and therefore the rate at which differential pressure increases, is proportional to the square root of time. (Fig 9-17)

The problem develops rapidly at first, continues to get worse but at a slower and slower rate, then plateaus once all the filtrate has drained out of the cake. The problem will continue to worsen if circulation is not re-established due to the growth of a static filter cake.

The quicker pipe movement is re-established, the less likely we are to get stuck. If we are stuck, we must quickly take the correct first action to prevent further sticking.

Side Loads

Doglegs through permeable sands should be avoided when possible, especially high in the open hole section. Long, heavy BHAs in high angles of inclination will cause high drag and impart a large side load on the low side of the hole. Good tripping practice calls for the pipe motion to be downward prior to setting the slips. This is partly to remove the excessive tension in the string that leads to higher side loads.

Warning Signs

Differential sticking begins the moment the lubricating layer has drained into the filter cake and differential pressure begins to develop. The friction force caused by the differential pressure will be small at first, but will increase with time until it reaches its maximum. If the pipe is moved before the sticking force becomes too great, it will be free. The lubricating layer will be replaced, and a differential pressure will no longer exist.

If the pipe was still long enough for the lubricating layer to drain away, the pipe will be stuck to the filter cake. Some force or torque must be applied to the pipe to free it. This force or torque is one of the early indicators that differential sticking is occurring.

An increase in torque or drag after the pipe has been motionless for any period of time is an indication of differential sticking. If this torque or drag disappears after the pipe has been moved, it is a strong indication of differential sticking. The sticking force due to differential sticking will disappear once the pipe is moved. There may still be drag due to settled cuttings, however.

There is no reduction in the annular clearance around pipe as it becomes differentially stuck. There is nothing that could cause an increase in pressure. Therefore, to confirm that we are experiencing differential sticking, we will check to see that there is no pressure increase accompanying the increase in over pulls.

The Differential Sticking Signature

The characteristic "signature" of differential sticking is an increase in drag or torque to initiate pipe movement after it has been motionless for any period of time. (Fig 9-22)

This torque or drag will disappear once pipe movement has been re-established. There will be no pressure increase to accompany the torque and overpull experienced.

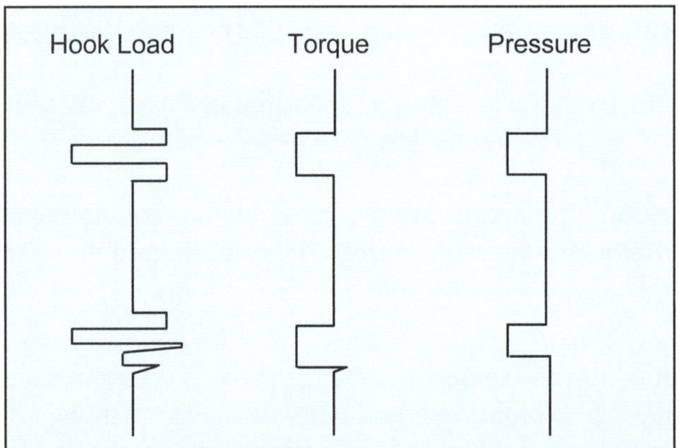

The characteristic "signature" of differential sticking on a Geolograph: No signature is present on the top connection. On the bottom connection drag and torque that disappears once motion is established is present. No change in pump pressure is observed.

Fig 9-22 Differential sticking "signature"

High Fluid Loss

In permeable formations high fluid loss may be experienced. This is a warning sign for thick permeable filter cakes. This leads to rapid compression of the cake and high wall contact.

Freeing Procedures

First Actions

If the string becomes differentially stuck, the first action should be to **circulate at as high a rate as possible while working maximum torque down the string**. The high flow rate will help erode the static filter cake. The annular friction losses will provide additional overbalance that may help compress the filter cake outside of the contact area.

Torque is far more effective than axial force in freeing differentially stuck pipe. Differential force creates a high resistance to rolling the pipe, and a very high friction force that resists axial movement. We are unlikely to move the pipe axially but we may be able to roll it off the wall.

The pipe should then be slumped to the maximum allowable set down weight, as quickly as possible.

The pipe should not be pulled upward! The pipe was static when it became stuck, which normally means the pipe was in tension across the high side of the hole when it became stuck. Pulling more tension on the pipe will only increase the side load, which pulls the pipe harder into the filter cake. The additional side load also increases the friction force, preventing movement. Another reason not to pull upward is that *tensional and torsional stresses are additive*. We want to apply as much torsion as the string can withstand, which does not leave much room for tension. Torsional stress and compression stress are *not* additive, so we can simultaneously put the full set-down and torsional limit on the string without twisting off.

If we have jars in the string we want to **jar down**. The circulation rate should be reduced just prior to the jars going off to minimize the pump open force acting on the jars, and thus maximize the jar blow. Reducing the flow rate will also reduce the additional overbalance from annular friction.

If we believe the pipe is stuck on the low side of the hole, and it has not become free from torquing and slumping, we may attempt to torque and pull within the design limitations of the drill string while considering secondary approaches.

Secondary Freeing Procedures

If torquing and slumping is unsuccessful, a number of alternative methods are available. However, the sticking force increases with the square root of time and these methods take time to apply. To prevent further sticking, torquing and slumping should continue while preparing and applying other techniques.

Reducing the overbalance by pumping a light-weight spacer into the annulus is a method that can be tried if the formations are competent enough to withstand it. (Fig 9-23) This is a common approach in older, more competent formations. In fact, in Northern New Mexico some operators have a policy of sending a nitrogen truck to blow all the mud out of the well when the string becomes differentially stuck. Well bore instability and well control are not concerns in fields where this policy is in place.

However, reducing the overbalance can be a dangerous approach to take. Many blowouts have destroyed rigs as a direct result of lowering the overbalance to become free from differential sticking. A large number of wells are also lost from well bore collapse as a direct result of this attempt.

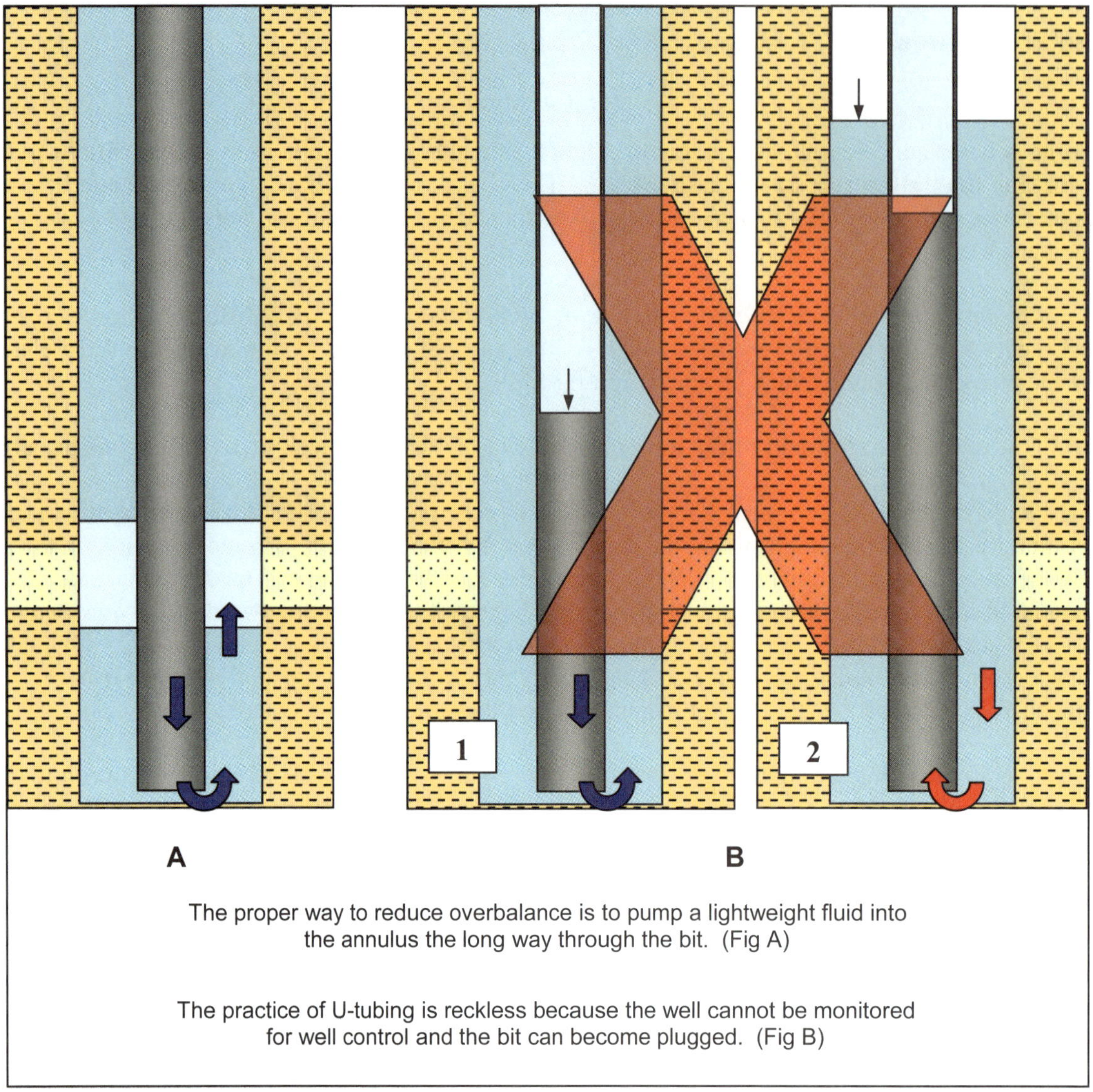

Fig 9-23 Reducing overbalance and "U-tubing"

If reducing overbalance is the desired secondary choice for attempting to free stuck pipe, then prudent practices should be employed. The maximum allowable reduction of overbalance should be established prior to getting stuck. The amount of light weight spacer should be carefully calculated and displaced into the annulus, above the stuck point. (Fig 9-23A)

The practice of "U-tubing" is a dangerous practice and should be avoided. "U-tubing" involves pumping a light weight spacer into the drill pipe and then allowing it to bleed back. (Fig 9-23B) This allows the level in the annulus to fall, and thus reduce the overbalance. However, we can no longer see the level of mud in the annulus and cannot monitor the well for well control. There is also the risk of plugging our nozzles, which will further complicate stuck pipe and well control operations.

Spotting Fluids and Techniques

When reducing the overbalance is too risky, spotting fluids become the secondary attempt. A combination of some reduction of overbalance in combination with spotting fluids might also be considered.

Spotting fluids work in part by attacking the filter cake. The spotting fluid blocks the pores in the cake to make it less permeable. This causes the cake to become compressed, just as the cake in the area of contact was when the pipe blocked the flow of filtrate into the filter cake. Compressing the cake reduces the contact angle and thus the area of contact. The shrinking filter cake may "crack" as the inside diameter increases in circumference. (Fig 9-24) This allows the spotting fluid to charge into the formation, and thus reduce the differential force across the filter cake and formation. The spotting fluid also attempts to "wet" the drill string between the steel and cake interface. If this occurs, the hydrostatic pressure will equalize around the pipe and the differential pressure will disappear.

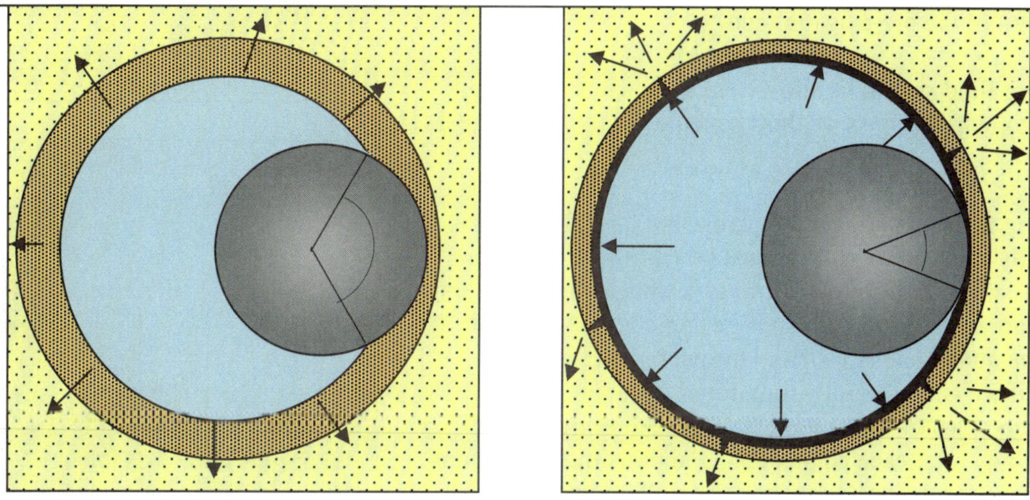

Spotting fluids slow fluid loss into the filter cake, which causes it to "shrink," just as it did in the area of contact. This reduces the area of contact. The shrinking filter cake "cracks" as the inner circumference stretches, allowing the fluid to charge into the formation and lower differential pressure.

Fig 9-24 Spotting Fluids

To be effective, the spotting fluid must displace the mud behind the drill pipe. The flow profile in an eccentric annulus tends to make the spotting fluid channel and prevent this. (Fig 9-25) A low viscosity calcium brine is generally pumped in front of the spotting fluid at the maximum possible flow rate to help displace this mud.[11] Calcium brine is used because it tends to flocculate the filter cake to aid in its removal. The spotting fluid and spacer should both be pumped at the same, maximum flow rate.

The spotting fluid should be at least as heavy as the mud weight in use to prevent channeling and migration. A slightly heavier spotting fluid will eliminate migration and maximize displacement of the pre-existing mud.

The spotting fluid should be allowed to soak while the string is worked to help the spotting fluid penetrate between the steel and cake.

The spotting fluid must be placed across the sand the string is differentially stuck against. We must be confident of where the string is stuck. Pipe stretch, torque readings, and free point indicators are methods to use. In high angle holes, drag will frustrate the quick pipe stretch and torque reading methods. However, if we have carefully monitored the drag trends, we may get some idea of where the pipe is stuck. If we re-start rotation slowly after each connection, and chart the revolutions against torque for the unstuck string, we will know if we have become stuck in the BHA or higher up in the string.

The farther the stuck point is from the bit, the greater the effect of channeling. Larger spacers and spots will likely have to be used as the open hole below the stuck point increases.

It may be necessary to blow a hole in the drill string just below the stuck point to improve the placement of the spotting fluid.

To be effective, the spotting fluid must displace the mud behind the drill pipe where the local annular velocity is lowest.

Fig 9-25 Placement of spotting fluid

The spotting fluid should be displaced a few feet up the well every 15 minutes or so by pumping as hard as possible for just a few strokes. This pumping action fluctuates the diameter of the pipe and helps the spotting agent wet the pipe and filter cake in the contact area.

Cathodic Currents

Another remedial attempt to free differentially stuck pipe is the induction of **cathodic currents.**[12]

Cathodic currents have been demonstrated to reduce the effects of bit balling. Bit balling is similar to differential sticking in that the ground-up cuttings stuck to the bit are (in part) differentially stuck to the bit, just as the filter cake is differentially stuck to the tubulars.

Cathodic currents are believed to reduce the coefficient of friction between the steel and cake interface by drawing water toward the steel by an electro-osmotic process. Brandon et al. suggests that hydrogen evolution at the cathode plays an even greater role of freeing stuck pipe than the buildup of a water film.

Brandon's work suggests that the coefficient of friction between cake and steel is reduced by half within two minutes of applying a cathodic current. The required torque to free the pipe may be reduced by 80% with a clay-base mud and 50% with a polymer-base mud.[12]

Low frequency resonance tools

Low frequency resonance tools are now commercially available.[13] These tools impart resonant standing-wave energy via wire line to the stuck points in the string. These vibrations break down and "fluidize" the rock and debris near the drill string. The drill string also dilates and contracts, which further reduces the friction forces.

Resonant pipe vibration can impart substantially more energy to the stuck point than any conventional mechanical means, such as jarring. Large chunks of debris or ledges are broken into small grains that are then "fluidized." When granular particles are excited by vibrational energy, they are transformed into a fluid-like material, which allows objects to pass through them as they would through a liquid.

The axial vibrations cause the pipe in the vicinity of the energy source to alternate between tensile and compressive stress. This, in turn, causes the diameter of the pipe to expand and contract. The pipe is thus moving both axially and radially along the wall. The friction force is reduced when the pipe is in motion because the dynamic coefficient of friction is lower than the static coefficient of friction. Also, portions of the pipe will have pulled away from the stuck point at times. The friction force is further reduced when the rock grains are fluidized, because they will move out of the way of tool joints rather than become wedged between the drill string and formation.

Resonant pipe vibration technology is more than 40 years old. It was tested in more than 70 wells between 1984 and 1986. Baker Hughes now offers a wire line low-frequency resonance tool called the "rattler."

Crippling the pumps to cause drill string vibration is an additional technique I have heard of, but I have no experience or documentation to support or disclaim the success of this technique.

Backing off

If the drilling jars don't fire, the pipe can be backed off above the stuck point so **fishing jars** can be installed. With additional heavy weight and jars just above the stuck point we greatly increase our chances of jarring free.

If jarring is unsuccessful, the pipe can be backed off above the stuck point and washed over with wash-over pipe. This is risky because the wash-over pipe is stiffer and may have more contact area than the drill string.

Test your knowledge of Differential Sticking

1. Why doesn't a rotating drill string become differentially stuck? (Fig 9-2, pg. 198)
2. What three functions does the lubricating film of fluid provide? (Pg. 197)
3. Name seven factors that contribute to differential sticking. (Fig 9-3, pg. 199)
4. Which 6 factors must be present to become differentially stuck? (Pg. 199)
5. Which factor has the most impact on differential sticking? (Pg. 201)
6. What is the difference between overbalance and differential pressure? (Pg. 193)
7. What happens to the filtercake once the lubricating layer has drained away? (Fig 9-4, pg. 202)
8. Why is differential sticking time dependent? (Fig 9-5, 9-17, pg. 203)
9. When is time most critical once the pipe becomes stuck? (Fig 9-5, 203)
10. What is the ideal filtercake for preventing differential sticking? (Pg. 196)
11. How does the filtercake affect differential sticking? (Fig 9-6, pg. 203)
12. What happens to the static filter cake once circulation is started again? (Pg. 205-206)
13. What factors influence the quality of the filtercake? (Pg. 208)
14. What are the three ideal properties of a filtercake?
15. How can solids be detrimental to the filtercake? (Pg. 208)
16. How do lubricants affect the filtercake? (Fig 9-11, pg. 209)
17. What are the three ideal properties of a filtercake? (Pg. 208)
18. What three things must a lubricant accomplish to be effective in preventing differential sticking? (Pg. 208)
19. Will lubricants work after the pipe has become stuck? (Pg. 208)
20. How can a lubricant reduce fluid loss? (Pg. 209)
21. How can a lubricant increase the amount of time that the pipe can remain static before becoming differentially stuck? (Fig 9-11, pg. 209)
22. How does overbalance affect a filter cake? (Pg. 210)
23. How does temperature affect a filter cake? (Pg. 210)
24. List 6 factors that can influence wall contact. (Fig 9-12 to 9-16, pg. 211)
25. How does side load affect differential sticking? (Fig 9-18, pg. 215)
26. Write an equation to find the differential sticking friction force. (Eq. 9.5, pg 217)
27. What causes the filtercake to adhere to the drill pipe? When is it a problem? (Fig 9-21, pg. 217)
28. List 10 conditions that should cause us to be alert for differential sticking? (Pg. 218)
29. What conditions do we have control over? (Pg. 218-220)
30. What is the classic differential sticking signature? (Fig 9-22, pg. 220)
31. What is the first action to take if we suspect we have become differentially stuck? (Pg. 221)
32. What secondary procedures are available to us? (Pg. 221-225)

Bibliography

1) Hayward, J.T.: "Cause and Cure of Frozen Drill Pipe and Casing," Drilling and Production Practice (1937)
2) Helmick, W.E. and Longley, J.: "Pressure-differential Sticking of Drill Pipe and How It Can Be Avoided or Relieved," Oil and Gas Journal (June 17, 1959)
3) Outmans, H.D., "Mechanics of Differential Sticking of Drill Collars," Trans AIME, Vol. 213 (1958)
4) Outmans, H.D., "Spot fluid quickly to free differentially stuck pipe," Oil and Gas Journal, July 17, 1974, 65-68
5) Clark, R.K. and Almquist, S. G., Shell Development Co.: "Evaluation of Spotting Fluids in a Full-Scale Differential Pressure Sticking Apparatus," paper SPE 22550 presented at the 66th Annual Technical Conference and Exhibition of the SPE, Dallas TX (Oct 1991)
6) Stewart, Maurice I. Jr., U.S. Minerals Management Service, Metaire, LA: "A Method of Selecting Casing Setting Depths to Prevent Differential-Pressure Pipe Sticking"
7) Herman, Eric: "The Entrapment Solution," Vac-Alert Suction Entrapment Archive, http://www.vac-alert.com/suction_entrapment_archive.htm
8) Gray, George R. & Darley, H. C. H.: "Composition and Properties of Oil Well Drilling Fluids" fourth edition, Gulf Publishing Company (1980)
9) Bushness-Watson, Y. M. and Panesar, S. S., BP Research Center, "Differential Sticking Laboratory Tests Can Improve Mud Design," paper SPE 22549 presented at the 66th Annual Technical Conference and Exhibition of the SPE, Dallas TX (Oct 1991)
10) Krol, David A., Gulf Research and Development Company: "An Evaluation of Drilling fluid Lubricants to Minimize Differential Pressure Sticking of Drill Pipe" Drilling Technology Conference Transactions (1984)
11) Fisk, J.V., Wood, R., and Kirsner, J., Baroid Drilling Fluids, Inc., "Development and Field Trial of Two Environmentally-Safe Water-Based Fluids Used Sequentially to Free Stuck Pipe," paper IADC/SPE 35060 presented at the 1996 IADC/SPE Drilling Conference, New Orleans, LA (March, 1996)
12) Brandon, N.P., Panesar, S.S., Bonanos, N., Fogarty, P.O., and Mamood, M.N., BP Research Center, Sunbury-on-Thames: "The Effect of Cathodic Currents on Friction and Stuck Pipe Release in Aqueous Drilling Muds" Journal of Petroleum Science and Engineering, 10 (1993)
13) Buck Bernat, Henry Bernat, Vibration Technology LLC Shreveport: "Mechanical Oscillator Frees Stuck Pipe Strings Using Resonance Technology" Oil and Gas Journal (Nov. 3, 1997)

Chapter 10
Well Bore Geometry

Introduction

Well bore geometry sticking occurs when there is a conflict between the shape of the BHA and the shape of the well bore. The BHA does not want to pass through this portion of the well. In order to become stuck, the BHA must be moved into that area where the conflict occurs. In other words, the drill string must be moving up or down to become stuck due to a well bore geometry issue.

Usually, there is no restriction in circulation because the cross sectional area of the annulus has not decreased. Thus, if the pipe was moving prior to becoming stuck, and there is no increase in pressure after getting stuck, then the sticking mechanism is likely to be well bore geometry.

Although most stuck pipe today is due to packoff and differential sticking, well bore geometry related sticking is still a serious problem. In the 1950s, keyseating was largely thought to be the number one cause of stuck pipe around the world. With the introduction of directional drilling, sticking began to occur when running in the hole with a stiffer assembly than the one used to build angle with. As directional drilling evolved to include mud motors and 90 foot stands, micro-dogleg sticking has become increasingly common.

Well bore geometry related sticking can be broken down into four main categories:

- Doglegs
- Ledges
- Squeezing formations
- Under-gauge hole

Doglegs

The majority of well bore geometry related trouble is due to doglegs. (Fig 10-1) Doglegs lead to keyseats, ledges, high side loads and torque, poor cementing around casing, trouble running casing and logs, drill string failure, production equipment failure, and casing wear while drilling. They also increase the risk of differential sticking and packoff.

The most notorious dogleg related problem is the keyseat, so we will discuss it first.

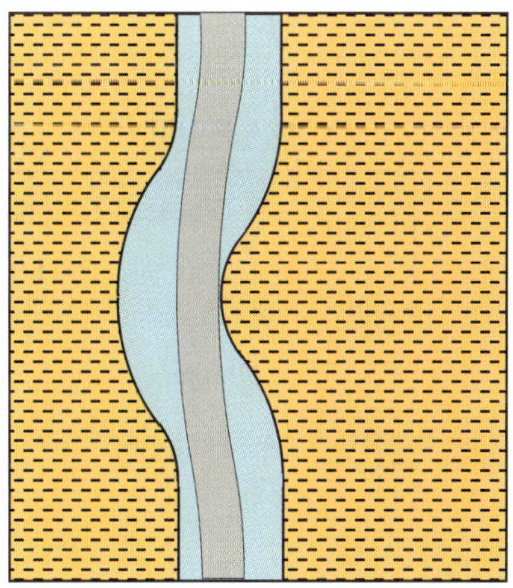

Fig 10-1 Dogleg

Keyseats

The earliest recognized well bore geometry related sticking problem is the *keyseat*. The keyseat gets its name from the keyhole shape it cuts in the well bore. (Fig 10-2) Rotating drill pipe against a dogleg cuts a slot into the formation that is smaller that the BHA. While pulling out of the hole, the drill pipe can pass through the keyseat, but the larger BHA becomes wedged in the small diameter slot and becomes stuck.

A lot of investigative work regarding keyseats was done in the 1950s while drilling deep vertical wells in the Anadarko basin. Many of these wells exceeded 25,000 ft TVD. It regularly took more than a year to drill these wells. The formations were hard, bits lasted only 8 hours, and there were frequent trips. Any dogleg in the upper part of the well led to drill string failure and keyseat formation. Thus, the emphasis on avoiding any dogleg whatsoever in the top portion of the well.

As directional drilling advanced in the 80s, many exasperated drillers demanded to know why a 1° dogleg was a problem in vertical wells, when a 90° dogleg was no problem in horizontal wells. The answer lies in the conditions that must be present to cut a keyseat.

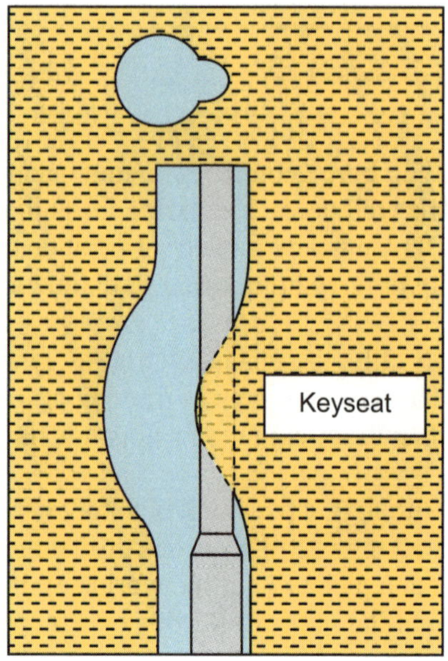

Fig 10-2 Keyseat

Factors Affecting Keyseat Formation

In order to cut a keyseat, the pipe must rotate against a dogleg with enough side load long enough for the keyseat to develop. (Fig 10-3) Thus, four conditions present themselves:

- A dogleg must be present for the "keyseat" to be cut into.
- A side load is necessary to force the rotating pipe against the formation.
- The pipe must be rotating to cut a keyseat.
- The pipe must rotate long enough to cut the keyseat into the formation.

Factors affecting Key Seat formation

The four factors affecting key seat formation are dogleg severity, tension in the pipe, pipe rotation, and rotating time.

All four of the factors must be present to cut a key seat.

An increase in any of the factors will increase the tendency to cut a key seat.

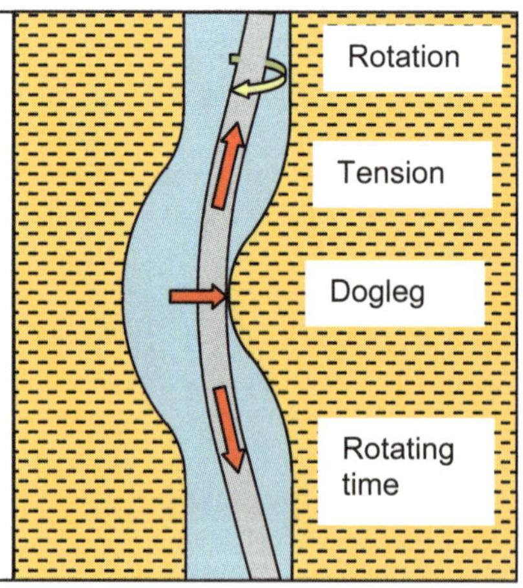

Fig 10-3 Factors affecting keyseat formation

Each of these four conditions must be present to cut a keyseat. If any one of the conditions is removed, a keyseat cannot be cut. The time required to cut a keyseat depends on how hard the formation is. Keyseats can be cut in extremely hard material if we have enough time and side load. Less time is required for soft formations, or if the other three conditions are more extreme.

Side load is an important factor. The harder the pipe is forced into the formation, the quicker a keyseat can be cut. The side load is dependent on the dogleg. The larger the dogleg, the higher the side load. (Fig 10-4) The side load is also dependent on how much tension is in the string. This is why sharp doglegs high in a deep vertical well are so harmful—there is more pipe below the dogleg to create tension in deep wells.

Deep vertical wells also tend to require heavy BHAs. Rotating off bottom with large BHAs greatly increases the side load against any doglegs. This is why it is considered poor practice to rotate off bottom in deep vertical wells.

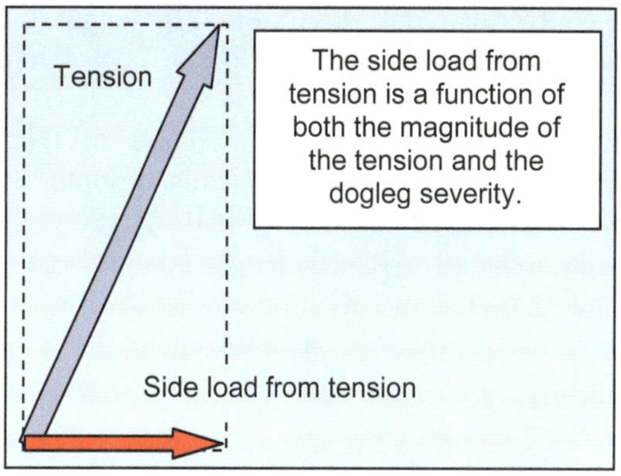

Fig 10-4 Side load and doglegs

A sharp dogleg high in the hole is a disastrous combination when drilling deep, vertical holes. When drilling horizontal wells however, we build a huge dogleg and rotate the pipe through it, but we don't create troublesome keyseats. This is because there isn't a high side load across the dogleg. (Fig 10-5) The side load comes from tension. The drill string is usually in compression while rotating across the dogleg at the bottom of the well. If we cut a keyseat at all, it will be on the low side of the dogleg. When we pick up and put the string in tension, the pipe moves out of the keyseat to the high side of the hole.

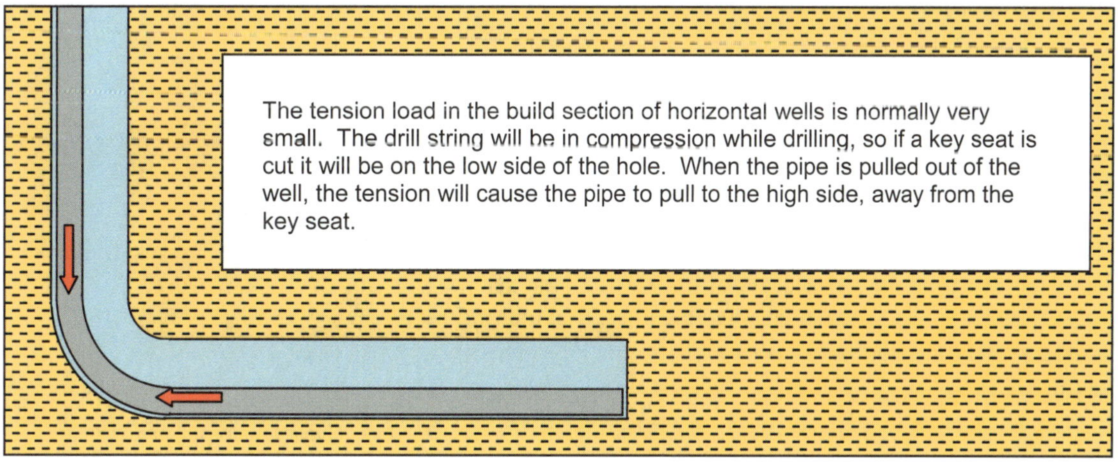

Fig 10-5 Horizontal wells

There is less keyseat sticking today than in the 50s and 60s due to smaller BHAs and better steering tools. Today's doglegs are generally less severe than those of the past, and the smaller BHAs provide less hook load and thus less side load against the doglegs. We also drill each section with fewer rotating hours, due to improved technology. With longer lasting bits, we also trip less frequently.

When to Expect Keyseats

Keyseats can be expected when there are high side loads and long rotating hours against a dogleg or ledge. High side loads exist when there are sharp doglegs near the surface and large BHAs. Rotating off bottom puts even more tension across the doglegs. If we have rotated off bottom for any length of time, we should be alert to keyseats.

Crooked hole country with hard and soft inter-bedding is notorious for keyseats. Inter-bedding leads to doglegs and ledges that keyseats can be cut into. (Fig 10-6) (See micro-doglegs.) Hole enlargement can expose ledges a keyseat can be cut into. Inter-bedding, even in very mild doglegs, leads to washouts that expose hard ledges, that can cause keyseating.

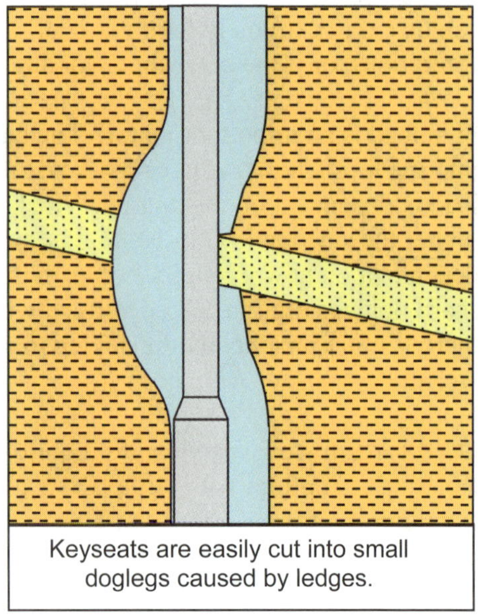

Fig 10-6 Keyseats in ledges

Warning Signs for Keyseats

Warning signs for keyseats include the conditions that lead to keyseats. If doglegs, side loads, and long rotating hours exist, then we are forewarned that a keyseat could exist. The drilling trends that warn us of doglegs include:

- Increase in torque and drag while tripping or drilling.
- Cyclic drag while tripping. There may be an overpull spike as tool joints are pulled through a keyseat. Short keyseats cut through hard ledges provide clear overpull spikes every 30 feet or so. (Fig 10-7) Longer keyseats cut through gradual doglegs may still provide cyclic drag patterns, but they will not be as obvious. Several tool joints will be in a long keyseat at the same time. Still, the drag may be higher in one section of the keyseat that sends a signal every 30 feet. Or, there may be some fluctuation in drag as the total number of tool joints in the keyseat fluctuates.

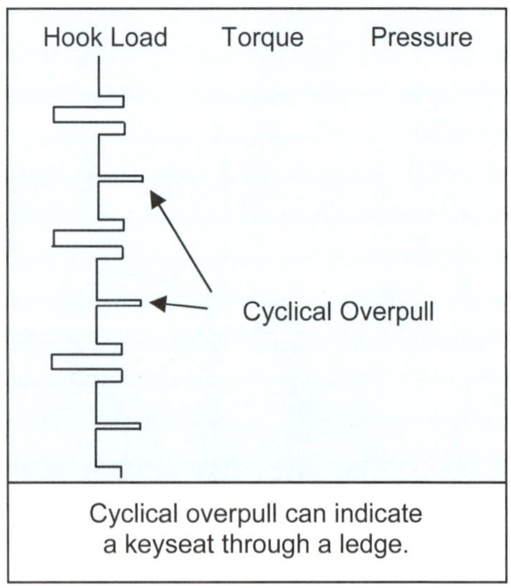

Fig 10-7 Cyclical overpull

- The traditional "signature" trend of keyseats is a progressive increase in overpull on subsequent trips, through the same section of hole. [1] We expect to see drag decrease, not increase, on subsequent trips, due to hole enlargement. But if a keyseat has been cut, the drag is likely to increase because the keyseat will be deeper each trip. (Fig 10-8) Note that we don't make as many trips between casing points as in the past, so we don't often have the luxury of being warned by this trend. When multiple trips are performed, this trend should be monitored.

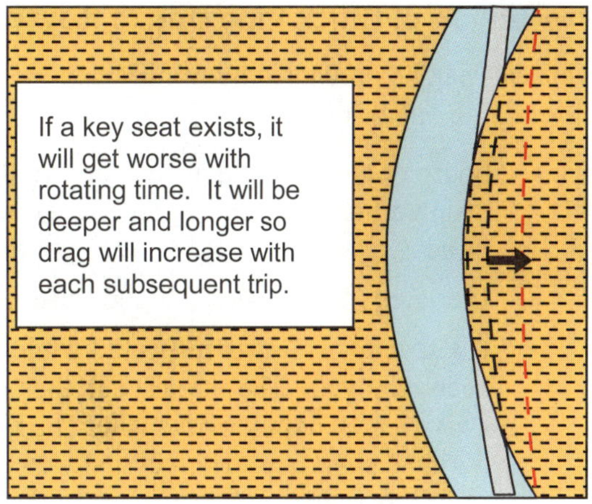

Fig 10-8 Keyseat trend

Many drillers have been surprised by becoming stuck in a keyseat, even though they were aware of the trends while tripping. In many wells, high, cyclic overpull was observed while POOH but downward motion was always possible to free the pipe. The tool joints eventually enlarged the keyseat, causing a reduction in overpull and a false sense that the problem had gone away. When the collars reached the keyseat, they became wedged and could not be freed with downward motion.

Keyseats are usually cut high in the hole. When first picking up off bottom, we have a lot of pipe below the keyseat with which to provide weight to move down, and out of the keyseat. By the time the collars stick in the keyseat, this weight has been lost and we have less chance of getting free. (Fig 10-9)

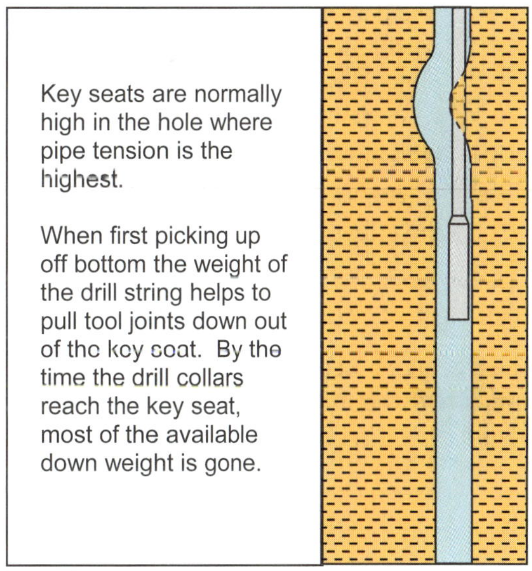

Fig 10-9 Free weight below keyseat

Prevention of Stuck Pipe Due to Keyseats

Keyseats can be prevented by minimizing the four conditions that contribute to forming a keyseat, or by eliminating one of them.

- Sharp doglegs should always be avoided high in the well, where drill string tension is highest. In a deep vertical well, the surface sections must be drilled cautiously, with the prevention of doglegs in mind.
- If doglegs are suspected while drilling, they should be reamed to wipe out the keyseat.
- If a severe keyseat is suspected while tripping, a keyseat wiper can be installed after enough pipe has been removed. The keyseat can then be wiped on a short trip, prior to pulling the BHA through it. This approach only works when the keyseat is high enough in the hole to install a keyseat wiper during a short trip. The keyseat wiper must reach the keyseat before the bit reaches TD. The practice of drilling with a keyseat wiper in the string may lead to more problems than it solves.
- Rotating time off bottom should be minimized, because the side load against a possible dogleg increases as hook load increases.
- As with any well bore geometry related sticking mechanism, **the driller must be aware of where the BHA is with respect to the well bore geometry** and pull through this section cautiously.
- The string can be rotated slowly while pulling the BHA through a suspected keyseat, such that the collars or stabilizers roll out of the keyseat.
- The use of heavy weight drill pipe provides more set-down weight above the stuck point to free ourselves. Generally, if the tool joints pass through the keyseat, then so should the heavy weight drill pipe. It will most likely be the collars, bit, or stabilizers that become stuck. We effectively lose the weight of the BHA below the stuck point, so if the stuck point occurs at the top of the collars, only the weight of the heavy weight pipe can be used for jarring.
- If inter-bedding and hole enlargement is expected, inhibitive mud should be considered to minimize the effect of proud ledges.

Freeing Procedures for Key Seats

The freeing procedure for any well bore geometry related sticking mechanism is to move and **jar in the opposite direction the pipe was moving before it became stuck**. When moving downward, maximum torque can be applied. When jarring up, torque must not be applied.

- In the case of keyseats, getting stuck always happens while moving up, so we need to **apply torque and jar downward** to get free.
- Once downward motion is possible, the pipe may be rotated and possibly pulled past the keyseat.

Secondary Freeing Procedures for Keyseats

If torque and slumping do not free the pipe, there are a couple of secondary techniques that have had some success.

- We would like to jar, if possible, but the jars are often below the stuck point. If the pipe cannot be freed by torquing and slumping, it may be necessary to back off above the keyseat (preferably in casing), and run in with fishing jars.
- If the keyseat is in a carbonate formation such as limestone, an acid pill may be spotted to dissolve and enlarge the keyseat.
- Spotting a lubricating pill may reduce the friction sufficiently for the pipe to be worked free.
- Wire line resonance tools may be able to free the pipe by lowering the effective friction between the pipe and the keyseat, or by "liquefying" the rock it is wedged against.
- Keyseat wipers may be used if downward motion is possible, but the pipe cannot be pulled past the keyseat with rotation. If the keyseat is high enough in the hole, the keyseat wiper may be installed at the rotary table and run down to the keyseat before the bit reaches bottom. If not, the pipe might be backed off in the casing and a wiper installed on a fishing run. This approach must be considered carefully, as it may not be possible to re-engage the fish with drill pipe.

Stiff Assembly

When directional drilling was introduced, a type of stuck pipe called "stiff assembly sticking" began to occur.[2] This type of sticking occurs when a stiff BHA is crammed into a dogleg. This would happen when a limber assembly was used to build up the angle, and then replaced with a stiff assembly to hold the angle. If the dogleg was more severe than anticipated, the stiffer assembly would not fit through it. Any assembly is more limber in compression than it is in tension, so the stiff assembly could be forced into the dogleg, but could not be retrieved. (Fig 10)

When a stiff assembly, such as casing, is forced into a dogleg that was drilled with a more limber assembly, high side loads are generated.

Fig 10-10 Stiff Assembly

This type of sticking is not as common as it once was. We have better steering equipment and don't build excessive doglegs as much as we did 20 or 30 years ago. We also don't use the big, stiff bottom-hole assemblies that were used in the '70s and '80s. We tend to hold angle with steering equipment rather than with big "locked" assemblies. The combination of smaller BHAs and better dogleg control has minimized this once common cause of stuck pipe. In areas where locked assemblies are used, or incomplete information about dogleg severity is available, this type of sticking is still likely to occur.

Casing is one of stiffest assemblies we run in the hole. Doglegs that aren't severe enough to stick the drill collars may still stick the casing.

When to Expect Stiff Assembly Sticking

Stiff assembly sticking occurs when a stiff assembly is forced into a dogleg. Traditionally, it occurs while running in the hole with a stiffer assembly than the one removed.

- If a dogleg exists, this type of sticking should be anticipated the first time a new bottom-hole assembly passes through it.
- Any time a stiffer assembly is installed, this type of sticking should be anticipated as the BHA enters known doglegs. Even though we may believe the build section does not contain severe doglegs, we could be surprised with sudden set down weights and stuck pipe while running in the hole.
- Casing is a very stiff assembly. This type of sticking can be anticipated when running casing past build sections or doglegs. This is especially true if there is little clearance between the casing and borehole wall.

Warning Signs for Stiff Assembly Sticking

The obvious warning signs for stiff assembly sticking are:

- Sudden set-down weight as the bottom-hole assembly enters a dogleg. The dogleg in this case may be the build section of a deviated well.
- Running in the hole with a stiffer bottom-hole assembly than the one most recently removed from the well.
- High torque while rotating down into a dogleg.
- Out of gauge stabilizers or other signs of wear on the bottom-hole assembly pulled from the well. This can indicate that severe doglegs may exist and the existing bottom-hole assembly has had difficulty rotating in them.
- High drag or torque pulling the bottom-hole assembly out of the well.

Preventing Stuck Pipe Due to Stiff Assembly Conflicts

- The best way to prevent this problem is to **limit dogleg severity**. Improved steering assemblies have helped limit the severity of doglegs in the build section of inclined wells. Gradual corrections vs. sudden corrections for directional control also help limit dogleg severity.

- **Be aware of any existing dogleg severity**. Simple inclination measurements with Totco surveys do not indicate how severe a dogleg is. We must know both direction and inclination to calculate dogleg severity. Consider a well that maintains the same inclination of 5° while changing direction 180° over a distance of 100 ft. The Totco survey would indicate zero dogleg severity, while in fact the true dogleg severity is 10° per 100 feet.

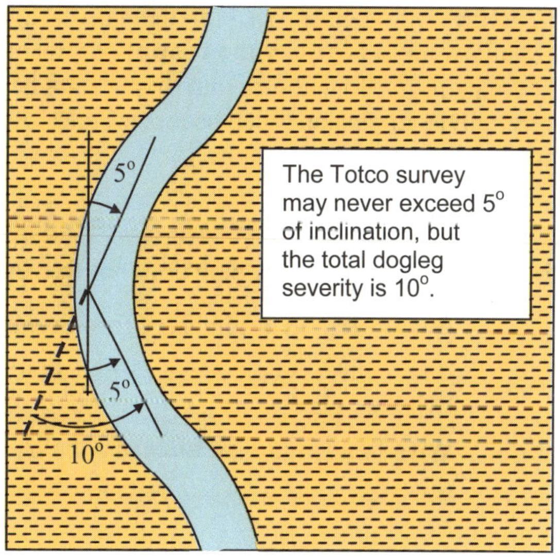

Fig 10-11 Measuring dogleg severity

- Caution must be exercised when changing bottom hole assemblies. The number of **bottom-hole assembly changes should be minimized**, especially those involving changes in stiffness. If changes must be made, careful consideration must be given to the stiffness of the new bottom-hole assembly. The stiffness of a drill collar is greatly influenced by its diameter. The bending strength is a function of diameter raised to the fourth power. If you double the diameter of the drill collar, its bending strength or stiffness is increased 16 fold.[3]
    - Stiffness = $\pi(D^4-d^4)/64$ eq. 10.1
    - The distance between stabilizers also affects stiffness. The closer the stabilizers, the stiffer the assembly.

- **Avoid cramming the collars into a dogleg** by limiting the set down weight. If a dogleg is offering too much resistance, it may be better to pull out of the hole and run a reaming trip rather than stick the pipe by forcing it.
- As with all well bore geometry related problems, the driller must **be aware of where the bottom-hole assembly is with respect to the well bore geometry at all times**. As the bottom-hole assembly approaches the dogleg, slow down and closely watch for excessive drag.
- **Severe doglegs should be reamed prior to picking up a stiffer assembly**. When severe doglegs are suspected, or when a stiffer assembly is being picked up, a wiper trip should be run to ream the suspect sections. A double hole opener makes a stiff reaming assembly, which can be used to test the well bore prior to running casing.

Freeing Procedures for Stiff Assembly Sticking

The freeing procedure for any well bore geometry related sticking mechanism is to **jar in the opposite direction the pipe was moving prior to getting stuck**. Note that tensile and torsion stresses are additive, so we *cannot* apply maximum tension and torsion simultaneously. We *can* apply the maximum set down limit with the maximum torsion limit simultaneously, because these stresses are not additive. Do not jar up while applying high torque or the string will part.

With the stiff assembly sticking the pipe usually becomes stuck while running in the hole. Therefore, we usually want to **jar up without applying torque**. However, the drill collars are more flexible in compression and may pass into a dogleg that they cannot be retrieved from. If the bottom-hole assembly has been forced through the dogleg and then becomes stuck while attempting to pull back into it, then we must jar downward while applying torque.

Secondary Freeing Techniques

If working and jarring the pipe does not free the pipe, several secondary attempts may be successful.

- A **spotting agent** to reduce friction can be placed across the dogleg while working the string.
- If the build section is in a carbonate formation, then **acid pills** may help dissolve the rock around the BHA to help free it.
- **Wire line resonance tools** may help liquefy the rock surrounding the stuck point.
- If the jar is not working, or is not installed in the string, a backoff may have to be run **to install a fishing jar**.

Micro-Doglegs

A small dogleg may not cause excessive trouble, but several small doglegs close together can cause what is known as "micro-dogleg sticking." (Fig 10-12)

Micro-doglegs reduce the effective diameter of the well. While the bottom-hole assembly is in compression, it is more limber and can pass through the smaller effective diameter caused by the micro-doglegs. When the string is put in tension, the bottom-hole assembly becomes more rigid and wedges against the successive doglegs.

Pipe usually becomes stuck in micro-doglegs while picking up, but stiffer assemblies and casing can become stuck while moving downward.

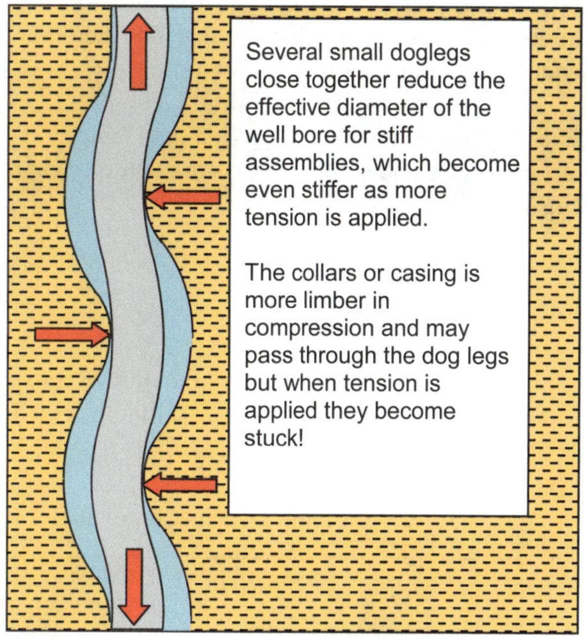

Fig 10-12 Micro-doglegs

Micro-doglegs are caused by frequent direction or angle changes in a directional well, or by the natural drilling tendencies in hard and soft inter-bedding. As the bit passes from a soft formation into a hard formation, there is a tendency for the bit to change direction. One side of the bit encounters the hard formation, while the other side continues to drill into the soft formation. This creates uneven forces, causing the bit to tip and drill up-dip or slide and drill down-dip. (Fig 10-13)

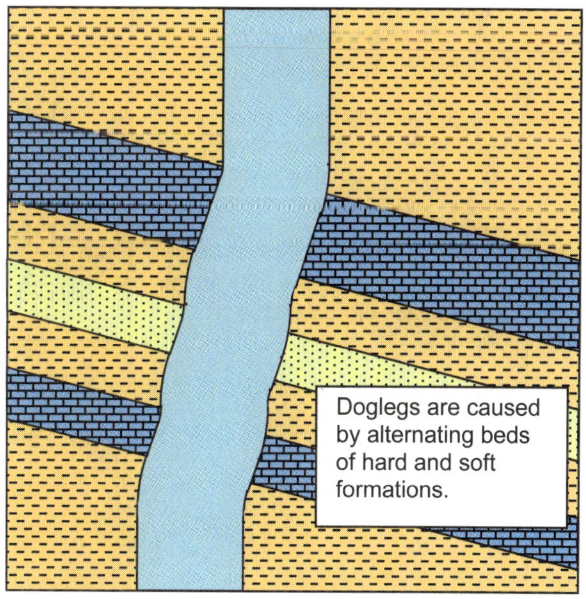

Fig 10-13 Alternating beds cause doglegs

If the formations are dipping less than 45° with respect to the bit, the tendency is to drill up-dip. If the formations are dipping more than 45° then the tendency is for the bit to slip down-dip.

This is easily demonstrated by tipping a table at various angles and setting a glass or cup down on it. (Fig 10-14)

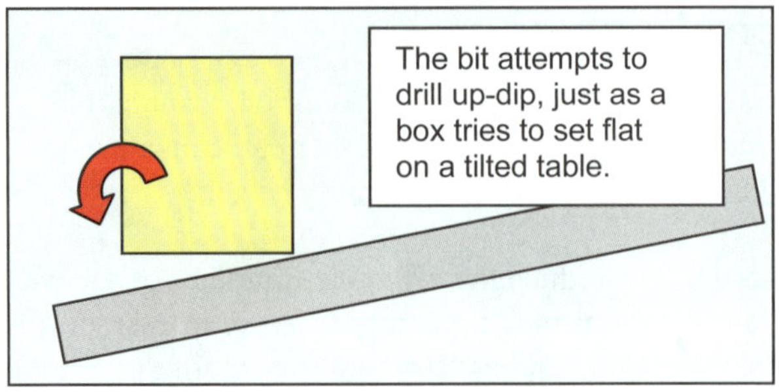

Fig 10-14 Bit deflection

Henry Woods and Arthur Lubinski identified this problem in 1954 and demonstrated that the diameter of the drill collar just above the bit controls the severity of the lateral bit movement that causes most micro-doglegs. (Fig 10-15)

$$\text{Minimum effective bit diameter} = \tfrac{1}{2}*(\text{Bit size} + \text{Drill Collar OD}) \qquad \text{eq. } 10.2$$

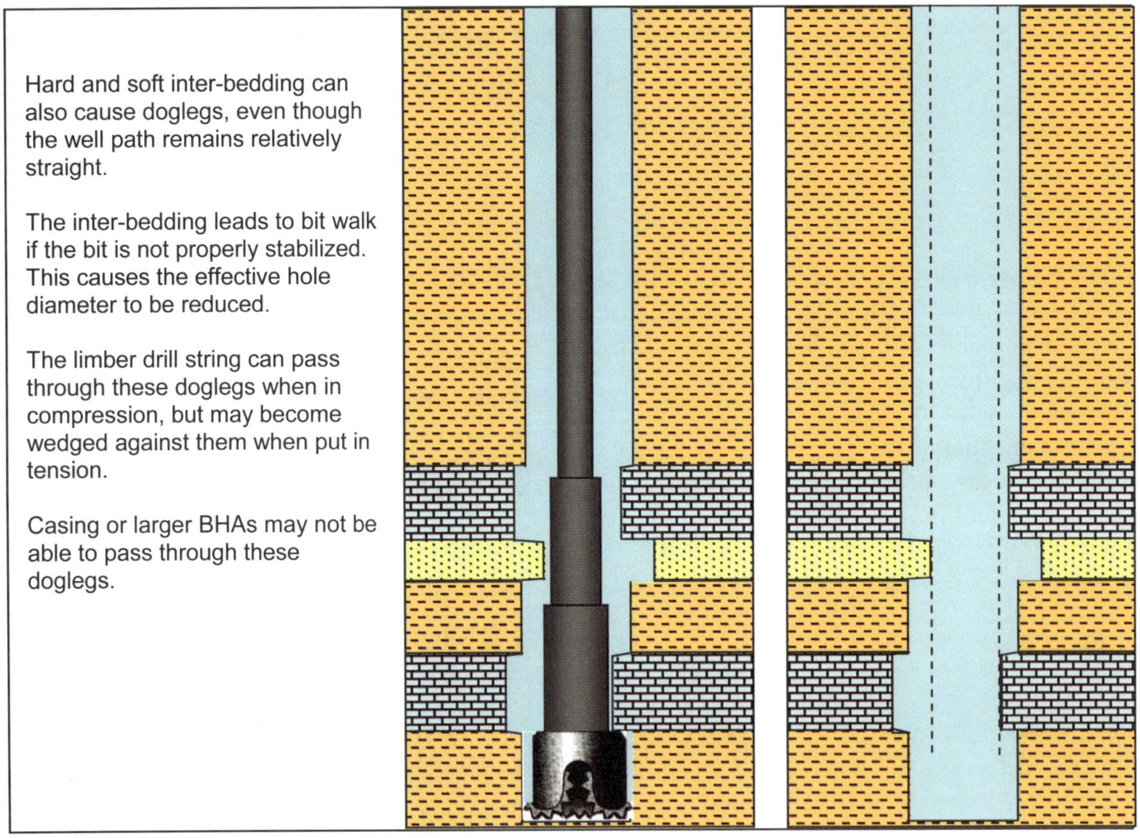

Fig 10-15 Bit walk

Micro-doglegs are becoming more common. This is due in part to the increased use of smaller bottom-hole assemblies. The smaller assemblies are more flexible and buckle easily while drilling. Also, the increased use of top drives lets us drill farther into trouble before picking up for a connection. Micro-dogleg sticking is common in areas with severe inter-bedding while drilling 90 ft stands with limber bottom-hole assemblies under heavy bit loads. The trend in the '90s has been to reduce the size of bottom-hole assemblies while drilling longer stands. Quite often, the problem isn't identified until becoming stuck while picking up for a connection.

Large bottom-hole assemblies prevent micro-doglegs, but can be a liability in high angle holes due to cutting beds and high wall drag. Smaller, un-stabilized BHAs allow bit walk that also leads to worn teeth and bearings. If the bit is not held on its true centerline while rotating, the teeth will scrape against the formation and wear out. If the collars are allowed to bend, the load will shift unevenly across the bearings. This causes the bearings and shirttail to fatigue and wear out quicker. (Fig 5-2) Large collars and stabilizers prevent these problems.[3]

When to Expect Sticking Due to Micro-Doglegs

Sticking in micro-doglegs generally occurs when picking up to make a connection, or when running in the hole with casing or stiffer assemblies. The formation of micro-doglegs can be expected when any of the following conditions are present:

- **Hard and soft inter-bedding has been encountered.** The bit will generally change direction as it passes from one formation into the next. This problem is exacerbated as the ratio of drill collar size to bit size decreases.
- **Limber or un-stabilized bottom-hole assemblies are used with high bit weight.** If excessive bit weight is used for the collar size, they will buckle and the bit will drill a corkscrew path.
- **Un-stabilized bit or a small drill collar size with respect to bit size is used.** If the natural tendency of the bit to move laterally is not constrained, the minimum effective well bore diameter will be smaller.
- **Making frequent directional changes** while directionally drilling.

Warning Signs for Micro-Doglegs

The warning signs for micro-dogleg sticking include the conditions that lead to their development:

- **A fluctuating rate of penetration** is an indication of inter-bedding.
- **An increase in torque and drag while picking up** for connections indicates that micro-doglegs may exist.
- **Increasing torque while drilling** can indicate severe micro-doglegging. Remember that the drill string is more rigid in tension than in compression. Any torque caused from micro-doglegs while the bit is loaded will increase when the bit is picked up off bottom. We will experience some torque from the bit while drilling and some from normal annular friction. The severity of micro-doglegs can be indicated by the torque trends with the bit both on and off bottom.
- **Making several direction changes or sliding with a motor**.

Prevention of Micro-Dogleg Sticking

The best way to avoid any dogleg-related problem is to avoid building doglegs in the first place.

- Directional changes should be minimized. Gradual course corrections can be done every other stand instead of every 30 feet.
- Avoid high bit weights with limber bottom-hole assemblies.
- Frequent back reaming while drilling through hard and soft inter-bedding will help wear the rough edges off any micro-doglegs that have been built. Suspected zones should be reamed prior to running a stiffer assembly or casing. Reaming helps to increase the minimum effective hole diameter.
- The driller must be aware of where the bottom-hole assembly is with respect to the anticipated doglegs, and move through them slowly and cautiously.

Freeing Procedures for Micro-Dogleg Sticking

The freeing procedure for any well bore geometry related sticking mechanism is to **jar in the opposite direction the pipe was moving prior to getting stuck.** Because the string is more rigid in tension than in compression, there is more tendency to get stuck in micro-doglegs while moving up. However, it is possible to get stuck moving downward—especially when a stiffer bottom-hole assembly or casing is being run.

Remember that torsion and tension stresses are additive. High Torque should only be applied while jarring down, not up.

Secondary Freeing Techniques

If working and jarring the pipe does not free the pipe, reducing friction with a **spotting agent** may be helpful.

If the suspect section is in a carbonate formation, then **acid pills** may help dissolve the rock around the bottom-hole assembly to help free it.

Wire line resonance tools may also help liquefy the rock surrounding the stuck point, or reduce the effective friction between the drill string and formation. (The dynamic coefficient of friction is less than the static coefficient of friction.)

Ledges

Doglegs can also lead to ledges when softer formation wears away to expose the harder formations. (Fig 10-16) Doglegs formed from inter-bedding are notorious for leading to ledges. However, doglegs are not required to develop a ledge. Ledges typically form in inter-bedded formations where the hard formations remain in gauge, while the softer formations break out to cause hole enlargement. They also form around fractures or faults. (Fig 10-17)

Ledges are most troublesome when running casing and logs because they prevent the casing and logs from getting to bottom. Often, there is little indication of trouble while tripping out to run casing, but because the casing is much larger and stiffer, it may not be able to negotiate the sharp ledge. (Fig 10-16)

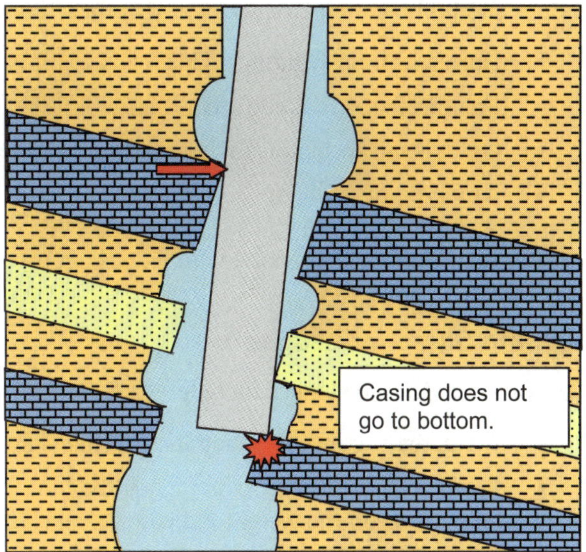

Ledges are easily formed as soft formations enlarge around firmer formations that hold their gauge.

Fig 10-16 Ledges

Stabilizer blades and sharp changes in drill sting component diameters may hang up on ledges while moving the string in either direction.

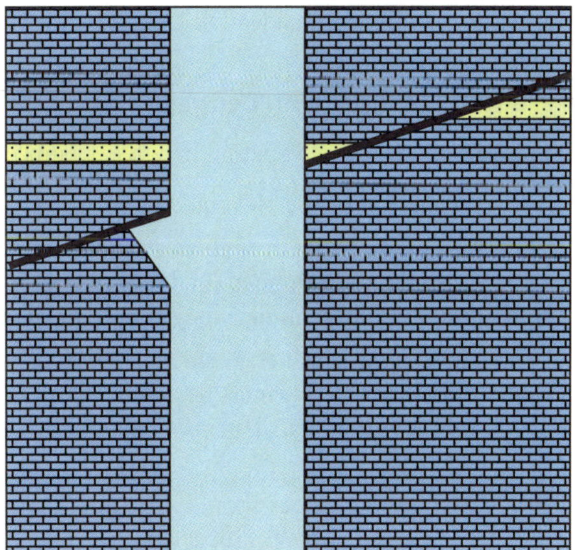

Ledges are also formed around faults when loose material breaks away.

Fig 10-17 Faulted formations

When to Expect Ledges

Ledges should be anticipated when any of these conditions are present:

- Hard soft inter-bedding. This leads to bit walk.
- Fractured and faulted formations.
- Graded salt formations where the salt dissolves at different rates.
- Any type of dogleg.

Most of the trouble experienced with ledges occurs while tripping pipe, or running casing or logs.

Warning Signs for Ledges

The warning signs for ledges include:

- A fluctuating rate of penetration. This is an indication of the inter-bedding that leads to ledges.
- Cyclic overpull while tripping. If the tool joints are forced against the ledge, some overpull ticks may be noticed on the weight indicator, every 30 feet or so. If the drag trends aren't too badly contaminated with hole cleaning trends, this can be a fairly easy trend to spot.
- Sudden and erratic overpull as the stabilizer blades or bit ram into the ledge.
- Sudden set down as the casing or drill string runs into the ledge.
- Known inter-bedding, fractured, or faulted formations.

If any of the conditions that lead to ledges are present, ledges may be present

Preventing Trouble with Ledges

The prevention of trouble relating to ledges begins by preventing the formation of ledges in the first place.

- Doglegs should be prevented.
- Hole enlargement of shale in sand and shale inter-bedding should be minimized with better mud programs. An inhibited mud program may not be called for to prevent well bore instability, but may be required to minimize ledges, so that casing can be safely run.
- When ledges can't be avoided, the driller must be careful as he passes by them with the bit or bottom-hole assembly. He must be aware of where the ledges are and where the drill string is likely to conflict with them. Trip speed should be slow as the bottom-hole assembly or casing approaches the ledges.
- Troublesome ledges should be reamed, but caution should be taken to avoid the bridging caused by knocking large blocks loose.

Freeing Procedures for Ledges

As with all well bore geometry sticking, the first action to become free should be to **jar in the opposite direction the pipe was moving prior to getting stuck**.

- Jarring should commence lightly at first, to avoid knocking debris off the wall. Then, it should become progressively more aggressive, if unsuccessful. Torque should only be used when jarring down.
- Often, downward motion is possible if the bottom-hole assembly became stuck pulling out of the hole, upward motion may be possible if casing becomes stuck going down. In these cases, it may be possible to rotate past the ledge.
- It may also be possible to "shake" the string past the ledge by sending standing waves down the string. A standing wave is similar to the "wave" you see in a hose or power cord as you whip and pull on it to pass it over an obstruction. Standing waves can be generated by lowering the string and suddenly stopping it with the brake. For example, imagine that a stabilizer blade is caught on a ledge and rotation is not possible or successful. Some of the overpull can be rapidly released to start a downward motion of the top of the drill string. When the downward motion is stopped suddenly, before all the overpull is released, the vibration from the sudden deceleration travels down the string. This vibration may be enough to cause the stabilizer to slip past the ledge. We must be careful, however, to avoid surging and swabbing when sensitive formations are exposed.
- Acid pills may be successful in carbonate formations (in some instances).

Squeezing Formations

Some formations such as salt, marls, and young plastic shale may squeeze or creep into the well bore, causing its diameter to shrink. The squeezing is caused by stresses from overburden, tectonic forces, or hydrational swelling.

Little problem is encountered while drilling the formation, but when pulling out of the hole the stabilizers and bit may become wedged in the smaller diameter hole. (Fig 10-18)

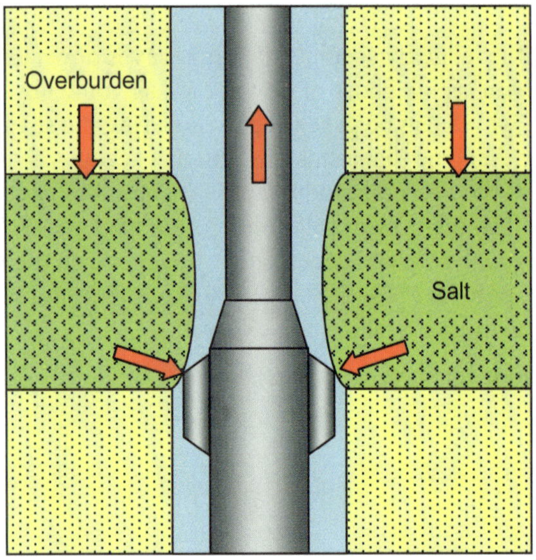

Fig 10-18 Squeezing salt

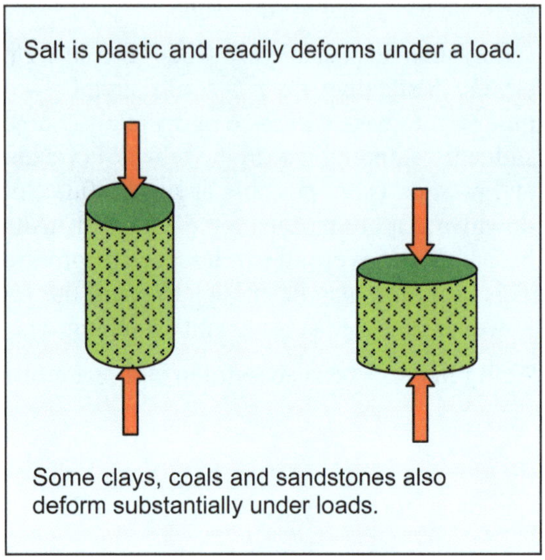

Fig 10-19 Plasticity of salt

Salt is notorious for squeezing, because it is very plastic. The drilled cuttings may appear hard, but a core sample of salt subjected to stress in a hydraulic press will readily change its shape. (Fig 10-19)

Overburden stress causes the salt to expand laterally into the well bore. The diameter of the well bore continuously decreases, from the moment the salt is first exposed. With enough time, the well may be squeezed completely shut.

No two salt formations behave the same. Some squeeze very rapidly, while others may not squeeze at all. Some salts are very pure, while others are heavily contaminated with other sediments. Still other salt formations were formed when the sea completely evaporated, leaving layers of different types of salt, which dissolve at different rates. (Fig 10-20) These "graded" salt formations are particularly troublesome because a salt saturated water-base mud may not prevent every layer of salt from dissolving.

A mud saturated with sodium chloride will prevent the sodium chloride layer from dissolving, but more soluble salts, such as magnesium chloride, may still dissolve, leaving ledges that can break off and fall into the well. To maintain gauge hole in a graded salt, an oil base mud is often used.

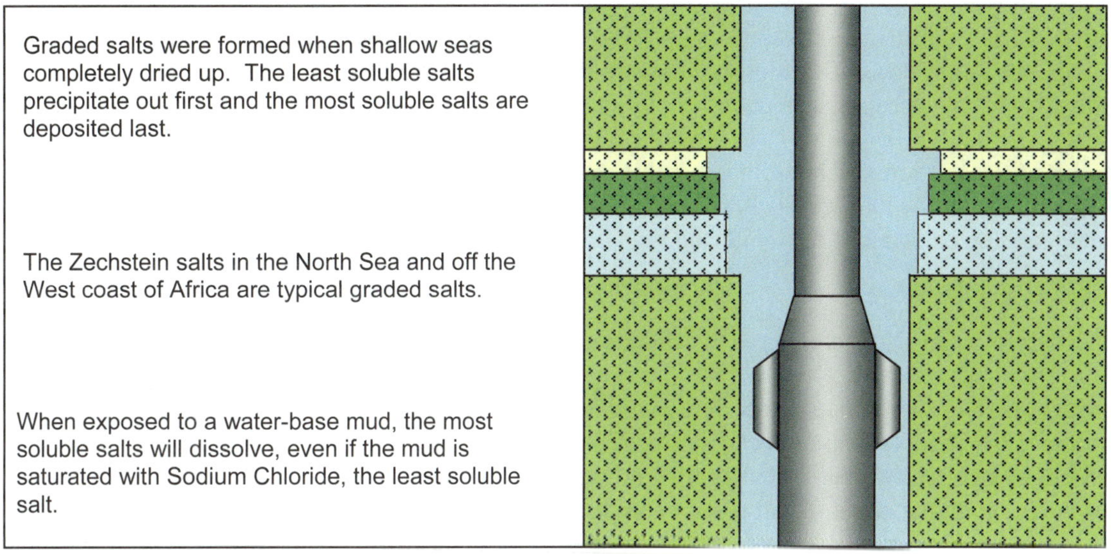

Fig 10-20 Graded salt formation

To better understand how much a salt formation might squeeze, we must understand the factors that affect salt creep.

Factors Affecting Salt Deformation or "Creep"

Several factors affect the rate of creep:

- Overburden and Tectonic stresses
- Purity and thickness of the salt
- Temperature
- Mud weight and mud type

Overburden has a substantial affect on creep. The deeper the salt is encountered, the more likely it is to creep. Salt has a high Poisson's ratio, so the overburden stress is readily transferred to horizontal stresses. **Tectonic stresses** have a similar impact.

The **purity** of the salt formation also has a significant effect. Impurities, such as sand and shale inter-bedded or mixed with the salt, will lower its Poisson's ratio or make it less plastic. The same is true for gumbo clays. We can make clay balls out of bentonite and then observe how impurities such as sand will make them less plastic.

The tendency to creep also increases as the **thickness** of the bed increases.

The **temperature** of the formation also influences its plasticity. All materials are more plastic at higher temperatures. The deeper the salt is encountered, the higher its temperature is likely to be. This and overburden are two reasons that salt at depth tends to be more plastic. Temperature also affects the saturation point or the maximum amount of salt that can be dissolved and carried in solution.

Mud weight provides a radial confining pressure to hold back the salt. As the radial stress increases, the hoop stress, and thus the shear stress, decrease. As confining pressure increases, the formation actually becomes more plastic, but the shear stress that leads to deformation is reduced, so there will be less creep.

The **mud type and salt saturation** affects the rate at which the salt formation will dissolve. It is possible to design a water-base mud that will dissolve the salt formation at the same rate that it is creeping in.

When to Expect Stuck Pipe Due to Creeping Formations

Creeping formations can be expected when factors that contribute to creep exist. Thick beds of pure salt are more likely to creep than thin, inter-bedded formations. Salt formations found at depth, or in regions of high tectonic stress, will creep at a faster rate than salts at shallow depths and low stress.

Salt formations are more likely to creep when high temperatures and high overbalances exist.

Warning Signs

The most common warning signs that creep may be taking place are:

- The presence of salt or coal
- Salt or coal across the shale shakers
- A lack of cuttings
- An increase in chloride concentration
- An increase in the rate of penetration
- Increasing torque and drag

The **existence of salt or coal formations** is the earliest warning sign of creep. If creeping salt formations or coal are prognosed, then creep should be expected.

Salt across the shale shakers or a lack of cuttings can indicate salt formations have been drilled. Water-base mud can completely dissolve the drilled salt cuttings, so **a lack of cuttings** could indicate that the cuttings were salt and have gone into solution. This can be confirmed with a mud check.

A **high concentration of chlorides** in the mud can indicate that salt, or salt water flows, have been encountered. Salt is highly impermeable, so salt water is often trapped within a salt deposit. Because salt is so plastic, its pore pressure can be much closer to the overburden pressure of rocks encountered at similar depth. (Fig 10-21) It may be very hard or impractical to stop some salt water flows.

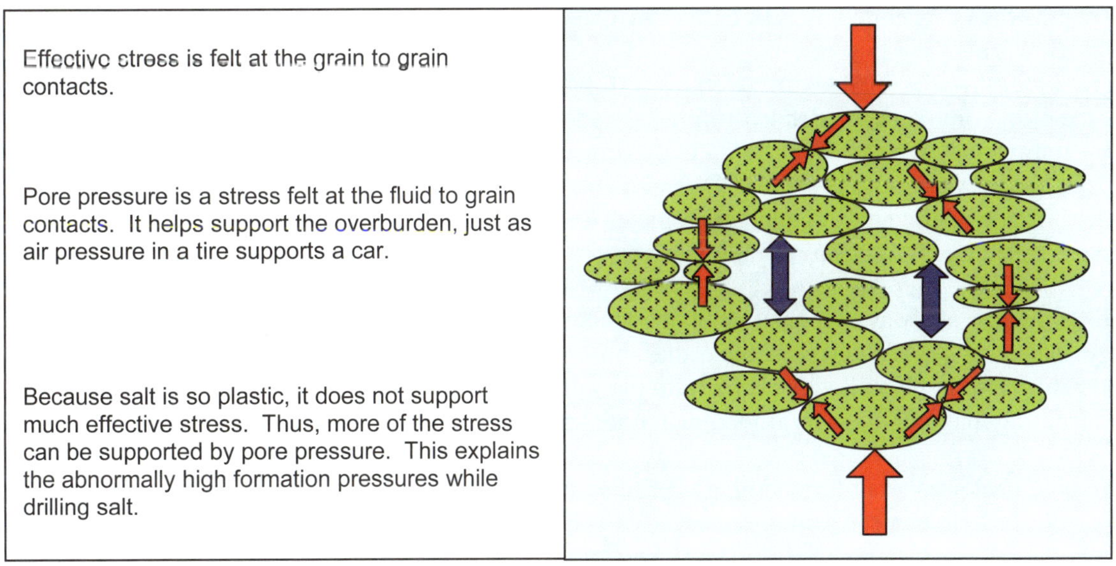

Fig 10-21 Pore pressure in salt formations

High penetration rates and **low bit vibration** are common in thick salt beds, and can thus indicate that salt may have been encountered.

High torque and drag while reaming or tripping can indicate that the well bore diameter has crept inwards.

Preventing Stuck Pipe from Creeping Formations

The most common approaches to preventing stuck pipe from creeping formations is:

- Avoiding the salt or creeping formations
- Raising the mud weight
- Using undersaturated muds
- Regular reaming of the salt interval
- Bi-center bits

The best prevention for becoming stuck in salt is to avoid salt in the first place. If the well plan calls for salt to be drilled, then the factors affecting creep should be avoided (or minimized) as much as possible.

A high mud weight can hold back the salt by increasing the radial stress against the well bore wall and minimizing the shear stress that leads to deformation. A mud weight high enough to prevent creep may not be practical, but any increase in mud weight will slow down the rate of creep.

An under-saturated, water-base mud can be designed to dissolve the salt at the same rate it is creeping inward. This is an effective approach when the rate of creep is predictable and not too excessive. This approach is not as effective in graded salts.

Salt-saturated, water-base mud is used to prevent excessive hole enlargement from dissolving the salt faster than it creeps inward. Hole enlargement from salt going into solution is more likely to be encountered at shallow depths than deeper depths. Oil-base or synthetic mud is often used to maintain in-gauge hole while drilling shallow salts.

Perhaps the most common approach to drilling salt is to ream the salt formations on a regular basis. The idea here is to continuously remove the material that is creeping inward.

Off-center bits can be used to drill a larger diameter well than the diameter of the bit and stabilizers. This allows more time to drill between reaming intervals.

The drill string almost always becomes stuck while attempting to pull or ream through a salt bed. As in all geometry related causes of stuck pipe, the driller must always know where the BHA is with respect to the well bore geometry. He should pull very slowly into the salt formations and avoid excess overpull. Fresh water pills and reaming may be necessary if too much time has elapsed between reaming intervals.

When tripping back into a salt formation, it is prudent to ream through it, unless we have enough information to know that it is no longer squeezing.

Freeing Procedures

As with any well bore geometry related sticking problem, the first action should be to jar in the opposite direction the string was moving prior to getting stuck. In salt formations, the string almost always gets stuck on the way out, so jarring will most likely be downward.

If the pipe does not become free from torque, downward movement, and jarring, a fresh water pill can be spotted across the stuck point to dissolve the salt. Assuming, of course, that circulation is possible. We must be careful not to pull so hard into the salt that we pack off completely and lose circulation.

Under-gauge Hole

The bit and stabilizers can become stuck if they are wedged into an under-gauge section of hole. An under gauge hole is typically the result of a worn bit. The diameter of the hole is reduced as the bit gauge becomes worn.

It is the drill crew's responsibility to measure the gauge of the bit and stabilizers before they are put into the well and as they are removed. Therefore, the driller should be aware of an under-gauge hole if it exists. If the driller is careless when running into the hole after a bit or BHA change, he may cram the bit or stabilizers into the under-gauge section of hole and wedge them so tightly that the drill string becomes stuck.

When to Expect It

An under-gauge hole can be expected when a bit or stabilizers are pulled under gauge. This usually occurs while drilling abrasive sands and/or when the bit quits drilling.

Over-gauge stabilizers or bits can also become wedged in full-gauge sections of hole. This sometimes occurs with retipped bits and rebuilt stabilizers. This is one reason it is considered good practice to gauge bits and stabilizers while going in the hole, as well as coming out.

An under-gauge hole can also be expected with squeezing formations. Sandstones subject to tectonic stress can squeeze in on one axis, thus reducing the effective diameter. Plastic formations such as salt, coal, and gumbos can squeeze in on both axis.

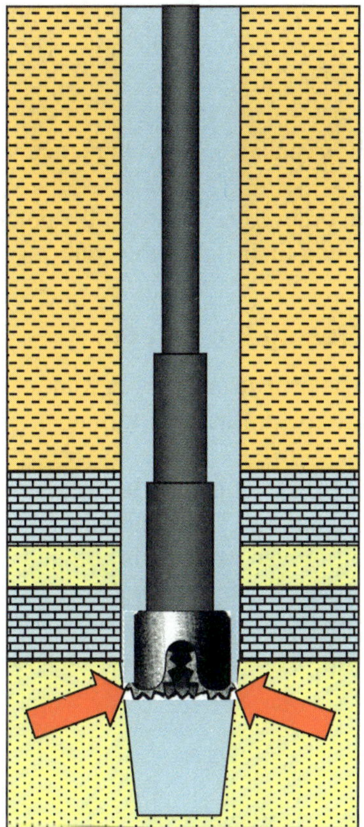

The bit or stabilizers can become wedged into an under-gauge hole while tripping in with a new bit.

Fig 10-22 Under-gauge hole

Hard/slow drilling formations tend to hold their gauge more than softer ones, and are the likely locations to become stuck if an over-gauge bit or stabilizer is run. The last few feet drilled are always suspect if an under-gauge bit is pulled.

Warning Signs for Under-gauge Hole

The earliest warning sign of an under-gauge hole is a change in penetration rate. In the classical case, the bit wears out and the penetration rate decreases. In the case of squeezing salts, the penetration rate increases. In either case, the driller is alerted to carefully check the gauge of the bit on the trip out and to note any tight spots that might indicate squeezing formations.

Slow-drilling formations tend to be hard formations that hold their gauge. Thus, slow drilling formations can warn us of full-gauge sections that a slightly over-gauge bit can become stuck in.

Preventing Stuck Pipe Due to Under-gauge Hole

To avoid becoming stuck in under-gauge hole, the driller must be aware of where the bit and stabilizers are while tripping in the hole. He should slow down and carefully ream through any sections suspected of being under gauge.

It is considered good practice to always ream the last couple of stands to bottom.

In order to be aware of the potential of under-gauge holes, the drill crew should always gauge every tool going in or out of the well.

Freeing Procedures

The drill string only becomes stuck in under-gauge hole while moving down. Thus, to become free we must jar upward without torque.

If the string cannot be freed with jarring, there are a number of secondary freeing procedures that can be attempted. These include *acid pills in carbonates* and *fresh water pills in salts* to dissolve the rock around the bit. Fresh water pills pumped at high rates might also erode the material around a bit. *Spotting agents* to reduce the friction between the bit and formation may also be useful.

A low frequency vibration tool can also be effective if it can be lowered close enough to the bit.

Bibliography

1) Bill Murchinson, Murchinson Drilling Schools: "Drilling Practices Course" Albuquerque, New Mexico
2) BP Amoco Training to Prevent Unscheduled Events Course, 1996
3) Bill Garrett, & Gerald Wilson; "How To Drill A Useable Hole" World Oil (August 1, 1976)

Chapter 11
Drilling Trends and Recorders

Introduction

This chapter focuses on "listening to the well" to stay ahead of trouble. We listen to the well by monitoring the drilling and tripping trends. The chapter begins by defining the word "trend" and listing the various parameters that we monitor. Some of the subtleties in trends that help us distinguish between one type of problem and another are also addressed. We start by looking at some easily recognized trends and then get into "masked" or hidden trends.

A comparison between mechanical and electronic chart recorders is made toward the end of this chapter. The emphasis here is on recognizing the strengths and weaknesses of each approach.

Trends

One key to successful drilling is careful monitoring of the drilling and tripping trends. *A trend is defined as the direction of change of one parameter with respect to another.* There are many parameters to consider, but the most important parameters are:

- Hook load, drag, and weight on bit
- Block height
- Standpipe pressure
- Bottom hole pressure
- Strokes per minute
- Rotary torque
- Rotary speed
- Rate of penetration
- Mud weight in and mud weight out
- Mud properties
- Cuttings and cavings at the shale shaker
- Pit volume
- Flow rate
- Gas shows
- Depth
- Time

Note that it is not so much the value of any of these particular parameters that matters. It is the direction of change in one of these parameters to any other parameter that matters. In others words, *it is the trend that counts, not the instantaneous value.*

A single parameter cannot be monitored by itself. By definition, a trend must involve at least two parameters. For example, an increase in pump pressure is meaningless without comparing it to pump strokes. If pump strokes increase, we expect the pump pressure to increase according to equation 11.1

$$P_2 = P_1 (SPM_2/SPM_1)^2 \qquad \text{eq. 11.1}$$

If our pressure is gradually decreasing while pump strokes are held constant, we are alerted to a potential washout. If our pressure is increasing while holding strokes constant, we are alerted to a possible packoff or mud property change. To look at the pump pressure gauge and notice that the pressure is 3,100 psi is meaningless unless we are comparing it to previously recorded pressures. Similarly, we must compare the change (or lack of change), in pressure to at least one other parameter, such as depth, time, strokes per minute, rate of penetration, etc.

The success of trend monitoring is largely based on how diligently parameters are observed and recorded. One means is to regularly record the parameters in a logbook and look for trends in the numbers. Another method is to install a chart recorder that continuously records the parameters and displays trends in the slopes of it lines. A "Geolograph" chart is a typical chart recorder. (Fig 11-1)

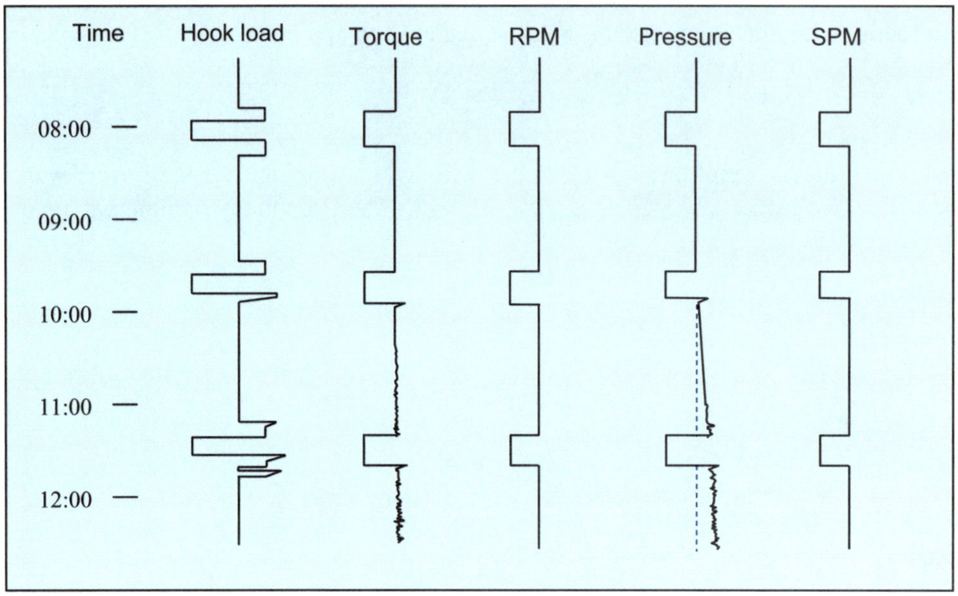

Fig 11-1 Geolograph chart

Notice in Figure 11-1 that the pump pressure is increasing with time and depth, while strokes per minute remain constant. Pump pressure has steadily increased over a two-hour interval. From Figure 7-59 we see that this could be an indication of potential pack off due to poor hole cleaning.

Drilling and tripping trends alert us to potential problems. We must analyze the subtleties of the trends to identify the cause of the problem. For example, the pressure in Figure 11-1 is building linearly. This is typical of cuttings loading a vertical annulus. If the pressure were building exponentially, it would indicate swelling clays. The annulus constricts gradually at first, and then more rapidly with time. Cuttings begin sticking together, and to the well bore and pipe above the swelling formation.

The value of the trend chart is that it gives a quick, visual indication of slight changes to the trained eye. The drilling crew must learn to recognize subtle changes and trends on the chart to get any benefit from them. The trick is pattern recognition. There are certain trends that we must learn to recognize and look for. With practice, even the most subtle trends will stand out with just a glance at the chart.

Professional chess players often drill before a tournament with a book of chess puzzles. The puzzles are known as "combinations." They include tricks, such as knight forks, pins, and skewers. Each page contains pictures of a chessboard and pieces in a partially played game. The student must analyze the position on the board and recognize that if his opponent's pieces were positioned just a little differently, the knight could "fork" or attack the opponent's king and queen simultaneously. His opponent can only move one piece at a time, so he can now capture the other piece. The setup requires luring his opponent into the necessary fork with a sacrifice. With practice, the chess player instantly recognizes potential combinations, many moves ahead. It is just a matter of *pattern recognition.*

Many drilling and tripping trends are flagrantly obvious, but some are extremely subtle and sometimes masked. A masked trend is a trend that is hidden, but can be identified by other trends. For example, look at the chart in Fig 11-2.

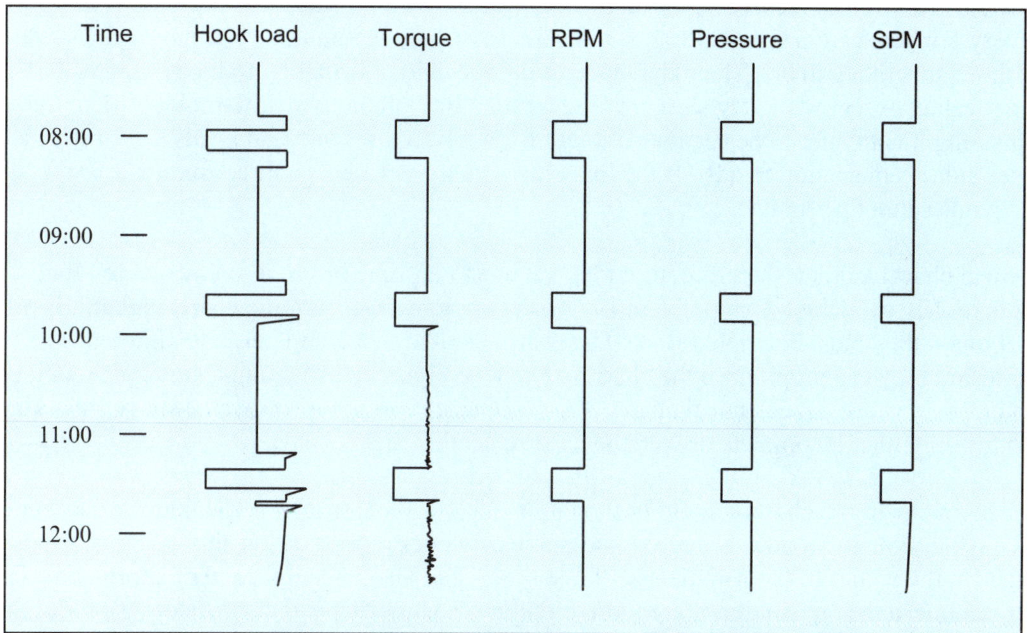

The annulus is beginning to pack off but pump pressure is masked.

Fig 11-2 Annulus packoff on Geolograph chart

Can you see that the annulus is packing off at 12:30? It is difficult to see because the pump pressure is masked by allowing the pump strokes to reduce. When the pump speed is reduced, we expect to see the standpipe pressure decrease. Less mud is being pumped through the jets, and this is where most of the pressure loss is occurring. The pressure should have reduced according to equation 11.1. The pressure that would have been lost by slowing down the pumps was replaced by packoff pressure in the annulus. In this case, the driller kept pressure constant by adjusting pump strokes.

Note also that as annular pressure increases, it causes a pistoning effect on the drill string. The hook load decreases as annular pressure increases. There are two trends indicating an increase in annular pressure, even though standpipe pressure has remained constant.

Mechanical vs. Computerized Charts

Twelve and twenty-four hour mechanical chart recorders have been around a long time. Over the past 25 years, electronic recorders have become more popular. The current trend is to replace the old mechanical charts with the new computerized charts. There are advantages and disadvantages to each type. Lets review them.

A mechanical chart rolls on a drum that is wound up like a clock. The chart is synchronized to this clock. Pressure signals are sent to the chart via diaphragms and coils and plotted real time. Electronic charts are not really real time. Pressure signals are converted to electrical signals that are then sent to a computer that sends a message to a printer or plotter to plot the information.

The main difference between the two is that the computer can only receive so many bits of information per second. The electronic chart recorder plots an average value, not an instantaneous value. The values may be averaged every few tenths of a second, or in some cases, every three minutes. The newer the system, the more "real time" it is likely to be. The older systems do not send information to the computer fast enough to be of any real value, except with very slow moving trends. Pit volume is a slow-moving trend that doesn't need a high strength computer to be accurate enough for our needs. Hook load needs to be refreshed fast enough to see subtle connection trends. If the line on the chart represents a three-minute average, then connection trends cannot be spotted.

The mechanical chart recorder, therefore, provides the most real time information available. It also produces a chart that is readily available. One problem I often encounter with electronic charts is that it is difficult to get a hold of one—they must be printed first. The chart is stored in a computer, and although everyone says it is available, I find they cannot produce the chart for me when I am investigating a stuck pipe incident. It can also be difficult to retrieve a mechanical chart. Many of these chart recorders are not in full working order, and their charts are either incomplete or were never produced.

The advantage an electronic chart recorder has over a mechanical chart recorder is that the electronic chart is *scaleable*. It is possible to broaden or narrow the range of values plotted. This allows one to focus in on very small changes in the values of the parameters. Time scales can also be lengthened or shortened. This makes the slope of the line stand out, so trends are more obvious. The mechanical chart recorder produces only one scale. Trends can be very small and hard to see. However, we do get a full tour or day on a single chart, every time.

The best advantage of electronic datais that it can be sent over the Internet to operation support centers. This allows experts far from the rig location to analyze the trends as they are occurring. Data can also be stored electronically, retrieved, and formatted to any scale desired, and then printed for analysis or projected for presentations.

In the 50s through the 70s, drillers monitored drilling trends themselves. In the 80's, mud logging crews began to set up more sophisticated equipment for monitoring trends and began taking this task away from the drillers. Fewer drillers remained devoted to trend analysis, and fewer still were learning the art of trend analysis.

Today, we enjoy the most sophisticated information gathering and data management systems the industry has ever seen. Unfortunately, too few are taking advantage of it. Fewer and fewer rig personnel are monitoring drilling trends. As I conduct on the rig training sessions in trend recognition, I find few drillers, and even fewer mud loggers, who can spot even the most flagrant stuck pipe trends!

The old Geolograph systems are disappearing. If they remain on the rig, they are most likely not in complete working order. Many rig crews go through the motions of sticking the charts on the drums, but don't write any pertinent data on them, such as the date, well name, depth, etc. Mud loggers are evaluated by how well they keep their equipment running, and by the reports they turn in, so they do not pursue trend analysis. All this equipment is worthless if it isn't used. The industry, in general, does a poor job of using this equipment to its potential.

Operation Support Centers

Electronic data communications have advanced to point that it is relatively easy to transmit real time trend data to a central location at company headquarters. This has lead to the development of "Operation Support Centers" where data is monitored and analyzed. The goal is to provide additional support to the drill crews. The support center analysts can become experts at pattern recognition and alert the drill crews to trends they might otherwise have missed. Eventually, these centers will develop algorithms to recognize trends automatically so we can set alarms.

It will become important for drillers to standardize their connection and tripping practices, and to become as consistent as possible. It is far easier to spot trends when the drillers actions are consistent, and when all drillers behave similarly. For example, when making a connection the driller should always start and stop his pumps at the same rate. He should pick up off bottom and out of the slips at the same speed so acceleration spikes don't confuse hole cleaning trends. The trend watcher is often looking for deviations from an established pattern. If the drillers actions aren't consistent it's much harder to see down hole influences.

Trend Analysis and Pattern Recognition

Trend analysis and pattern recognition are a key to stuck pipe prevention. The drilling crews should practice and "drill" with trend recognition charts until the common patterns associated with stuck pipe are easily recognized. It seems difficult at first, but the significant trends tend to follow common patterns. With time and practice, these patterns present themselves effortlessly. The packoff at 12:30 depicted in Fig 11-2 would literally jump out to anyone who has had any practice analyzing trend charts.

We often hear of how the driller has to "listen to the hole talking to him." It is with trend analysis that we can hear what the well is saying.

Most of the stuck pipe cases my colleagues and I have investigated had obvious trends warning of the impending danger—hours before the pipe became stuck. The drilling crews would argue that they were completely surprised by the incident and had no warning. However, when we showed them how to read a trend chart, they usually agreed that they should have seen it coming.

Chapter 12
Problems Associated with Stuck Pipe

Introduction

One problem often leads to another problem. Just as lost circulation can lead to well bore collapse or a blowout, tight hole and stuck pipe bring new problems to deal with. The biggest problems that commonly occur as a result of tight hole and stuck pipe are:

- Induced kicks
- Lost circulation
- Drill string and equipment failure
- Personal injury

Well Control Issues

Many blowouts begin with stuck pipe and tight hole. This is partly because we become so engrossed with attempting to get free or to avoid getting stuck that we miss the warning signs of an influx. It is also because the mechanics of the tight hole problem can also cause an induced kick.

If we are fighting a potential pack off due to swelling clays or cutting beds, we will probably find ourselves working tight hole. The collars and stabilizers may be balled up and acting as a piston or plunger. This causes severe surging and swabbing.

The filter cake on the well bore wall can act as a one-way check valve, allowing us to swab an influx into the well, but preventing it from being pumped back into formation when the pipe is surged downward. (Fig. 12-1)

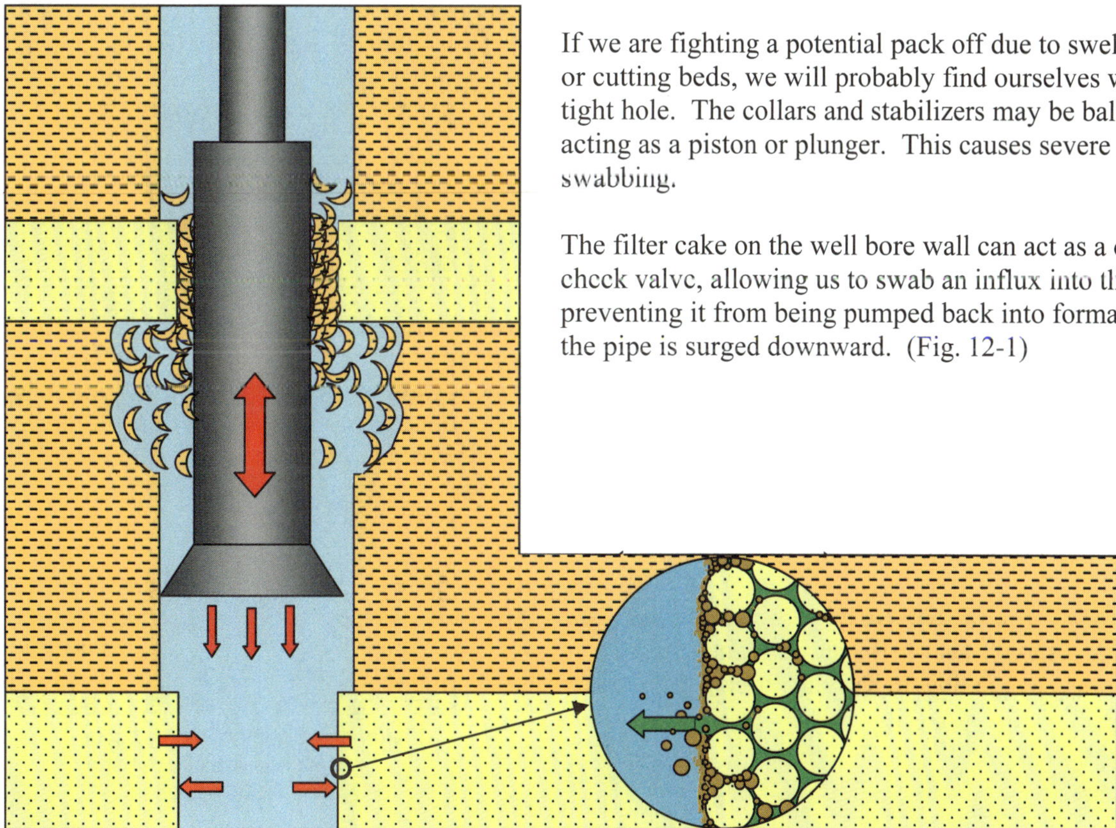

Filter cake can act like a check valve while surging and swabbing tight hole.

Fig 12-1 Swabbing tight hole

When working tight hole, we must monitor flow and pit volume closely. We do not want to take a kick that forces us to shut in and attempt a well control method with a partial or complete packoff! If circulation is lost due to surging, we must be certain to keep the hole full of mud.

We are especially vulnerable while tripping without a top drive. If an influx has been swabbed in below the bit and the annulus is packed off, the path of least resistance is through the drill pipe. As Murphy's Law would have it, when the well begins flowing, it often flows through the drill pipe when the tool joint is 10 feet above the floor. This makes it extremely difficult to stab and make up the full opening safety valve. We may want to consider making up a safety valve prior to working the string.

If an influx is trapped beneath a packoff, it may have an impact as it migrates up to the point of the packoff. *If the influx is oil*, and it migrates to packoff, it may lubricate the debris in the packoff, helping to free the pipe. *If the influx is gas*, it will bring higher pressures with it as it migrates to the packoff. This may piston the drill string into the packoff, making it more severe. The packoff may also suddenly break free, allowing the gas to rapidly expand and quickly blow a lot of mud out of the well.

Differential Sticking and Well Control

Many rigs have been lost by going under-balanced to get free from differential sticking. Going under-balanced is a secondary freeing procedure that should be used only if necessary, and only when it is safe to do so. Generally, we don't need to go completely under-balanced to get free. If necessary, we might try reducing the overbalance, but we don't want to go under-balanced if a hydrocarbon-bearing reservoir is open.

One common mistake is to rotate the drill string while circulating out a kick to prevent differential sticking during a well kill procedure. In fact, many well control instructors instruct their students to always rotate the pipe while circulating out a kick!? This is absurd!

If we do decide to move the pipe, which is strongly discouraged, *it should be reciprocated, not rotated!* If the pipe is rotated, the heat built up by friction will have nowhere to go and the elastomer seal will be quickly lost.

In most cases, the pipe won't become differentially stuck while circulating out a kick, because we were under balanced when we took the kick.

If there is low-pressure sand above the reservoir, we may be under-balanced at the reservoir but overbalanced against the higher formation. It may be possible to become differentially stuck in this case. (Fig 12-2)

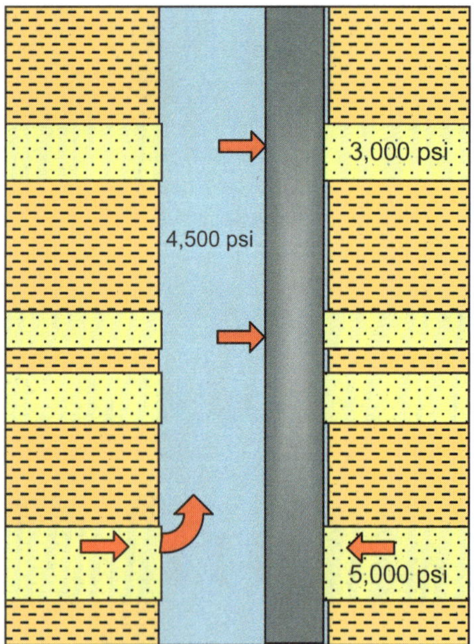

Differential sticking can only occur if an overbalance exists against a permeable formation.

Fig 12-2 Differential sticking and well control

Before risking reciprocating the drill string while circulating out a kick, we should carefully assess the needs and consequences of doing so. If the chance and/or cost of getting stuck is high, and the possibility of getting free is low, then we may feel compelled to move the pipe. But, we may decide to do so only if the risk associated with losing the rams is low. If the reservoir has low pressure and low permeability, and there is plenty of backup BOP, (and an escape route if necessary), then the risk from losing a pipe ram is relatively low. If high casing pressures are anticipated, or there is high permeability or H_2S, the risk may be too great.

As you evaluate any situation, keep Murphy's Law in mind. "Whatever can go wrong, will go wrong." The back up rams may fail. Replacement rams may be the wrong size or type, or simply fail to make a seal. The moving pipe may dislodge cavings that cause a pack off. It is generally more prudent to deal with the well control issue first, and then worry about freeing stuck pipe.

Barite Sag

A final point to consider is **barite sag**. When a weighted mud sits for a long time, especially in an inclined well bore, the barite can settle out. This reduces the mud weight and can cause us to go under-balanced. Even if the well was static at the time we got stuck, slowly *migrating gas bubbles* and *settling barite* could cause the well to go under-balanced.

The driller's primary responsibility is well control. He must never lower his guard, no matter how intense the battle with stuck pipe is raging.

Lost Circulation

The surging and swabbing pressures that occur in tight hole due to packoffs can lead to lost circulation. Lost circulation can lead to kicks and well bore instability.

The balled collars and bit that caused swabbing in Fig 12-1 will also cause surging as the pipe is reciprocated. If we are not pumping and have a check valve in the bit, we will not be able to determine how much pressure is caused by lowering the drill string through the packoff. What appears as down drag on the weight indicator might actually be a pistoning effect. The increase in pressure due to surging may try to pump the drill string back out of the well. If this pressure is higher than the fracture pressure, the formation will break down and we will lose whole mud into the formation.

The equations that engineers use to calculate surge and swab pressures don't really apply in a packoff because the flow through the packoff is likely to be turbulent flow, not laminar flow. We do not know the cross sectional area of the annulus around the packoff, so we cannot predict the fluid velocity.

If we are pumping as we pick up into a packoff, not all of the mud leaving the bit will be able to get past the packoff. If the formations are permeable, we may not see much, if any, increase in pressure. The mud may just move into the formation. *This is one way to spot a packoff.*

If the mud return drops when the pipe is at one elevation, but returns when the pipe is moved to a different elevation, there is a packoff somewhere in the string. If the formations that are open below the packoff are not permeable, pressure is likely to increase rapidly until the fracture pressure is reached.

We must be careful when reciprocating the string, especially when bottom-hole pressure cannot be read. We must also be extra careful when moving the string through a tight spot while pumping. If not, we will be fighting tight hole *and* lost circulation. Well control problems may turn up as well.

Drill String and Equipment Failure

When the string becomes stuck, it might be pulled or torqued it to its theoretical limit, and beyond. The string may fail immediately when subjected to this stress, or it may weaken to the point that it fails from fatigue in the near future.

The drill crew needs to know the current operating limits of each member in the drill string. The maximum tensile and torsion strengths are reduced as the pipe diameter is worn. If the pipe is pulled through a sharp dogleg, the bending stress must also be accounted for. Pipe that is tensioned across a dogleg will fail with a lower tension load than one that is hanging straight.

Some drill crews are not aware that *tensile and torsion stresses are additive.* If the drill pipe is pulled to its tensile limit, there is no room for torsion. Similarly, if the pipe is torqued to its limit, there is no room for tension. The combined loading formula in appendix B, or the combined tension torsion loading charts in the Standard DS-1[tm] manual depict the safe operating limits for most drill pipe.

The drill string isn't the only equipment that may be worked to the limit when we are stuck. The top drive and rotary tables may be worked hard as well. Pressure surges can cause Kelly hoses and fittings to burst.

Jarring fatigues the drill string, the top drive, (if in use), and the mast and sub-structure. It is prudent to inspect all of this equipment after jarring for any length of time. Remember, the equipment may not fail at the time it is severely loaded, but it will have a shorter fatigue life.

Personal Injury

When the equipment is operated near its operating limit, things can break and release stored energy. The pipe may part, a drill line could part, and so on. It is easy for someone to get hit with flying debris, or to get caught between a rock and a hard place. With good operating practices, no one should be in a position to get hurt if any equipment should fail.

One common type of accident occurs while rotating with slips. The drill pipe slips are not designed to transmit torque to the drill string. Yet when we are differentially stuck, or have the Kelly set back, we often use slips to rotate the drill string. With a high set down load on the slips, there may be enough friction between the bowl and the slips, and between the slip dies and the pipe. But if the load is decreased, the friction will decrease, and the surfaces may slip. The trapped torsional energy of the string will cause the slips to rotate backward violently, and possibly be thrown out of the table. If a man is hit with the spinning handles or with the falling slips, he may be very seriously injured or killed. Rotating with the slips is considered bad practice by every authority to have ever discussed it.

We should not use the slips to rotate the string. However, desperate people sometimes do desperate things. A driller should always hold a short safety meeting with his crew to inform them of potential dangers and keep them away from the rig floor when the loads are high.

Chapter 13
Tripping Practices

Introduction

A high percentage of stuck pipe and well control incidents occur while tripping. Most studies on well control suggest that most kicks occur while tripping. It is also safe to say that at least half of all stuck pipe incidents occur while tripping.

Poor tripping practices also damage formations, lose circulation, cause drill string and bit failures, and injure the crew. For these reasons, the drilling crews and rig site management must stay focused on all aspects of the trip.

This chapter begins with the importance of planning of the trip, then moves into preparations to make before the trip. A discussion of well control is included because most kicks and blowouts occur while tripping. The importance of using trip sheets and trip tanks while pulling out, *and running in*, the hole is addressed. The correct procedures for conducting trip drills and pumping weighted slugs is also addressed.

A discussion about pumping out of the well is included to warn the reader of common pit falls often encountered. Other topics include planning wiper trips, wellbore stability issues and differential sticking.

The trip is one of the most important parts of drilling the well. Not only is a significant percentage of all down hole problems caused by poor tripping practices, but this is where the drill crew can best asses the condition of the well through tripping trends.

Planning the Trip

The trip begins in the planning stage. Trips can be detrimental to the well. Some shale formations are severely damaged by trips. The surging and swabbing cause violent fluctuations in the radial and hoop stresses around the well bore. Temperature fluctuations when the well is static can also cause dramatic changes in the well bore stresses. The side loads imparted by the drill string and drill collars are also higher when the bit is off bottom.

Unnecessary trips waste valuable time and extend open hole exposure time. For this reason, we do not trip pipe indiscriminately. We also cannot postpone wiper or bit trips if they are required. Stuck pipe or lost cones may be the result.

We cannot trip the pipe until we are ready. The hole and mud must be conditioned, a trip sheet prepared, and the necessary well control and pipe handling equipment must be in good working order. A heavy slug should also be planned. The size and weight of the slug should be calculated so we know exactly how far the mud will fall in the drill pipe. (Fig 13-5)

Circulating times, maximum overpull, tripping speed, and anticipated or potential problem areas should be communicated in writing to the driller.

Ton-miles should be calculated for the trip to determine if, and when, a slip and cut is required.

The next bit and BHA or casing program should be determined and made ready prior to pulling out of the hole.

Preparations for the Trip

Once a trip is planned, the first step is to **condition the hole and the mud** during the **clean up cycle**. Cuttings and cutting beds must be completely removed before pulling the pipe. The plastic viscosity should be made as low as possible to minimize surge and swab pressures, and to reduce the thixotropic tendencies of the mud.

Insufficient hole cleaning prior to a trip is responsible for a large percentage of stuck pipe. Even if we don't suffer stuck pipe due to pack off, we will probably end up fighting tight hole somewhere on the way out. The time saved by circulating less may be lost several times over fighting our way past tight spots. These tight spots also lead to severe surging and swabbing, which leads to well bore instability, lost circulation, and induced kicks.

We must circulate at least bottoms up to ensure there is no gas in the well. Migrating gas will expand on its way up, causing a reduction in bottom-hole pressure. The clean up cycle should not end until the pressure, torque, and drag trends have stabilized. No matter how low the values have dropped, the clean up cycle is not over if the values are still dropping.

Torque, drag, and pressure **trends should be carefully monitored** while conditioning the hole. The shaker trends, gas shows, and mud property trends should also be monitored and recorded. Just prior to pulling out of the hole, several baseline measurements need to be documented in case problems are encountered later on:

- Up and down drag – with and without rotation at expected trip speeds
- Off bottom torque
- Free rotating weight
- Full flow pump pressure off bottom and total strokes per minute
- Slow pump rate pressures
- Mud weight in and out
- Mud temperature in and out
- Flow rate out

We use the up and down drag as a baseline with which to monitor overpull. The full flow pump pressure is a baseline to monitor packoff, in case we need to circulate on the way out.

The best time to take slow pump rates is just before taking a kick. Because most kicks are taken while tripping out, it makes good sense to take slow pump rates just before starting out of the well. The slow pump rates are taken on bottom because we will strip back to bottom to circulate the kick out.

A trip sheet must be prepared while circulating the well. There should be a trip sheet for tripping out and for tripping in. The mud tanks should be properly isolated so pit gain can accurately monitored. A trip tank is necessary for accurate pit gain measurements. The practice of counting pump strokes is reckless, and it can't be used on the way in.

A working mud bucket is required in order to manage and account for all the mud. If some mud escapes the trip tank, then we aren't practicing good well control methods. The mud bucket is not there to keep the rig floor clean—it is there to contain the mud, so that hole fill can be accurately monitored!

I also recommend preparing a kill sheet for a 10 bbl kick. A kick occurring during a trip will be swabbed in, so the kill weight mud is already in the well. By preparing a kill sheet for the depth at which the trip starts, we will have already calculated strokes to bit and bottoms up strokes. We can do this leisurely while circulating the hole clean, so if a kick does occur, we will have already prepared for it, and won't have to prepare one under pressure. (We will want to check our work, of course.)

A heavy slug is normally pumped after five stands are pulled, and the well takes the correct amount of mud to replace pipe displacement. We must calculate how far the slug will fall in the drill string, so we know how much mud to get back. This must be factored into the trip sheet. (Equation 13.1 and Fig 13-5)

If possible, the trip should be planned to avoid a crew change, or a need to relieve the driller for lunch or dinner. Most stuck pipe incidents occur during such crew changes. Poor handovers are largely responsible for these incidents. If a driller is going to be relieved for dinner, or by the next crew, thorough handovers are required. The toolpusher and foreman should be on the floor long before, and after, the crew change.

A lithographic chart with drilling trends should be printed and made available on the rig floor. A paper model of the bottom-hole assembly should be prepared to the same scale as the trend chart. The driller and drilling supervisors can then anticipate and monitor problem areas by dragging the model along the chart as the bit is pulled. (Fig 13-1)

Most packoffs and well bore geometry problems occur around the bottom-hole assembly. Tight spots are recorded at the bit, but it isn't always the bit that causes the tight spot. It could be a stabilizer, any other change in diameter, or just a stiff collar pressed against a dogleg. If there is one troublesome ledge, it is likely to show up as four tight spots if we have a BHA with three stabilizers. By holding the bit of the model at the appropriate depth on the trend chart, we can anticipate when the collars and stabilizers enter a suspected trouble spot.

The model can also help determine what is causing the problem. If the problem is related to well bore geometry, it will show up quite clearly as the stabilizers pass through it. If the problem is due to cutting beds, the tight spot will not be fixed—it will move as the cutting bed is moved. This will also be quite apparent with the use of the model.

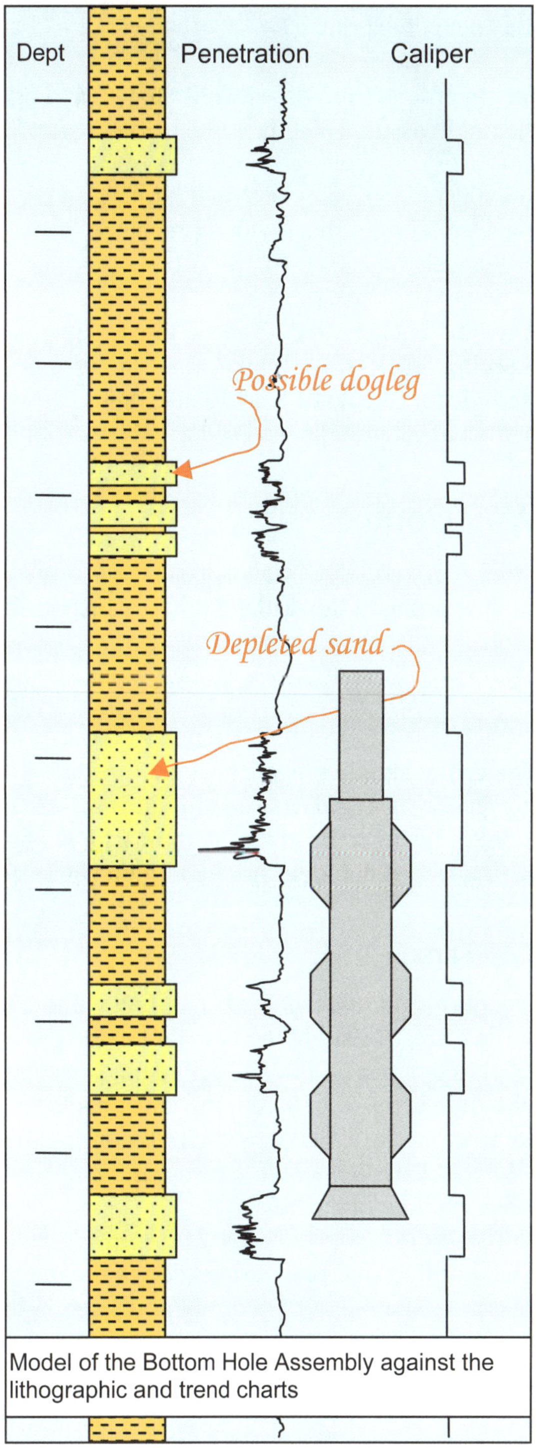

Fig 13-1 BHA model for trips

The charts and model take some time to prepare so we need to notify the mud loggers of the trip early and check up on them to ensure they will have the chart prepared in time.

Ideally, the chart will show hole diameter and formation type against depth. If caliper logs are not available, we can use offset well information and/or rate of penetration to estimate well bore diameter. Usually, the well bore breaks away in the softer formations that drill faster. Shale tends to grow in diameter, whereas sands may actually shrink.

If caliper logs from offset wells show that a particular formation holds its gauge while another does not, then we might assume that the same is true in our well.

We must identify any potential hazards on the chart and any tight spots noticed as we come out. Careful and consistent use of this tool has prevented many stuck pipe incidents and helped crews diagnose the condition of the well.

The mud bucket, pipe spinners, tongs, slips, full opening safety valve, and inside BOP must be in good working order. If we think we need to change the tong or slip dies, it should be done while circulating the well clean. We don't want to interrupt the trip to repair equipment, open another bucket of pipe dope, pump hydraulic oil into the load cells, and so on.

Well Control

More than half of all kicks occur while tripping pipe. Many result in blowouts, lost rigs, or lost wells. Well control is one of the driller's primary responsibilities, and he must pay particularly close attention to well control while tripping the pipe.

Trip Sheets

The driller should prepare his own trip sheet while circulating the well. With today's modern tools, mud loggers or rig engineers can provide him with one to work with. However, the driller *must* prepare and monitor his own trip sheet. **A man who is not mathematically strong enough to prepare, and keep up with a trip sheet, does not belong on the brake!**

Trip Tanks

The rig must use a trip tank! Drilling without a trip tank is reckless! A trip tank is used because it has a small surface area. A slight increase in volume produces a noticeable change in fluid level in a trip tank. The fluid level in a typical, 2,000 bbl active mud system will probably increase only ⅝ inch for each 10 bbls of influx. The fluid level in a 100 bbl trip tank should increase about 12 ½" for the same 10 bbl influx. Some contractors argue that they can monitor hole fill with pump strokes. This is not true. *A trip tank must be used when running in the hole, as well as out of the well.* Pump strokes have nothing to do with pipe displacement when running in the hole.

Many drillers and tool pushers do not see the need for running in on a trip tank. This behavior demonstrates a basic weakness in well control knowledge. Many blowouts are the result of the lengthening and artificial migration of gas as an influx is displaced up and around the drill collars. (Fig 13-2)

Artificial Migration

The lengthening of the influx around the collars reduces bottom-hole pressure in vertical wells. The pressure can be further reduced as the gas is displaced up the well where it can expand. The problem of "artificial migration" becomes even more important in directional wells. A large influx will not reduce bottom-hole pressure while it is in the horizontal section, because the height of the influx is very small. (Fig 13-2) In the example in fig 13-2 the height of the influx is only 12¼" regardless of how long the influx is. However, as the influx is displaced into the vertical section, its length is converted to height and bottom-hole pressure is reduced. As the pipe continues farther into the hole, the influx is displaced up the well. This is often referred to as artificial migration.

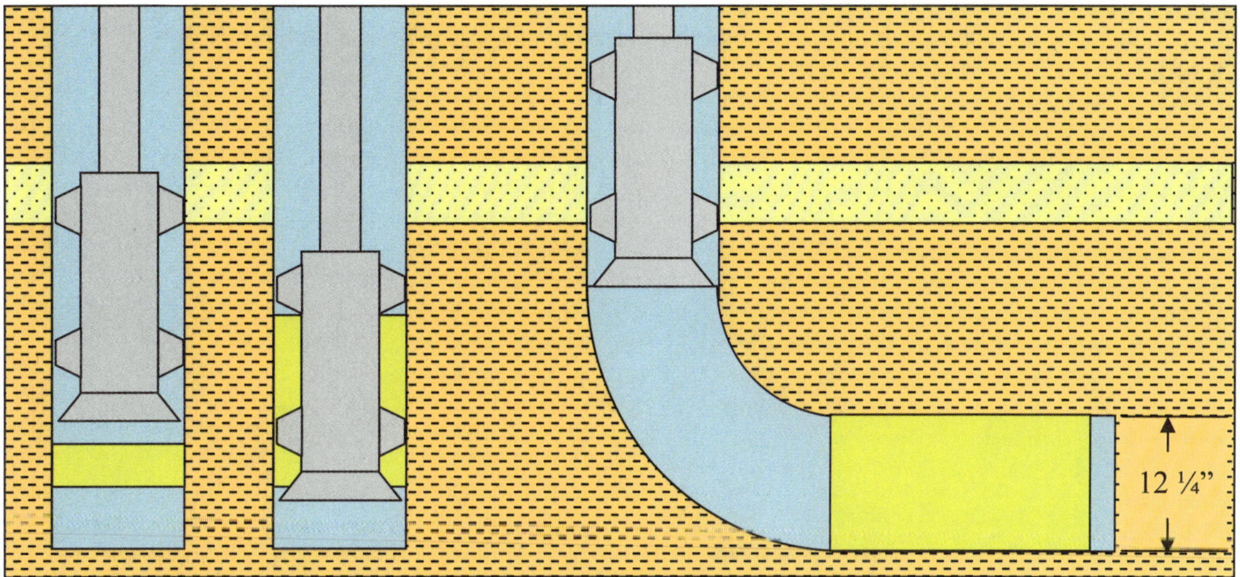

When the collars penetrate an influx, its length increases. In a vertical well, this increases influx height and reduces bottom hole pressure. In a horizontal well, the influx height does not increase until the influx is in the vertical section of the well.

Fig 13-2 Artificial migration

Mud Management

Trip sheets are prepared in order to manage the displacement of mud into, and out of, the well. This means that all of the mud that enters or leaves the well must be accounted for, even on wet trips. This implies that *the mud bucket is a well control tool.* Good well control is not being practiced if we trip wet with a mud bucket that leaks! The mud that is captured by the mud bucket must be directed to the trip tank.

It is not good practice to transfer mud while tripping, unless we can accurately monitor the total mud volume in all pits and tanks involved in the transfer.

BOPE

A working, full opening **safety valve** and **inside BOP** are required on the rig floor at all times. We must be able to stop a flow through the drill string *and* be able to strip back to bottom to circulate the influx out of the well. This means that both the full opening safety valve and inside BOP must be small enough to enter the smallest hole section. If we have a combination string of 5" and 3½" drill pipe, a mere crossover sub will only be sufficient if the 5" inside BOP and safety valve can be stripped into the smaller liner or open hole.

While tripping, we must always be able to stab and make up the full opening safety valve. If we are fighting tight hole, we do not want to get the pipe stuck and take a kick when the tool joint is 15 feet above the rig floor. It may be prudent make up the full opening safety valve and pull singles until past the tight spot. This is especially true if we are swabbing in the tight hole section and or if there is a high probability of H_2S gas. (see fig 12-1)

Trip Drills

A trip drill should be conducted on every trip. A drill is performed to train the crew to *react to a crisis* in a predetermined manner. The best possible plan of action is thought out prior to the crisis, and practiced until our actions become reactions that require no thought or analysis.

We must inspect what we expect. If we expect our drill crew to react a certain way to a crisis, then we must perform drills to ensure they actually perform that way. Well control drills must be conducted in earnest. We cannot simply pretend to have a drill and then report it in the morning report as an actual drill.

The Drilling Foreman and Tool Pusher must inspect the drillers' response to pit gains and flow in the flow and pit drills. Many supervisors allow the driller to initiate these drills. This is inadequate. *We must inspect what we expect.* If we expect the driller and crew to monitor pit gain and flow, we must pull surprise drills to inspect the drillers' and mud loggers' responses.

Surging and swabbing

One of the biggest concerns while tripping, is swabbing in a kick. This can occur from pulling the pipe too fast or from the syringe effect when the bit and BHA are pulled through a tight spot. Swabbing becomes more likely as the pipe is pulled faster, with small well bores, and/or with big pipe. Usually, swabbing can be recognized and monitored with a trip tank. However, it is possible to swab an influx in from one formation, and inject mud into another, while the pipe is surged downward. This will mask the total amount of influx swabbed in while working tight hole. *We cannot safely assume that a lack of pit gain ensures no influx while working tight hole!* It is more prudent to trip back to bottom and circulate bottoms up after a short trip through tight hole.

Surging causes lost circulation. With lost circulation, we can lose our hydrostatic overbalance and take a kick. This is another reason for tripping in on a trip tank. If we are tripping too fast, it can be spotted on the trip tank with insufficient displacement.

Surging can be disguised as down drag on the weight indicator as we pass through tight hole. The plunger and piston effect causes a reduction in hook load. This piston effect disappears as the pressure beneath the bit reduces when downward pipe motion is slowed or stopped. (Fig 13-3)

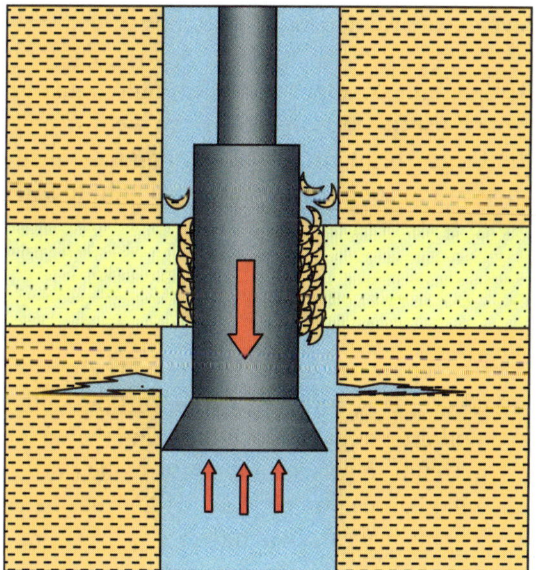

The pressure beneath the bit caused by surging will reduce the hook load. On the weight indicator, this appears as down drag.

Fig 13-3 Surging disguised as down drag

Backreaming

Pumping out is a good way to avoid the trouble with swabbing but it must never be done without rotation. It is too easy to pump ourselves into a packoff while backreaming, and its even more likely if we are pumping without rotation!

Packoffs are most likely going to occur while pulling the pipe upward. The risk is highest while pumping out of the hole through inter-bedded shale and sandstone sequences. (Fig 13-4) The drill collars tend to pack off where the change in diameter in the drill string first contacts the change in diameter of the well bore. Cavings tend to hold up in the enlarged sections of the well bore and are dragged into the restricted diameters with the drill collars, causing a packoff.

As the well begins to pack off, the pump pressure increases and the drill string is "pumped" into the packoff. *This masks the overpull trend.* If a driller is watching only his weight indicator, the overpull will not increase substantially and may even decrease due to the pistoning effect.

If tight hole becomes severe enough to justify back reaming the pipe must first be run back in the well several stands below the first sign of trouble. Back reaming must then proceed very slowly and should not be stopped until the BHA is above the build section in cased hole.

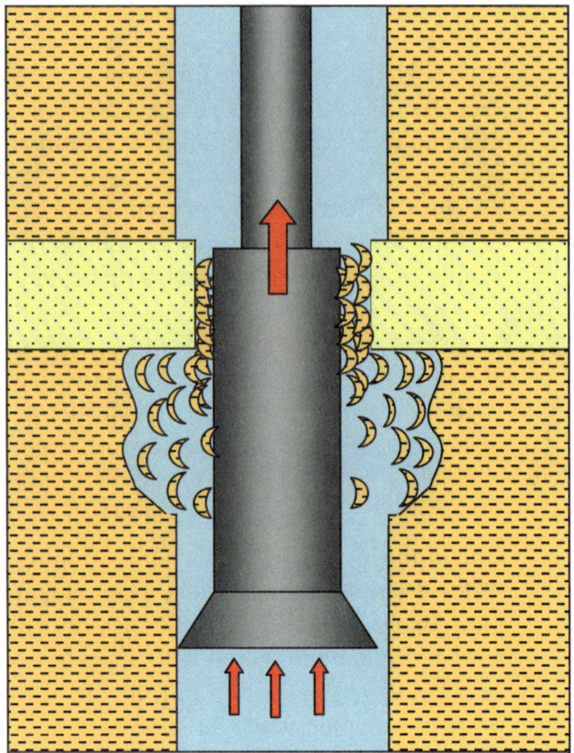

Packoffs tend to occur where the BHA is pulled into full gauge sections of the hole just above hole enlargements. If the string is being pumped out a pistoning effect will mask the overpull trend.

Fig 13-4 Pumping the BHA into packoffs

One exception to the above rule is when back reaming is initiated immediately upon reaching total depth to avoid dropping angle. If we are drilling with a mud motor in friable formations there is a strong possibility we will drop angle while circulating the well before tripping begins. The slow reciprocation with a mud motor tends to carve out the lower side of the wellbore and cause angle to drop. In this case very slow back reaming begins immediately. If the shakers come clean before the BHA reaches any wellbore geometry or tight spots, we can switch from back reaming to conventional tripping.

This is not back reaming through tight hole. It is just circulating the well bore clean while occasionally removing stands. This prevents the bit from dropping angle by wearing in the same spot for several hours. The back reaming will continue only as long as necessary to clean the well. Note that although this is also a good method to avoid swabbing, careful well control vigilance must still be maintained. This is especially true when conventional tripping resumes.

When pumping into the hole, we must monitor pressure more closely than hook load in order to avoid severe surging that could lead to lost circulation and/or a packoff.

It is best to avoid circulating while running into a directional well. There will most likely be cuttings beds comprised of cavings that were produced while pulling out of the hole. These beds will be disturbed as we run in the hole. If we begin circulating we will have to continue circulating until those cuttings are completely removed from the well. A common mistake is to circulate past the cuttings bed and then stop to continue running in. *If cuttings are circulated above the 65° angle they can slide back down the well and packoff if circulation is stopped prematurely.*

Circulating off bottom is never a good idea because the high jet velocity can wash out that part of the hole. If it becomes necessary to circulate off bottom we must keep the pipe moving so the jets are not at the same depth very long.

For these reasons, if circulation is ever started while running in, we may be committed to continue circulating to total depth. *If circulation is started we must circulate with a sufficient flow rate to clean the hole*, usually the same flow rate we drilled the well with.

Barite Slugs

If we are pumping out, we won't pump a barite slug until after we have reached a casing shoe and circulated the well clean.

Barite slugs should not be pumped until we know the well is taking the correct amount of fluid. This means that the first five stands will be pulled wet and slow, while carefully monitoring hole fill. When it is safe to pump the slug, the premixed slug is pumped into the string and chased with either clear water or normally weighted mud.

We must work the string until the heavy slug has settled and the fluid levels have equalized inside the drill string and annulus. We should have already calculated how far the fluid level will drop inside the drill string, and thus how much mud should come back to the trip tank. (Fig 13-5)

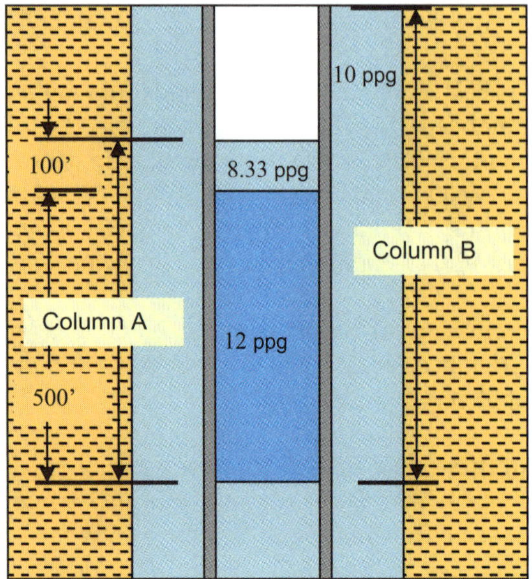

The volume of mud displaced by a weighted slug is equal to the void space inside the pipe.

Fig 13-5 Weighted slug

The hydrostatic column of fluid inside the drill string must equal the hydrostatic column in the annulus. If we pump a known quantity of weighted mud and of chase fluid we will know the total hydrostatic head at the bottom of the weighted slug in the pipe. (Column A of figure 13-5.) The hydrostatic head due to the height of fluid in the annulus, at the same depth, must be equal to that inside the pipe. (Column B of Figure 13-5)

The hydrostatic pressure of Column A = the hydrostatic pressure of Column B eq. 13.1

Example 17.1

Assume an original mud weight of 10 ppg. A volume of 12 ppg mud is calculated to give 500 feet of height inside the drill pipe. This slug is chased with 100 feet of 8.33 ppg water. The hydrostatic head inside the pipe must equal the hydrostatic head outside the pipe.

(12 ppg mud column) + (8.33 ppg mud column) = (10 ppg column outside pipe)
(500 ft)(12 ppg)(.052) + (100 ft)(8.33 ppg)(.052) = (X ft)(10 ppg)(.052)

[(500 ft)(12 ppg) + (100 ft)(8.33 ppg)] = (X ft)(10 ppg)

6833 = 10X

683 = X

The height of column B = 683 ft
The void space inside the pipe equals 683 ft – 600 ft = 83 ft

The volume of mud required to fill 83 ft of pipe is the volume of excess mud we expect to see come back to the trip tank.

Wiper Trips

Wiper trips are often required to condition the well, but unnecessary trips can be detrimental to well bore stability. Wiper trips are required to wipe away thick filter cakes and swelling clays. They are also useful for reaming away ledges and taking the rough edges off of small doglegs. Wiper trips can also tell us a lot about the condition of hole that was recently drilled. For these reasons, rig hands like to make regular wiper trips.

Wiper trips take time and can be detrimental. Therefore, design engineers and drilling optimizers tend to discourage them. A balance must be struck. The hole can dictate when a wiper trip is required, but only to experienced drilling hands who recognize the trends and have regional experience.

Attempting to save time on a well by minimizing wiper trips can result in stuck pipe. Likewise, unnecessary trips can result in stuck pipe.

Well Bore Instability

Well bore stability is compromised while tripping due to the fluctuating stress distribution around the well bore caused by:

- High side loads imparted by the drill pipe across doglegs when the weight on bit goes to zero
- The temperature fluctuations when circulation is stopped
- Surging and swabbing
- The axial loads caused by the drill string dragging along the wall
- Fluid loss in shale due to longer open hole exposure time

Well bore stability can be monitored with pressure and drag trends while tripping. If the well is relatively unstable, it may take longer to make a wiper trip than it took to drill the section. Patience is required. Ideally, as little energy as possible will be imparted to the well bore walls while tripping out. We must also get back to bottom and circulate quickly in order to minimize open whole exposure time and temperature fluctuations.

On deeper wells, it may be necessary to stage our way into the well in order to minimize pressure and temperature surges. The thixotropic qualities of the mud cause it to thicken when it is static. A pressure surge occurs when circulation is started. The pressure drops as the viscosity quickly returns to normal.

The lower portion of the well will heat up when cool mud is no longer circulated across it, and the upper portion of a deep well may cool down when warm mud is no longer circulated past it. The mud in the tanks also cools while circulation is stopped. Then when circulation is started again, the cool mud sweeps across the warmer formations, and hot mud sweeps past the cooler formations at the top. It takes time to get the formations back to normal circulating temperatures. Some well bore caving or lost circulation may occur during these temperature changes. (See Fig 8-106)

Differential Sticking

If differential sticking is a concern, we must pay very close attention to the first few stands stood back. Contrary to popular belief, most differential sticking occurs in the drill pipe high in the hole, not around the collars down lower in the hole.[1] This is largely due to the high side load the drill pipe applies to the well bore wall when the bit is off bottom. It may also be due to high wall contact due to minor keyseating of the pipe against the high side of the hole. (See chapter 9, Figures 9-13 and 9-18.)

If we have doglegs through an under-balanced sand, we should be particularly careful. We must keep the pipe moving as much as possible. We may need to condition the mud with lubricants prior to tripping out.

When tripping back in, we may not notice much trouble with differential sticking as the collars first pass the sand. However, as more and more weight hangs beneath the pipe against the sand, the side load against it increases. We may become stuck just a few stands off bottom, even though the collars are not against a sand formation.

Note also that the static filter cake will be very thick when we are tripping back into the well. It may be prudent to stop and circulate and rotate to condition the filter cake before continuing to trip to bottom.

Circulating After the Trip

The trip begins when drilling has stopped and circulation begins to condition the well and mud. The trip doesn't end until the bit or casing is on bottom and the well has been circulated to condition the well and mud.

Circulation to condition the well after the bit is back on bottom cannot be taken lightly. Trip gas will come to surface sometime before bottoms up strokes. The lag time between maximum trip gas and bottoms up stroke can provide valuable insight into migration speeds and/or the origin of the gas. The gas expands as it nears the surface and can cause the well to go under-balanced. It may be necessary to circulate the last half (or fourth) of the bottoms up on the choke.

As well bore temperatures stabilize, some caving, partial losses and ballooning can occur. We don't want to start drilling or cementing casing until these problems have stabilized.

Bibliography

1) Stewart, Maurice I. Jr., U.S. Minerals Management Service, Metaire, LA: "A Method of Selecting Casing Setting Depths to Prevent Differential-Pressure Pipe Sticking"

Conclusion

The title of this book is Trouble-free Drilling. As the name implies, it is intended to be a guide for drilling trouble-free.

The focus of this book has been primarily on the mechanics of stuck pipe. This is because stuck pipe accounts for the vast majority of down hole problems. There are other problems however, such as drill string failure, lost circulation, slow drilling, directional control, and well control. Recently, we have had to add "deep water" problems and high pressure/high temperature issues as well.

It was my intention to address those issues in this book. However, I feel the need to get the material on stuck pipe to the presses now. I intend to add the other information later, in additional volumes.

One point I want to stress is that **it is the men on the rig who drill the well and are best positioned to monitor and deal with the problems they encounter**. They must be knowledgeable about these problems and properly trained to deal with them if we are to drill trouble-free. I cannot say this often enough, "all the understanding the drilling industry has acquired over the past century is of no use if it does not reach the man on the rig."

This book was written for the drillers, tool pushers, drilling foreman, and engineers who actually drill the wells. Hopefully, this manual will find its way into their hands, where they can use it to increase their understanding of down hole mechanics.

I am currently preparing material for a second edition, and additional volumes of this manual. I welcome comments, challenges, criticism, and questions pertaining to any of the material presented thus far. I also welcome additional source material and comments for existing or future material presented in the second edition.

I can be contacted at my international drilling school and consulting firm at;
www.Drilbert.com.

I will end this book with one final thought;

> Just as in the game of chess, one can improve his game substantially by reading and studying books on the subject. But studying alone does not help a chess player become great—he needs practical, on the board, experience. Then he must carefully analyze this experience with his peers and trained instructors. Without professional training and formal endgame analysis, a chess player never moves beyond the rank of novice. The same is true for drillers and drilling supervisors.

Regards,

John Mitchell
President
Drilbert Engineering Inc.

Appendix A Hole Cleaning Charts (for wells with full pipe rotation)

Procedures for using the hole cleaning charts from IADC/SPE paper 27486 "Simple Charts to Determine Hole Cleaning Requirements"

1. Select one of the three hole sizes:
   - 17 ½"
   - 12 ¼"
   - 8 ½"
2. Enter the appropriate Rheology Factor chart. (the ones on the left) Using Plastic Viscosity and Yield Point values, read off the value of the rheology factor, RF.
3. Get the Angle Factor, AF, from Table 1.

Table 1 Angle Factor for Deviated Holes

Hole Angle (deg)	Angle Factors
25	1.51
30	1.39
35	1.31
40	1.24
45	1.18
50	1.14
55	1.10
60	1.07
65	1.05
70-80	1.02
80-90	1.0

© SPE

4. Calculate the Transport Index, TI, using the following equation:

   TI = RF x AF x MW

   Where: TI = Transport Ratio
   RF = Rheology Factor from the hole cleaning charts
   MW = Mud weight in specific gravity

   Note: Specific gravity refers to the weight of mud relative to the weight of fresh water. To convert mud weight to specific gravity, divide it by 8.33 lbs/bbl, or use the following equation:

   TI = RF x AF x MW ÷ 8.33

   Where: TI = Transport Ratio
   RF = Rheology Factor from the hole cleaning charts
   MW = Mud weight in lbs/gallon

5. Use the calculated Transport Index in the charts on the right to find either the minimum flow rate for a desired penetration rate, or the maximum rate of penetration for a particular flow rate.

Hole Cleaning Charts for 17 ½" Holes

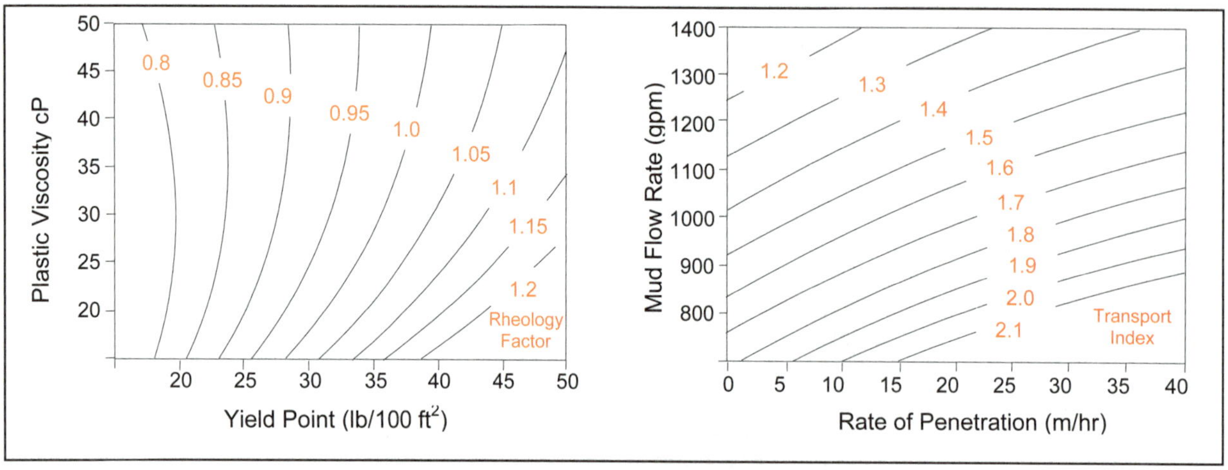

Hole Cleaning Charts for 12 ¼" Holes

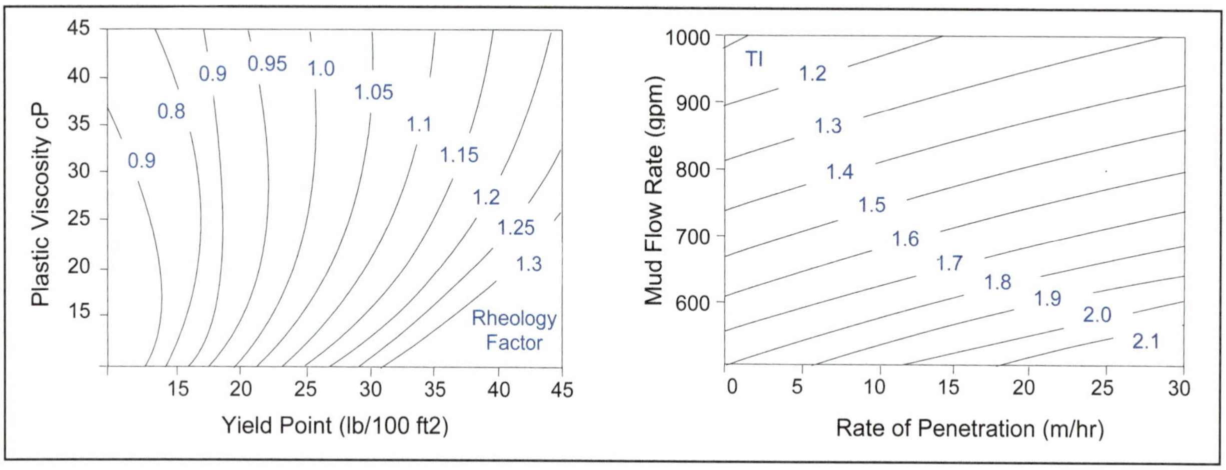

Hole Cleaning Charts for 8 ½" Holes

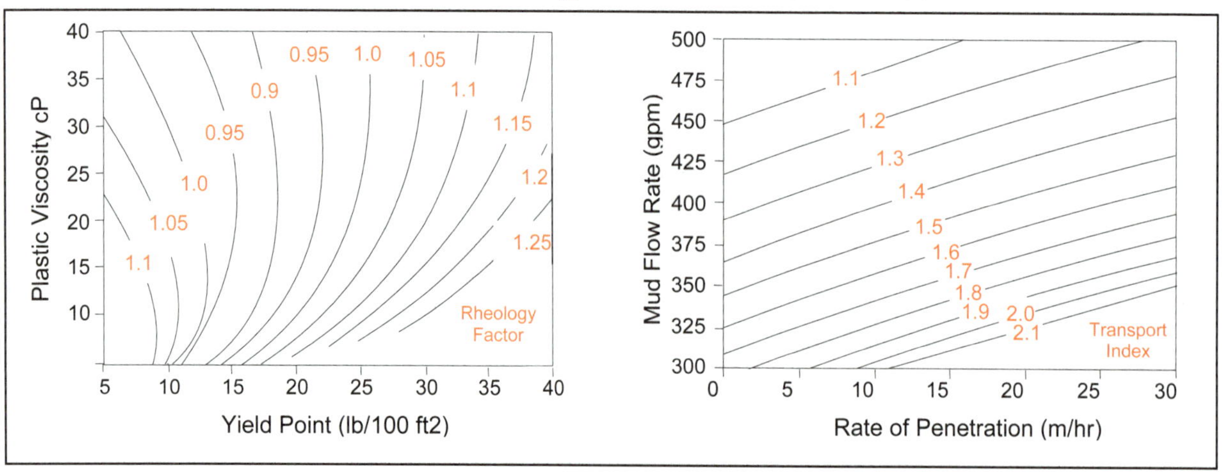

Example:
Assume a horizontal 8 ½" hole with 12 ppg mud, a PV of 25 cP, and a YP of 18 lb/100 ft^2.

Question:
What is the maximum rate of penetration we can achieve with 450 gpm flow rate?

Answer:
From the RF chart, for 8½" hole, we find that the RF = 0.91
From Table 1, we find the Angle Factor, AF = 1.0
The Transport Index, TI, is calculated to be:

$$TI = RF \times AF \times MW \div 8.33$$

$$TI = 0.91 \times 1.0 \times 12 \div 8.33 = 1.31$$

From the ROP chart, for 8½" hole, at a TI of 1.31 and a flow rate of 450 gpm we can have a maximum ROP of about 21 meters per hour.

Question:
What is the minimum flow rate we need to drill at a rate of penetration of 20 meters per hour ?

Answer:
From the ROP chart, for 8 ½" hole, at the TI of 1.31 calculated above, and a ROP of 20 m/hr, we find we need a flow rate of 440 gpm.

Appendix B Equations

Kirsch produced equations for determining the stress distribution around a circular tunnel in 1898. A full discussion on the derivation of these equations is given by Jaeger and Cook.[5] The equations are summarized below.

Kirsch Equations

Stress components at point (r,θ)

Radial $\sigma_r = \frac{1}{2}\sigma_z\{(1 + k)(1 - a^2/r^2) + (1 - k)(1 - 4a^2/r^2 + 3a^4/r^4)\text{Cos}2\theta\}$

Hoop $\sigma_\theta = \frac{1}{2}\sigma_z\{(1 + k)(1 + a^2/r^2) - (1 - k)(1 + 3a^4/r^4)\text{Cos}2\theta\}$

Shear $\tau_{r\theta} = \frac{1}{2}\sigma_z\{-(1 - k)(1 + 2a^2/r^2) - 3a^4/r^4)\text{Sin}2\theta\}$

Principal stresses in plane of paper at point (r,θ)

Maximum $\sigma_1 = \frac{1}{2}(\sigma_r + \sigma_\theta) + (\frac{1}{4}(\sigma_r - \sigma_\theta)^2 + \tau_{r\theta}^2)^{1/2}$

Minimum $\sigma_2 = \frac{1}{2}(\sigma_r + \sigma_\theta) - (\frac{1}{4}(\sigma_r - \sigma_\theta)^2 + \tau_{r\theta}^2)^{1/2}$

Kirsch Equations

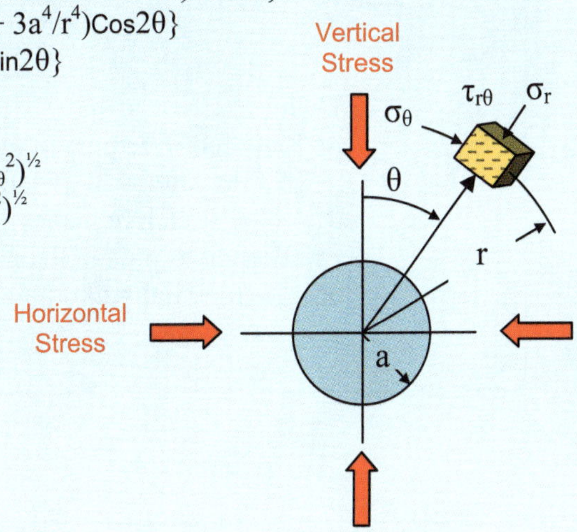

Fig B-1 Kirsch equations for wellbore stress

Note: These equations are for a horizontal tunnel. They work exactly as printed for horizontal wells. For vertical wells, the vertical stress must be substituted for the major horizontal stress, and the horizontal stress must be substituted for the minor horizontal stress.

Pipe Stretch or Free Point Formula (English units)

$$L = \frac{E \cdot \Delta L \cdot w}{40.8 \cdot \Delta P} \qquad \text{eq. B.2}$$

Where:
L = Length of free pipe in feet
E = 29,000,000 lbs/in^2 (Modulus of elasticity)
ΔL = Pipe stretch in inches
w = Weight of pipe in lbs/ft
ΔP = Differential pull on pipe in pounds

Pipe Stretch or Free Point Formula (metric units)

$$L = \frac{127.5 \cdot E \cdot \Delta L \cdot w}{\Delta P} \qquad \text{eq. B.3}$$

Where:
L = Length of free pipe in meters
E = 20390 kg/mm^2
ΔL = Pipe stretch in millimeters
w = Weight of pipe in kg/m
ΔP = Differential pull on pipe in kg

Combined Tension and Torsion Loading (English)

$$T_{max} = \left(\frac{0.096167 \bullet J}{OD} \right) \sqrt{Y_m^{\,2} - \frac{P^2}{A^2}} \qquad \text{eq. B.4}$$

where:

T_{max} = Maximum torque at a given tension P (ft-lbs)
J = Polar moment of inertia = $(D^4 - d^4)/10.19$ (in^4)
D = Outside diameter in inches
D = Inside diameter in inches
Y_m = Minimum yield stress in lbs/in^2
P = Total tension load in pounds
A = Cross sectional area (in^2)

Combined Tension and Torsion Loading (Metric)

$$N_{max} = \left(\frac{100 \bullet L \bullet S}{\pi \bullet D \bullet G} \right) \sqrt{1 - \frac{100\, T^2}{3 \bullet A^2 \bullet S^2}}$$

where:

N_{max} = Maximum number of turns at a given tension T
L = Length of drill pipe (m)
D = Outside diameter of drill pipe (cm)
S = Maximum shear stress: 57.7% of minimum yield stress
G = Modulus of elasticity: 84,000 MPa
T = Total tension load (daN)
A = Cross sectional area (mm^2)

Appendix C Volume of fluid in a horizontal tank

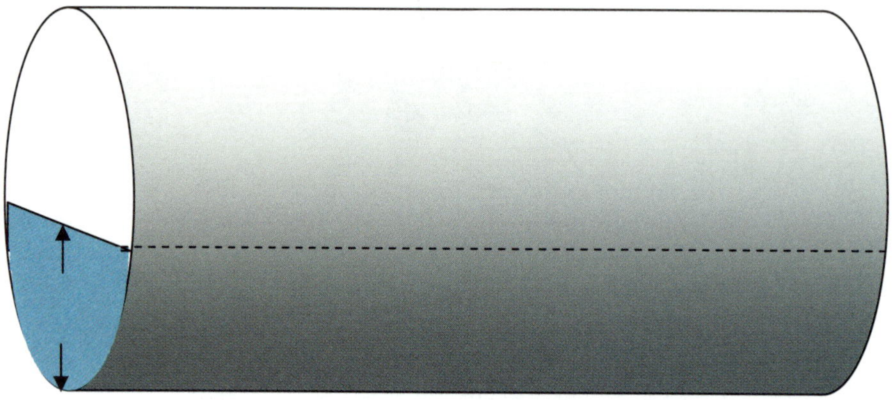

$$\text{Vol. (ft}^3\text{)} = \text{Length}\left[\frac{\pi\, r^2 (180 - 2h \sin\left(\frac{r-h}{r}\right)}{360} - \frac{r^2 \sin(180 - 2h \sin\left(\frac{r-h}{r}\right)}{2}\right]$$

Where:

Vol. = Volume of fluid in the tank in cubic feet

Length = Length of the tank in feet

h = Height of fluid in the tank in feet

r = Radius of the tank in feet

or

$$\text{Vol. (bbls)} = \frac{\text{Length}}{5.6146}\left[\frac{\pi\, r^2 (180 - 2h \sin\left(\frac{r-h}{r}\right)}{360} - \frac{r^2 \sin(180 - 2h \sin\left(\frac{r-h}{r}\right)}{2}\right]$$

Where:

Vol. = Volume of fluid in the tank in barrels

Length = Length of the tank in feet

h = Height of fluid in the tank in feet

r = Radius of the tank in feet

Note: The top equation will give the volume of the tank in cubic meters if length, height, and radius are in meters.
Cubic meters can then be converted to barrels by multiplying the answer by 6.2899 bbls/m^3.

Appendix D Conversion Factors

1 Acre	=	43560 Square feet (ft^2)
1 Acre	=	4046.8 Square meters (m^2)
1 Atmosphere (atm)	=	14.691 Pounds per square inch (lbs/in^2)
1 Atmosphere (atm)	=	29.92 inches of mercury (inHg)
1 Atmosphere (atm)	=	101325 Pascals (Pa)
1 Atmosphere (atm)	=	1.01325 Bar (bar)
1Barrel (bbl)	=	42 Gallons (gal)
1Barrel (bbl)	=	34.9726 Imperial Gallons (gal UK)
1Barrel (bbl)	=	158.984 Liters (l)
1Barrel (bbl)	=	0.158984 cubic meters (m^3)
1Barrel (bbl)	=	5.6146 cubic feet (ft^3)
1 Bar (bar)	=	14.5038 Pounds per square inch (lbs/in^2) or (psi)
1 Bar (bar)	=	100 Kilopascals (kPa)
1 Centimeter (cm)	=	0.3937 Inches (in)
1 Centipoise (cP)	=	0.001 Pascal-seconds (Pa-s)
1 Decanewton (daN)	=	1.019716 Kilograms (kg)
1 Decanewton (daN)	=	2.2482 Pounds (lbs)
1 Fathom (fth)	=	6 Feet (ft)
1 Foot (ft)	=	0.3048 Meters (m)
1 Foot (ft)	=	12 Inches (in)
1 Cubic foot (ft^3)	=	28.302 Liters (l)
1 Cubic foot (ft^3)	=	7.4805 Gallons (gal)
1 Cubic foot (ft^3)	=	0.17811 Barrels (bbl)
1 Gallon imperial	=	1.200912 Gallons U.S. (gal)
1 Gallon U.S (gal)	=	128 Ounces (oz)
1 Gallon U.S (gal)	=	4 Quarts (qt)
1 Gallon U.S (gal)	=	8 Pints
1 Gallon U.S (gal)	=	231 Cubic inches (in^3)
1 Gallon U.S (gal)	=	3.7854 Liters (l)
1 Gram (g)	=	15.4324 Grains (gr)
1 Gram (g)	=	0.035274 ounces (oz)
1 Hectares (ha)	=	10000 Square meters (m^2)
1 Hectares (ha)	=	2.47105 Acres
1 Horsepower (hp)	=	0.7457 Kilowatts (KW)
1 Horsepower (hp)	=	745.701 Watts (W)
1 Horsepower (hp)	=	550 foot-pounds per second (ft-lbs/sec)
1 Inch (in)	=	2.54 Centimeters (cm)
1 Cubic inch (in^3)	=	0.5541 Ounces (oz)
1 Joule (J)	=	0.737561 Foot-pounds (ft-lbs)
1 Kilogram (Kg)	=	2.204586 Pounds (lbs)
1 Kilogram (Kg)	=	9.81 Newtons (N)
1 Kilogram/cubic meter	=	0.3505 Pounds per barrel (lbs/bbl)
1 Kilogram/cubic meter	=	.062427 Pounds per cubic foot (lbs/ft^3)
1 Kilometer (km)	=	0.621373 Miles (mile)
1 Kilometer (km)	=	0.539957 Nautical miles
1 Kilopascal (kPa)	=	0.145038 Pounds per square inch (lbs/in^2)
1 Kilowatt (kW)	=	1.34102 Horsepower

1 Liter (l)	=	61.025844 Cubic inches (in^3)
1 Liter (l)	=	0.264178 gallon (gal)
1 Liter (l)	=	0.219976 Imperial gallons (gal UK)
1 Liter (l)	=	1000 Cubic centimeters (cm^3)
1 Liter (l)	=	1.0567 Quarts (qt)
1 Liter (l)	=	0.001 Cubic Meters (m^3)
1 Megapascal (Mpa)	=	10 Bars (bar)
1 Megapascal (Mpa)	=	145.038 Pounds per square inch (lbs/in^2)
1 Megapascal (Mpa)	=	9.87 Atmosheres (atm)
1 Meter	=	3.28084 (ft)
1 Cubic meter (m^3)	=	6.28994 Barrels (bbl)
1 Cubic meter (m^3)	=	1000 Liters (l)
1 Cubic meter (m^3)	=	264.17 Gallons (gal)
1 Mile (mile)	=	5280 feet (ft)
1 Mile (mile)	=	1760 Yards (yd)
1 Mile (mile)	=	1609.3 meters (m)
1 Mile (mile)	=	0.8684 Nautical miles
1 Square mile (sq mile)	=	641.025 Acres
1 Newton (N)	=	0.101972 Kilograms (kg)
1 Newton (N)	=	0.224809 Pounds (lbs)
1 Ounce (oz)	=	1.804728 Cubic Inches (in^3)
1 Pound (lb)	=	4.44822 Newtons (N)
1 Pound/inch^2 (lbs/in^2)	=	0.06894745 Bars (bar)
1 Pound/inch^2 (lbs/in^2)	=	6.894745 Kilopascals (kPa)
1 Pound/inch^2 (lbs/in^2)	=	2.03583 Inches of mercury (inHg)
1 Pounds/foot^3 (lbs/ft^3)	=	7.4805 Pounds/gallon (lbs/gal)
1 Pounds/foot^3 (lbs/ft^3)	=	0.0160185 Kilogram/liter (kg/l)
1 Pound/gallon (lb/gal)	=	0.1336809 Pounds/foot^3 (lbs/ft^3)
1 Pound/gallon (lb/gal)	=	0.1198 Kilogram/liter (kg/l)
1 Pound/gallon (lb/gal)	=	119.8288 Kilogram/meter^3 (kg/m^3)
1 Quart (qt)	=	0.94634239 Liters (l)
1 Short ton	=	2000 Pounds (lbs)
1 Long ton	=	2240 Pounds (lbs)
1 Long ton	=	1000 Kilograms (kg)
1 Yard (yd)	=	3 Feet (ft)
1 Cubic Yard (yd^3)	=	764.6 Liters (l)
1 Watt-hour (Wh)	=	3600 Joules (J)

Table of Figures

Index

First Action Freeing procedures

Once the sticking mechanism is identified, the correct first action to free ourselves may be taken. The recommended first actions to take for each mechanism are outlined below.

Pack off and Bridge

1. Relieve trapped pressure and apply a low pressure. (Too much pressure will piston the BHA into the packoff. A little pressure is required to help establish circulation if any movement is attained)
2. Apply torque and jar down. If jars are not in place, or not functioning, then apply torque and apply maximum set down weight. (We want to move the pipe in the opposite direction is was moving when we packed off. If we pull into the packoff we will only packoff harder. The goal is to establish pipe movement and then circulation in an attempt to disturb the debris and wash it up the well) (Note that if we are moving <u>down</u> in a high angle well as the packoff occurred we will want to pull or jar <u>up</u> without torque instead)
3. If any circulation can eventually be established increase the flow rate to the maximum amount without losing circulation. Circulate until the well is cleaned.
4. Ream the section and eventually return to bottom to circulate before running casing or logs.

Differential Sticking

1. Immediately apply maximum torque and work the torque to the stuck point.
2. Circulate at the maximum allowable rate. (This should be done simultaneously with the working of the torque.) (If jars are in the string reduce the pump rate to minimize the pump open force while jarring down)
3. Slump the pipe to the maximum set down weight while holding torque in the string. Never pull up! (Pulling up only pulls the pipe harder into the stuck point and tension reduces the amount of torque we can safely apply to the drill string)
4. Jar down if jars are in the string. (Remember to reduce pump pressure while jarring down to minimize the "pump open" effect on the jars)

Wellbore Geometry

1. Jar in the opposite direction the pipe was moving prior to sticking. Apply torque while jarring down but never apply torque when jarring up.
2. Do not forget to take the pump open force into account while cocking or firing the jars. Pump pressure increases the jar blow up and decreases the jar blow down. It also resists cocking the jars for an up blow and helps cock the jars for a down blow.

Secondary freeing procedures can be prepared while continuing with the first action for any of the stuck pipe mechanisms.